FIDIC 2017: The Contract Manager's Handbook

emerald PUBLISHING
ice Publishing

FIDIC 2017: The Contract Manager's Handbook

Geoffrey Smith
BSc, CDipAF, LLDip, CEng, FICE, FCIArb
PS Consulting

Published by Emerald Publishing Ltd, Floor 5, Northspring, 21-23 Wellington Street, Leeds, LS1 4DL, UK.

ICE Publishing is an imprint of Emerald Publishing Limited

Other ICE Publishing titles:
FIDIC Green Book: A companion to the 2021 Short Form of Contract
Jakob B Sørensen. ISBN 978-0-7277-6629-8
FIDIC 2017: A definitive guide to claims and disputes
Nicholas Alexander Brown. ISBN 978-0-7277-6531-4
FIDIC 2017 Contracts Companion – three-volume set
Jakob B Sørensen. ISBN 978-0-7277-6448-5

A catalogue record for this book is available from the British Library.

ISBN 978-0-7277-6652-6

Cover photo: Bridge close-up. hallojulie/Shutterstock

Commissioning Editor: Michael Fenton
Assistant Editor: Cathy Sellars
Production Editor: Sirli Manitski

Typeset by Manila Typesetting Company
Index created by Michael Allerton
Printed and bound by CPI Group (UK) Ltd, Croydon, CR0 4YY

Contents

Preface

My experience of FIDIC contracts dates back to 1981. Since then, I have had occasional spells of working with other standard forms of contract, such as the JCT suite and the ENAA forms, but I estimate that 85–90% of the contracts with which I have worked, initially as contract manager, then expert, dispute board member and arbitrator, have been FIDIC based.

For more than ten years now, I have spent a significant proportion of my time teaching and training people all around the world on FIDIC contracts, contracts management, claims and disputes. The interaction with students and trainees has confirmed the view which I had already formed from my many years of using FIDIC contracts, which is that a large proportion of disputes arise from poor preparation of the Particular Conditions and/or Specifications and Employer's Requirements, and even more so, from misunderstandings about how the General Conditions are to be operated.

The FIDIC 2017 suite of Contracts is still young, and relatively few people have had the opportunity to work with them. Being much more detailed than the earlier versions, particularly with respect to procedures, the scope for misunderstanding is magnified. Therefore, when I was approached by ICE Publishing, I seized the opportunity to compile a book, which I hope will serve as a practical guide to help Contract Managers.

This book would not have been possible without the agreement of FIDIC for which I am grateful. I am also grateful to the World Bank for allowing me to refer to their Standard Procurement Documents.

On a more personal note, I wish to thank my colleagues and friends, Jim Perry, Guillaume Sauvaget and Dick Appuhn, for their constant encouragement since the outset of this venture. I also thank Michael Fenton of ICE Publishing for his support. Above all I thank my wife, Marie-France, for being so tolerant and understanding throughout the many precious hours that I diverted from our family life to spend on drafting.

About the author

Geoffrey Smith, BSc, CDipAF, LLDip, CEng, FICE, FCIArb, Barrister, Accredited Conciliator and Mediator, FIDIC Certified Adjudicator, Member of FIDIC President's List of Disputes Adjudicators, has more than 40 years' practical experience in the use of FIDIC contracts throughout Africa, Asia and Europe as Contract Manager, Expert, Counsel, Dispute Board Member and Arbitrator.

After gaining initial experience in the United Kingdom, his international career began in the Middle East in 1981 preparing and negotiating claims on behalf of a contractor on a billion-dollar project. Several projects later, he moved to France in 1988, where he opened an office for a UK firm of contract managers, advising European contractors working in Africa and Asia.

After qualifying as a barrister in 1995, Geoffrey moved towards dispute resolution, through arbitration and mediation. He was admitted to the FIDIC President's List of Disputes Adjudicators in 2012 and since then has dedicated the majority of his time to Dispute Boards. As a Dispute Board member, he is known to be a keen advocate and practitioner of dispute avoidance.

Projects have included tunnels, dams, ports and marine works, power stations, electricity distribution, water treatment and distribution, sewerage, roads, bridges, railways, petrochemical and industrial installations, as well as buildings. Since 2008, he has trained large numbers of people in FIDIC contracts and Contracts Management for FIDIC, World Bank, Asian Development Bank, Islamic Development Bank, Japanese International Cooperation Agency, the European Union, the OECD and ILO. He also lectures at Université Paris II–Panthéon Assas, the École Supérieure des Ponts et Chaussées and the University of Turin.

He is a past Director of the Dispute Resolution Board Foundation (Region 2) and from 2012 to 2022, he chaired its Bank Liaison Committee developing and maintaining links between the Foundation and the Multilateral Development Banks and bilateral funding agencies. He advised the World Bank with respect to its Special Provisions for use with FIDIC 2017 Contracts, many of which have been adopted by other banks. He has also frequently advised the Asian Development Bank and the Japanese International Cooperation Agency on contractual matters.

He was a 'friendly reviewer' of the FIDIC Green Book 2021 (2nd Edition) and reviewed the French translations of FIDIC 2017 Contracts and the FIDIC White Book 2017. He chairs FIDIC Task Group 20 which is preparing a Joint Venture agreement for Contractors. He is one of the assessors of candidates aiming to be FIDIC Certified Adjudicators.

In 2021, he was the winner of the 'Al Mathews Award for Dispute Board Excellence' from the Dispute Resolution Board Foundation and received its Distinguished Service Award in 2022.

Glossary

Accepted Contract Amount	The amount accepted in the Letter of Acceptance for the execution of the Works in accordance with the Contract.
Advance Payment Certificate	The Payment Certificate issued by the Engineer for advance payment under Sub-Clause 14.2.2 [*Advance Payment Certificate*].
Advance Payment Guarantee	The guarantee provided to the Employer by the Contractor under Sub-Clause 14.2.1 [*Advance Payment Guarantee*] as security for the advance payment.
Base Date	The date 28 days before the latest date for submission of the Tender.
Bill of Quantities	The document entitled bill of quantities (if any) included in the Schedules, containing the estimated quantities of work together with the Contractor's rates and prices.
Claim	A request or assertion by one Party to the other Party for an entitlement or relief under any Clause of these Conditions or otherwise in connection with, or arising out of, the Contract or the execution of the Works.
Commencement Date	The date as stated in the Engineer's Notice issued under Sub-Clause 8.1 [*Commencement of Works*] at which the Time for Completion begins.
Compliance Verification System	The compliance verification system to be prepared and implemented by the Contractor for the Works in accordance with Sub-Clause 4.9.2 [*Compliance Verification System*].
Conditions of Contract	General Conditions as amended by the Particular Conditions.
Contract (Red Book)	The Contract Agreement, the Letter of Acceptance, the Letter of Tender, any addenda referred to in the Contract Agreement, these Conditions, the Specification, the Drawings, the Schedules, the JV Undertaking (if applicable) and the further documents (if any) which are listed in the Contract Agreement or in the Letter of Acceptance.

Contract (Silver Book)	The Contract Agreement, any addenda referred to in the Contract Agreement, these Conditions, the Employer's Requirements, the Schedules, the Tender, the JV Undertaking (if applicable) and the further documents (if any) which are listed in the Contract Agreement.
Contract (Yellow Book)	The Contract Agreement, the Letter of Acceptance, the Letter of Tender, any addenda referred to in the Contract Agreement, these Conditions, the Employer's Requirements, the Schedules, the Contractor's Proposal, the JV Undertaking (if applicable) and the further documents (if any) which are listed in the Contract Agreement or in the Letter of Acceptance.
Contract Agreement	The agreement entered into by both Parties in accordance with Sub-Clause 1.6 [*Contract Agreement*].
Contract Data	The pages, entitled contract data which constitute Part A of the Particular Conditions, which provide the data necessary for the operation of some provisions of the Conditions of Contract.
Contract Price	The price defined in Sub-Clause 14.1 [*The Contract Price*], including Variations and Adjustments.
Contractor	The person(s) named as contractor in the Letter of Tender accepted by the Employer and the legal successors in title of such person(s).
Contractor's Documents	The documents prepared by the Contractor as described in Sub-Clause 4.4 [*Contractor's Documents*], including calculations, digital files, computer programs and other software, drawings, manuals, models, specifications and other documents of a technical nature.
Contractor's Equipment	All apparatus, equipment, machinery, construction plant, vehicles and other items required by the Contractor for the execution of the Works. Contractor's Equipment excludes Temporary Works, Plant, Materials and any other things intended to form or forming part of the Permanent Works.

Contractor's Personnel
The Contractor's Representative and all personnel whom the Contractor utilises on Site or other places where the Works are being carried out, including the staff, labour and other employees of the Contractor and of each Subcontractor; and any other personnel assisting the Contractor in the execution of the Works.

Contractor's Representative
The natural person named by the Contractor in the Contract or appointed by the Contractor under Sub-Clause 4.3 [*Contractor's Representative*], who acts on behalf of the Contractor.

Cost
All expenditure reasonably incurred (or to be incurred) by the Contractor in performing the Contract, whether on or off the Site, including taxes, overheads and similar charges, but not including profit. Where the Contractor is entitled under a Sub-Clause of these Conditions to payment of Cost, it shall be added to the Contract Price.

Cost Plus Profit
Cost plus the applicable percentage for profit stated in the Contract Data (if not stated, five percent (5%)). Such percentage shall only be added to Cost, and Cost Plus Profit shall only be added to the Contract Price, where the Contractor is entitled under a Sub-Clause of these Conditions to payment of Cost Plus Profit.

Country
The country in which the Site (or most of it) is located, where the Permanent Works are to be executed.

DAAB Agreement
The agreement signed or deemed to have been signed by both Parties and the sole member or each of the three members (as the case may be) of the DAAB in accordance with Sub-Clause 21.1 [*Constitution of the DAAB*] or Sub-Clause 21.2 [*Failure to Appoint DAAB Member(s)*], incorporating by reference the General Conditions of Dispute Avoidance/Adjudication Agreement contained in the Appendix to these General Conditions with such amendments as are agreed.

Date of Completion	The date stated in the Taking-Over Certificate issued by the Engineer; or, if the last paragraph of Sub-Clause 10.1 [*Taking Over the Works and Sections*] applies, the date on which the Works or Section are deemed to have been completed in accordance with the Contract; or, if Sub-Clause 10.2 [*Taking Over Parts*] or Sub-Clause 10.3 [*Interference with Tests on Completion*] applies, the date on which the Works or Section or Part are deemed to have been taken over by the Employer.
day	A calendar day.
Daywork Schedule	The document entitled daywork schedule (if any) included in the Contract, showing the amounts and manner of payments to be made to the Contractor for labour, materials and equipment used for daywork under Sub-Clause 13.5 [*Daywork*].
Defects Notification Period	The period for notifying defects and/or damage in the Works or a Section or a Part (as the case may be) under Sub-Clause 11.1 [*Completion of Outstanding Work and Remedying Defects*], as stated in the Contract Data (if not stated, one year), and as may be extended under Sub-Clause 11.3 [*Extension of Defects Notification Period*]. This period is calculated from the Date of Completion of the Works or Section or Part.
Delay Damages	The damages for which the Contractor shall be liable under Sub-Clause 8.8 [*Delay Damages*] for failure to comply with Sub-Clause 8.2 [*Time for Completion*].
Dispute	Any situation where: (*a*) one Party makes a claim against the other Party (which may be a Claim, as defined in these Conditions, or a matter to be determined by the Engineer under these Conditions, or otherwise); (*b*) the other Party (or the Engineer under Sub-Clause 3.7.2 [*Engineer's Determination*]) rejects the claim in whole or in part; and

(*c*) the first Party does not acquiesce (by giving a NOD under Sub-Clause 3.7.5 [*Dissatisfaction with Engineer's determination*] or otherwise),

provided however that a failure by the other Party (or the Engineer) to oppose or respond to the claim, in whole or in part, may constitute a rejection if, in the circumstances, the DAAB or the arbitrator(s), as the case may be, deem it reasonable for it to do so.

Dispute Avoidance/ Adjudication Board The sole member or three members (as the case may be) so named in the Contract or appointed under Sub-Clause 21.1 [*Constitution of the DAAB*] or Sub-Clause 21.2 [*Failure to Appoint DAAB Member(s)*].

Drawings The drawings of the Works included in the Contract and any additional and modified drawings issued by (or on behalf of) the Employer in accordance with the Contract.

Employer The person named as the employer in the Contract Data and the legal successors in title to this person.

Employer's Equipment The apparatus, equipment, machinery, construction plant and/or vehicles (if any) to be made available by the Employer for the use of the Contractor under Sub-Clause 2.6 [*Employer-Supplied Materials and Employer's Equipment*], but not including Plant which has not been taken over under Clause 10 [*Employer's Taking Over*].

Employer's Personnel The Engineer, the Engineer's Representative (if appointed), the assistants described in Sub-Clause 3.4 [*Delegation by the Engineer*] and all other staff, labour and other employees of the Engineer and of the Employer engaged in fulfilling the Employer's obligations under the Contract; and any other personnel identified as Employer's Personnel, by a Notice from the Employer or the Engineer to the Contractor.

Employer's Representative	The person named by the Employer in the Contract Data appointed by the Employer for the purposes of the Contract, or any replacement appointed under Sub-Clause 3.1 [*The Employer's Representative*].
Employer's Requirements	The document entitled employer's requirements, as included in the Contract, and any additions and modifications to such document in accordance with the Contract. Such document describes the purpose(s) for which the Works are intended, and specifies Key Personnel (if any), the scope, and/or design and/or other performance, technical and evaluation criteria, for the Works.
Employer-Supplied Materials	The materials (if any) to be supplied by the Employer to the Contractor under Sub-Clause 2.6 [*Employer-Supplied Materials and Employer's Equipment*].
Engineer	The person named in the Contract Data appointed by the Employer to act as the Engineer for the purposes of the Contract, or any replacement appointed under Sub-Clause 3.6 [*Replacement of the Engineer*].
Engineer's Representative	The natural person who may be appointed by the Engineer under Sub-Clause 3.3 [*Engineer's Representative*].
Exceptional Event	An event or circumstance as defined in Sub-Clause 18.1 [*Exceptional Events*].
Extension of Time	An extension of the Time for Completion under Sub-Clause 8.5 [*Extension of Time for Completion*].
Final Payment Certificate	The payment certificate issued by the Engineer under Sub-Clause 14.13 [*Issue of FPC*].
Final Statement	The Statement defined in Sub-Clause 14.11.2 [*Agreed Final Statement*].
Foreign Currency	A currency in which part (or all) of the Contract Price is payable, but not the Local Currency.
General Conditions	The document entitled '*Conditions of Contract for Construction for Building and Engineering Works designed by the*

Employer', or alternatively '*Conditions of Contract for Conditions of Contract for Plant and Design-Build*', or alternatively '*Conditions of Contract for EPC/Turnkey Projects*', all as published by FIDIC.

Goods

Contractor's Equipment, Materials, Plant and Temporary Works, or any of them as appropriate.

Interim Payment Certificate

A Payment Certificate issued by the Engineer for an interim payment under Sub-Clause 14.6 [*Issue of IPC*].

Joint Venture

A joint venture, association, consortium or other unincorporated grouping of two or more persons, whether in the form of a partnership or otherwise.

JV Undertaking

The letter provided to the Employer as part of the Tender setting out the legal undertaking between the two or more persons constituting the Contractor as a JV. This letter shall be signed by all the persons who are members of the JV, shall be addressed to the Employer and shall include:

(*a*) each such member's undertaking to be jointly and severally liable to the Employer for the performance of the Contractor's obligations under the Contract;
(*b*) identification and authorisation of the leader of the JV; and
(*c*) identification of the separate scope or part of the Works (if any) to be carried out by each member of the JV.

Key Personnel

The positions (if any) of the Contractor's Personnel, other than the Contractor's Representative, that are stated in the Specification or Employer's Requirements.

Laws

All national (or state or provincial) legislation, statutes, acts, decrees, rules, ordinances, orders, treaties, international law and other laws, and regulations and by-laws of any legally constituted public authority.

Letter of Acceptance	The letter of formal acceptance, signed by the Employer, of the Letter of Tender, including any annexed memoranda comprising agreements between and signed by both Parties. If there is no such letter of acceptance, the expression '*Letter of Acceptance*' means the Contract Agreement, and the date of issuing or receiving the Letter of Acceptance means the date of signing the Contract Agreement.
Letter of Tender	The letter of tender, signed by the Contractor, stating the Contractor's offer to the Employer for the execution of the Works.
Local Currency	The currency of the Country.
Materials	Things of all kinds (other than Plant), whether on the Site or otherwise allocated to the Contract and intended to form or forming part of the Permanent Works, including the supply-only materials (if any) to be supplied by the Contractor under the Contract.
month	A calendar month (according to the Gregorian calendar).
No-objection	The Engineer/Employer's Representative has no objection to the Contractor's Documents, or other documents submitted by the Contractor under these Conditions, and such Contractor's Documents or other documents may be used for the Works.
Notice	A written communication identified as a Notice and issued in accordance with Sub-Clause 1.3 [*Notices and Other Communications*].
Notice of Dissatisfaction	The Notice one Party may give to the other Party if it is dissatisfied, either with an Engineer's determination under Sub-Clause 3.7 [*Agreement or Determination*] or with a DAAB's decision under Sub-Clause 21.4 [*Obtaining DAAB's Decision*].
Part	A part of the Works or part of a Section (as the case may be) which is used by the Employer and deemed to have been taken over under Sub-Clause 10.2 [*Taking Over Parts*].

Particular Conditions	The document entitled particular conditions of contract included in the Contract, which consists of Part A – Contract Data and Part B – Special Provisions.
Party	The Employer or the Contractor, as the context requires.
Parties	Both the Employer and the Contractor.
Payment Certificate	A payment certificate issued by the Engineer under Clause 14 [*Contract Price and Payment*].
Performance Certificate	The certificate issued by the Engineer (or deemed to be issued) under Sub-Clause 11.9 [*Performance Certificate*].
Performance Damages	The damages to be paid by the Contractor to the Employer for the failure to achieve the guaranteed performance of the Plant and/or the Works or any part of the Works (as the case may be), as set out in the Schedule of Performance Guarantees.
Performance Security	The security provided to the Employer by the Contractor under Sub-Clause 4.2 [*Performance Security*].
Permanent Works	The works of a permanent nature which are to be executed by the Contractor under the Contract.
Pink Book	'*Conditions of Contract for Construction for Building and Engineering Works designed by the Employer, MDB Harmonized Edition*', as published by FIDIC.
Plant	The apparatus, equipment, machinery and vehicles (including any components) whether on the Site or otherwise allocated to the Contract and intended to form or forming part of the Permanent Works.
Programme	A detailed time programme prepared and submitted by the Contractor to which the Engineer has given (or is deemed to have given) a Notice of No-objection under Sub-Clause 8.3 [*Programme*].

Provisional Sum	A sum (if any) which is specified in the Contract by the Employer as a provisional sum, for the execution of any part of the Works or for the supply of Plant, Materials or services under Sub-Clause 13.4 [*Provisional Sums*].
QM System	The Contractor's quality management system (as may be updated and/or revised from time to time) in accordance with Sub-Clause 4.9.1 [*Quality Management System*].
Red Book	'*Conditions of Contract for Construction for Building and Engineering Works designed by the Employer*', as published by FIDIC.
Retention Money	The accumulated retention moneys which the Employer retains under Sub-Clause 14.3 [*Application for Interim Payment*] and pays under Sub-Clause 14.9 [*Release of Retention Money*].
Review	Examination and consideration by the Engineer of a Contractor's submission to assess whether (and to what extent) it complies with the Contract and/or with the Contractor's obligations under or in connection with the Contract.
Schedule of Payments	The document(s) entitled schedule of payments (if any) in the Schedules showing the amounts and manner of payments to be made to the Contractor.
Schedule of Performance Guarantees	The document(s) entitled schedule of performance guarantees (if any) in the Schedules showing the guarantees required by the Employer for performance of the Works and/or the Plant or any part of the Works (as the case may be) and stating the applicable Performance Damages payable in the event of failure to attain any of the guaranteed performance(s).
Schedule of Rates and Prices	The document(s) entitled schedule of rates and prices (if any) in the Schedules.
Section	A part of the Works specified in the Contract Data as a Section (if any).

Schedules	The document(s) entitled schedules prepared by the Employer and completed by the Contractor, as attached to the Letter of Tender and included in the Contract. Such document(s) may include data, lists and schedules of payments and/or rates and prices, and guarantees.
Silver Book	'*Conditions of Contract for EPC/Turnkey Projects*', as published by FIDIC.
Site	The places where the Permanent Works are to be executed and to which Plant and Materials are to be delivered, and any other places specified in the Contract as forming part of the Site.
Special Provisions	The document (if any), entitled special provisions which constitutes Part B of the Particular Conditions.
Specification	The document entitled specification included in the Contract, and any additions and modifications to the specification in accordance with the Contract. Such document specifies the Works to be executed.
Statement	A statement submitted by the Contractor as part of an application for a Payment Certificate under Sub-Clause 14.3 [*Application for Interim Payment*], Sub-Clause 14.10 [*Statement at Completion*] or Sub-Clause 14.11 [*Final Statement*].
Subcontractor	Any person named in the Contract as a subcontractor, or any person appointed by the Contractor as a subcontractor or designer, for a part of the Works; and the legal successors in title to each of these persons.
Taking-Over Certificate	A certificate issued (or deemed to be issued) by the Engineer in accordance with Clause 10 [*Employer's Taking Over*].
Temporary Works	All temporary works of every kind (other than Contractor's Equipment) required on Site for the execution of the Works.
Tender	The Letter of Tender, the Contractor's Proposal, the JV Undertaking (if applicable), and all other documents which the Contractor submitted with the Letter of Tender, as included in the Contract.

Tests after Completion	The tests (if any) which are stated in the Specification/Employer's Requirements, and which are carried out in accordance with the Special Provisions after the Works or a Section (as the case may be) are taken over under Clause 10 [*Employer's Taking Over*].
Tests on Completion	The tests which are specified in the Contract or agreed by both Parties or instructed as a Variation, and which are carried out under Clause 9 [*Tests on Completion*] before the Works or a Section (as the case may be) are taken over under Clause 10 [*Employer's Taking Over*].
Time for Completion	The time for completing the Works or a Section (as the case may be) under Sub-Clause 8.2 [*Time for Completion*], as stated in the Contract Data as may be extended under Sub-Clause 8.5 [*Extension of Time for Completion*], calculated from the Commencement Date.
Unforeseeable	Not reasonably foreseeable by an experienced contractor by the Base Date.
Variation	Any change to the Works, which is instructed as a variation under Clause 13 [*Variations and Adjustments*].
Works	The Permanent Works and the Temporary Works, or either of them as appropriate.
year	365 days.
Yellow Book	'*Conditions of Contract for Conditions of Contract for Plant and Design-Build*', as published by FIDIC.

Abbreviations

ADB	Asian Development Bank
AfDB	African Development Bank
AIIB	Asian Infrastructure Investment Bank
BOQ	Bill of Quantities
DAAB	Dispute Avoidance/Adjudication Board
DAB	Dispute Adjudication Board
DNP	Defects Notification Period
EOT	Extension of Time
FPC	Final Payment Certificate
GC	General Conditions
IADB	Inter-American Development Bank
IFI	International Financing Institution
IPC	Interim Payment Certificate
IsDB	Islamic Development Bank
JV	Joint Venture, Consortium or similar
MDB	Multilateral Development Bank
NOD	Notice of Dissatisfaction
PC	Particular Conditions
RB	Red Book
RB 1999	Red Book 1999 Edition
RB 2017	Red Book 2017 Edition
SB	Silver Book
SB 1999	Silver Book 1999 Edition
SB 2017	Silver Book 2017 Edition
TOC	Taking-Over Certificate
WB	World Bank
YB	Yellow Book
YB 1999	Yellow Book 1999
YB 2017	Yellow Book 2017

Smith G
ISBN 978-0-7277-6652-6
https://doi.org/10.1680/fcmh.66526.001

Chapter 1
Introduction

The author first worked with a FIDIC Contract in the Middle East in 1981. It was a Red Book 3rd Edition of 1977. It weighed 133g. This was followed by experience with the Red Book 4th Edition which weighed 258g. The new Red Book published in 1999 (now referred to as the 1st Edition) weighed 532g. The Red Book 2017 (now referred to as the 2nd Edition), which was reprinted in 2022 with amendments, weighs 1022g. It can be seen that each new version is approximately twice as heavy as its immediate predecessor.

Between 1977 and 2022, the rights and obligations of the Employer and the Contractor as stated in the contract, have changed little. The descriptions of the rights and obligations have perhaps become more complex, and a few new obligations have been inserted such as those dealing with labour. However, most of the increase in weight comes from the procedures.

Take as an example the provisions related to claims and disputes. The Red Book 3rd Edition of 1977 contained a single sub-clause (Sub-Clause 52.3) dealing with claims and was composed of two short paragraphs. Sub-Clause 67, composed of a single paragraph, dealt with Engineer's decisions and arbitration.

In the 4th Edition of 1987, claims became the subject of a separate clause (Clause 53), composed of five sub-clauses, each of a single paragraph. The Engineer's decisions and arbitration became the subject of four sub-clauses, totalling nine paragraphs.

By 1999, the corresponding claims provisions had expanded to nine paragraphs under Sub-Clause 20.1 [*Contractor's Claims*] and four paragraphs under Sub-Clause 2.5 [*Employer's Claims*]. In addition, Sub-Clause 3.5 [*Determinations*] contained two paragraphs dealing with the Engineer's processing of the claims and Sub-Clauses 20.2 to 20.8 dealt with Dispute Boards and arbitration (25 paragraphs), which referred to sample forms for the Dispute Adjudication Agreement, the General Conditions of Dispute Adjudication Agreement and the Procedural Rules.

In the reprint of Red Book 2017, Clause 20 [*Claims*] contains eight sub-clauses totalling 25 paragraphs which deal with Claims; Sub-Clause 3.7 [*Agreement or Determination*] dealing with the Engineer's handling of the Claims has been expanded to 17 paragraphs in five sub-clauses. Sub-Clause 21 [*Disputes and Arbitration*] stretches to almost seven pages plus 15 pages of General Conditions of DAAB Agreement and DAAB Procedural Rules.

A second example is the use of the word 'deemed' in Red Book 2017. It is used 120 times, mostly to bestow a right on one Party if the other Party (or the Engineer) fails to act within a specified period or to cause a Party to lose a right by failing to act within the specified period.

It is therefore essential for the Contract Manager to fully understand the detailed procedures set out in the FIDIC 2017 Contracts, to be aware of their strengths and weaknesses and to ensure that the appropriate action is taken at the required time.

This task of the Contract Manager is made more complicated by the existence of differences between the books which make up the 2017 suite, as well as other differences imposed by the Multilateral Development Banks (MDBs) and International Financing Institutions (IFIs) for use with FIDIC 2017 Contracts for projects which they finance. Finally, the situation is made even more complex by amendments introduced in the 2022 Reprints of the FIDIC 2017 Contracts.

It is widely recognised that proper implementation of the FIDIC 2017 Contracts will require more Contract Managers within the Contractor's team but also within the teams of the Engineer and the Employer.

This book is intended to offer practical guidance to all such Contract Managers working with FIDIC 2017 Contracts – in their 'pure' state or as amended by the MDB and IFI and/or by the 2022 Reprints.

It concentrates on the Red Book 2017 (RB 2017) or to use its formal title: *FIDIC Conditions of Contract for Construction for Building and Engineering Works designed by the Employer, Second Edition 2017.* However, it also addresses differences between RB 2017 and the Yellow Book 2017 (YB 2017), formally known as: *FIDIC Conditions of Contract for Plant and Design-Build for Electrical & Mechanical Plant, and for Building and Engineering Works, designed by the Contractor, Second Edition 2017* and the Silver Book 2017 (SB 2017), formally known as: *FIDIC Conditions of Contract for EPC/Turnkey Projects, Second Edition 2017.*

The structure of the book generally follows the structure of the Contracts, except that Chapter 2 sets out some pre-contract considerations. Chapter 3 covers the General Provisions (Clause 1 in FIDIC 2017). Chapters 4 and 5 correspond to Clauses 2 and 3 of FIDIC 2017 and address the roles of the Employer and the Engineer/Employer's Representative. Chapter 6 considers the Contractor's obligations under Clauses 4, 5 and 6 of RB 2017 (Clauses 4 and 6 of YB 2017 and SB 2017). Chapter 7 covers the Contractor's role as designer (Clause 5 of YB 2017 and SB 2017). Chapter 8 addresses all the steps to be taken prior to commencement of the Works. Chapter 9 deals with quality (Clause 7 in FIDIC 2017). Chapter 10 deals with timing (Clause 8 in FIDIC 2017).

Chapter 11 covers matters related to completion of the Works and corresponds to Clauses 9, 10 and 11 of FIDIC 2017. Chapters 12 and 13 deal with payments and Variations and Adjustments. Chapter 14 deals with termination, both by the Employer and by the Contractor (Clauses 15 and 16 under FIDIC 2017). Chapter 15 groups together the Contractor's obligations with respect to care of the Works, Employer's Risks and Exceptional Events (Clauses 17 and 18 in FIDIC 2017). Chapter 16 deals with insurances (Clause 19 of FIDIC 2017) and Chapters 17 and 18 cover Claims and Disputes respectively (Clauses 20 and 21 of FIDIC 2017).

A reprint of each of the FIDIC 2017 contracts was issued at the end of 2022 (2022 Reprint) which introduced modifications to several clauses. Chapter 19 explains these modifications.

Modifications to the standard provisions which are imposed by the MDB, are shown in boxes.

Hints from the author, specific points to be noted and examples are shown in shaded boxes.

A glossary is included on pp. xv–xxvi, and a list of abbreviations is to be found on p. xxvii.

Smith G
ISBN 978-0-7277-6652-6
https://doi.org/10.1680/fcmh.66526.005

Chapter 2
Setting the scene

2.1. Introduction

Most of this book addresses the General Conditions of the FIDIC 2017 contracts. However, experience shows that many of the challenges that arise when using FIDIC contracts do not arise from the General Conditions but from poorly prepared Particular Conditions. The seeds for future disputes are sown during the pre-contract phase and the preparation by the Employer of the bidding documents.

DAAB members and arbitrators are not unused to seeing projects for which the incorrect contract was chosen at the outset. In RB 2017, FIDIC warns:

> *'These Conditions allow for the possibility that the Contractor may be required to design a small proportion or a minor element of the Permanent Works, but they are not intended for use where significant design input by the Contractor is required or the Contractor is required to design a large proportion or any major elements of the Permanent Works. In this latter case, it is recommended that the Employer consider using FIDIC's Conditions of Contract for Plant and Design-Build, Second Edition 2017 (or, alternatively and if suitable for the circumstances of the project, FIDIC's Conditions of Contract for EPC/Turnkey Projects, Second Edition 2017).'*

Yet, notwithstanding this warning, some Employers make the fundamental error of giving the Contractor responsibility under RB 2017 for the detailed design of Works such as hydroelectric schemes and are then surprised when the Contractor 'over-designs' such that it requires payment for significantly increased quantities.

Other Employers choose to use SB 2017, and insist upon approving every Contractor's Document before it can be used, despite the warning in the introductory note:

> *'These Conditions of Contract for EPC/Turnkey Projects are not suitable for use in the following circumstances:*
>
> - *If there is insufficient time or information for tenderers to scrutinise and check the Employer's Requirements or for them to carry out their designs, risk assessment studies and estimating;*
> - *If construction will involve substantial work underground or work in other areas which tenderers cannot inspect, unless special provisions are provided to account for unforeseen conditions; or*
> - *If the Employer intends to supervise closely or control the Contractor's work, or to review most of the construction drawings.'*

The starting point for a successful project is therefore to choose the correct set of General Conditions.

2.2. Contract Data

Like previous editions of the Red, Yellow and Silver Books, the 2017 editions provide guidance for the preparation of the Particular Conditions, which are now composed of two parts: Part A – Contract Data (previously referred to as the 'Appendix to Tender') and Part B – Special Provisions. The Contract Data provides information needed for some of the General Conditions to be applied, such as the duration of the Time for Completion of the Works, or the amount of Delay Damages payable for each day of delay in completing the Works, or a Section. FIDIC provides a form listing the relevant sub-clauses and the description of the information to be provided. Some of the information is to be inserted by the Employer prior to issuing the bidding documents. Some is to be inserted by the bidders. In relation to one matter (the list of proposed members of the DAAB), input is required from both the Employer and the bidders.

In some cases, if the relevant information is not inserted, it will not be possible to apply the sub-clause. Examples include: the duration of the Time for Completion under Sub-Clause 1.1.84; the amount of the Performance Security under Sub-Clause 4.2 [*Performance Security*]; the daily rate for Delay Damages under Sub-Clause 8.8 [*Delay Damages*]; and the amount of the advance payment under Sub-Clause 14.2 [*Advance Payment*]. For some such provisions, it is expressly stated that the sub-clause will not apply if the information is not provided in the Contract Data, for example with respect to the advance payment under Sub-Clause 14.2 [*Advance Payment*]. For others, there is no such express statement, but it is established practice that the sub-clause will be inapplicable: for example, Delay Damages.

In other cases, if the relevant information is not inserted in the Contract Data, the default position described in the sub-clause will apply. Examples are: the duration of the Defects Notification Period (DNP) under Sub-Clause 1.1.27 and the payment periods under Sub-Clause 14.7 [*Payment*].

Most of the items listed by FIDIC in its template are uncomplicated and can be readily understood by reference to the relevant sub-clause. However, some are worthy of comment.

In its discussion with respect to the drafting of the Special Provisions (see below), FIDIC strongly recommends that drafters take due regard of FIDIC's five Golden Principles.[1] FIDIC explains that these Golden Principles are necessary to ensure that modifications to the General Conditions

- are limited to those necessary for the particular features of the Site and the project, or necessary to comply with the applicable law
- do not change the fair and balanced character of a FIDIC contract and
- do not make the Contract unrecognisable as a FIDIC contract.

When considering the application of these Golden Principles, the Employer should bear in mind that risk has a price and that a transfer of risk to the Contractor will lead to higher bid prices.

The Employer should also bear in mind that the Contract Data is one of the first documents that potential bidders review before deciding whether to submit a bid.

[1] https://fidic.org/sites/default/files/_golden_principles_1_12.pdf

The items which are particularly relevant in this respect, are the following:

Sub-Clause	Data to be given	Data
1.1.84	Time for Completion	. . . days
1.8	Total liability of the Contractor to the Employer under or in connection with the Contract	
4.2	Performance Security (as percentages of the Accepted Contract Amount in Currencies)	. . .. %
8.8	Delay Damages payable for each day of delay	
	Maximum amount of Delay Damages	
12.3	Percentage profit	. . .. %
14.2	Total amount of Advance Payment (as a percentage of Accepted Contract Amount)	. . .. %
14.7	Periods of payment	. . .. days
14.8	Financing charges for delayed payment (percentage points above the average bank short–term lending rate as referred to under sub-paragraph (a))	.. . . %
14.15 (f)	Rates of exchange	
17.2 (d)	Forces of nature, the risks of which are allocated to the Contractor	
21.6	Institution/Rules for administering arbitration and/or place of arbitration	

If the Time for Completion is too short, some bidders will add 10% to their bid to cover the liability for Delay Damages. Others will accept the risk but will be exceptionally aggressive from the outset with regard to Claims.

With respect to the amount of the Performance Security, some international contractors will not bid if the amount is to be in excess of 15%.

The position is similar with respect to the Delay Damages and/or Performance Damages.

With respect to the amount of the advance payment and the repayment conditions, bidders will plan to remain in a positive cash-flow position for much of the Time for Completion. If it is apparent that this will not be possible, they will add to their bids an allowance for financing costs or increase the percentage for risk.

Note:

The template for Contract Data provided with the FIDIC 2017 Contracts does not include any mention with respect to Sub-Clause 21.6 [*Arbitration*]. This is because the General Conditions state that the arbitration will be finally settled under the Rules of Arbitration of the International Chamber of Commerce (ICC) '*unless otherwise agreed by the Parties*'. By excluding any reference to Sub-Clause 21.6 in the Contract Data, FIDIC discourages the Parties from agreeing otherwise.

The template for the Contract Data used by many MDBs, allows the Parties to insert rules of arbitration other than those of the ICC and also the place of arbitration.

Hint:

Another item in the FIDIC template which merits comment is the list in Sub-Clause 21.1 [*Constitution of the DAAB*] of proposed members of the DAAB. The Employer is to provide three names in the bidding documents and each bidder is required to provide three names. This is regardless of whether the DAAB is to be composed of a sole member or three members. Thus, for a project for which six bids are submitted, a total of 21 names could be proposed for an appointment of a sole member. Thus, the probability of any named person being appointed is less than 5%. For a three-member DAAB, the probability of a candidate being appointed to the DAAB is slightly higher but still low.

Before adding names to the list, the Employer and bidders should check whether the potential candidates are interested, expect to be available and are free of any conflict of interest. Because of the low probability of being appointed and the uncertainty about when the project is likely to start, potential candidates often reply positively. However, the award of the contract is often delayed significantly and, in the absence of news, the potential candidates accept other appointments. So, when finally contacted by the Employer and Contractor, the chosen candidate(s) are no longer available. It is not unusual for two years or more to pass without news between when the potential candidate is first contacted and when the Parties agree to appoint the person – only to find that it is not possible.

To minimise this risk, it is recommended that the Employer and the bidders keep the potential candidates regularly informed of the position: for example, whether their names have been included in the bidding documents or bid, whether the bidding process has been delayed, whether a Letter of Acceptance has been issued, and so forth. It is also important and polite to inform candidates that have been unsuccessful.

2.3. Special Provisions

The Special Provisions delete, add to or otherwise modify provisions in the General Conditions, to suit the particularities of the project. This is because the FIDIC document containing the General Conditions cannot be modified without the agreement of FIDIC due to copyright. The Special Provisions, when attached to the General Conditions, implement the modifications which the Employer wishes to make. One reason for including the modifications in a separate document is to make them readily visible. If the wording of the General Conditions was itself changed, such modifications might not be identified by bidders, for example, if the word 'not' was inserted in the text, it might not be found by bidders during the short period available for preparing bids.

FIDIC suggests that the bidding documents should draw the attention of bidders to the fact that the Special Provisions (Particular Conditions – Part B) take priority over the corresponding sub-clauses in the General Conditions, and the provisions of the Contract Data (Particular Conditions – Part A) take priority over the Special Provisions (Particular Conditions – Part B).

One of the complexities of contracts based on the FIDIC model is that within the General Conditions there are many cross references to other sub-clauses, but also to other documents forming part of the Contract, such as the Contract Data, the Specification or Employer's Requirements, the Drawings or the Schedules (which include the Bills of Quantities (BOQ) or Schedule of Payments).

Thus, great care is needed when drafting Special Provisions to avoid the introduction of a conflict between provisions which can only be resolved (if at all) by reference to Sub-Clause 1.5 [*Priority of Documents*]. In some cases, a badly worded Special Provision can render a General Condition inoperable and have the opposite effect to that desired. To limit the possibility of such an occurrence, drafting should only be done by experienced personnel of diverse disciplines working together (e.g. procurement personnel, engineers, lawyers). All should have a deep and thorough understanding of the operation of FIDIC General Conditions.

On one occasion, when the author was training staff from an MDB, a person in the audience asked why they were being trained on FIDIC General Conditions as their role only involved reviewing Particular Conditions (Contract Data and Special Provisions)! Unless the drafters (and reviewers) of Special Provisions fully understand the functioning of FIDIC Contracts and the links between provisions and the other documents in the Contract package, they will be sowing the seeds of dispute.

Another aspect which may give rise to misunderstandings is where the Special Provisions use terms which are not defined. It is not uncommon to encounter Special Provisions which refer to, for example, '*Substantial Completion*', '*Practical Completion*' or '*Provisional Completion*', whereas the correct FIDIC expression is '*Taking Over*'.

To limit the possibility of future disputes, drafters of Special Provisions should take care to use only those terms which are defined in the General Conditions (usually commencing with an upper-case letter) or they should define precisely any new term which they consider to be necessary.

One other area of difficulty which unfortunately arises too often is that the Employer or the Employer's advisors decide to 'cut and paste' sub-clauses from one FIDIC Contract into another. Those encountered most frequently are Silver Book clauses such as Sub-Clause 4.12 [*Unforeseeable Difficulties*] which are inserted into Yellow Book contracts with the aim of transferring to the Contractor all risk associated with Unforeseeable adverse physical conditions.

Such practices substantially modify the risk allocation between the Parties and are in complete conflict with FIDIC Golden Principle No. 3.

The five Golden Principles are as follows:

> *GP.1: The duties, rights, obligations, roles and responsibilities of all the Contract Participants must be generally as implied in the General Conditions, and appropriate to the requirements of the project.*

GP.2: The Particular Conditions must be drafted clearly and unambiguously.
GP.3: The Particular Conditions must not change the balance of risk/reward allocation provided for in the General Conditions.
GP.4: All time periods specified in the Contract for Contract Participants to perform their obligations must be of reasonable duration.
GP.5: All formal disputes must be referred to a Dispute Avoidance/Adjudication Board (or a Dispute Adjudication Board, if applicable) for a provisionally binding decision as a condition precedent to arbitration.

Although FIDIC cannot prevent the Employer from breaching the Golden Principles, financiers such as the MDB will be careful to ensure that this does not happen and may refuse to provide finance for a project which is in breach of these principles.

For each of the FIDIC 2017 Contracts, the Notes on Preparation of Special Provisions provide alternative wording for many of the sub-clauses that are frequently amended by Employers. Some of these are relatively minor and evident, such as the need to provide interpreters for Key Personnel who do not speak the language being used for communications. Others, such as provisions related to intermediate milestones are more significant and are examined in detail at the relevant point in the following chapters.

The Notes on Preparation of Special Provisions provide not only alternative wording for some sub-clauses but also guidance with respect to the meaning of some General Conditions. For example: in relation to the meaning of '*consent*' under sub-paragraph (g) of Sub-Clause 1.2 [*Interpretation*], FIDIC states that this '*does not mean 'approve' or 'approval' which, under some legal jurisdictions, may be interpreted as accepting or acceptance that the requested matter is wholly satisfactory – following which the requesting party may no longer have any responsibility or liability for it*'.

Another example relates to Sub-Clause 3.1 [*The Engineer*]. In this respect, FIDIC states: '*It is recommended that the engineer who was appointed by the Employer to carry out the Employer's design of the Works is retained by the Employer to act as the Engineer under the Contract for services during construction of the Works.*'

Note:

While these comments provide additional information, particularly with respect to the thinking of the drafters of the FIDIC 2017 Contracts which could be useful to the drafters of Special Provisions, they are by no means binding. The MDBs, for example, do not allow the same consulting firm which designed the Works to be the Engineer under RB 2017 or YB 2017 responsible for supervision of the execution of the Works.

2.4. Specifications, Drawings and Employer's Requirements

Other documents to be prepared during the pre-contract phase which must be read in conjunction with the General Conditions and the Particular Conditions include the Specifications, Drawings and Schedules under RB 2017 and the Employer's Requirements and Schedules under YB 2017 and SB 2017.

The General Conditions frequently indicate: 'as stated in the Specification' or 'unless otherwise stated in the Specification' or 'unless otherwise stated in the Employer's Requirements'.

Sub-clauses which specifically refer to the Specification/Employer's Requirements in each of the FIDIC 2017 Contracts are the following:

Sub-Clause	Topic	RB 2017	YB 2017	SB 2017
1.8	Care and Supply of Documents	X	X	X
1.13	Compliance with Laws	X	X	X
2.1	Right of Access to the Site	X	X	X
2.5	Site Data and Items of Reference	X	X	X
2.6	Employer-Supplied Materials and Employer's Equipment	X	X	X
4.1	Contractor's General Obligations	X	X	X
4.4	Contractor's Documents	X		
4.5	Training	X		
4.5	Nominated Subcontractors		X	X
4.6	Cooperation	X	X	X
4.8	Health and Safety Obligations	X	X	X
4.9	Quality Management and Compliance Verification Systems	X	X	X
4.16	Transport of Goods	X	X	X
4.18	Protection of the Environment	X	X	X
4.19	Temporary Utilities	X	X	X
4.20	Progress Reports	X	X	X
5.1	General Design Obligations		X	X
5.2	Nominated Subcontractors	X		
5.2	Contractor's Documents		X	X
5.4	Technical Standards & Regulations		X	X
5.5	Training		X	X
5.6	As-Built Records		X	X
5.7	Operation & Maintenance Manuals		X	X
6.1	Engagement of Staff and Labour	X	X	X
6.6	Facilities for Staff and Labour		X	X
6.7	Health and Safety of Personnel	X	X	X
6.12	Key Personnel	X	X	X
7.3	Inspection	X	X	X
7.4	Testing by Contractor	X	X	X

Sub-Clause	Topic	RB 2017	YB 2017	SB 2017
7.8	Royalties	X	X	X
8.3	Programme	X	X	X
9.1	Contractor's Obligations	X	X	X
10.2	Taking Over Parts	X	X	X
11.11	Clearance of Site	X	X	X
12.1	Procedure for Tests after Completion		X	X

Drafters of the Contract documents should check that the relevant information is provided in the correct document, with respect to all the above sub-clauses.

Hint:

When inserting information into the Specification, Drawings or Employer's Requirements, it should be remembered that these are lower in the order of priority under Sub-Clause 1.5 [*Priority of Documents*] than the General Conditions and the Particular Conditions. Therefore, in the event of conflict between documents, the Particular Conditions (and General Conditions) will override the Specification, Drawings or Employer's Requirements.

Smith G
ISBN 978-0-7277-6652-6
https://doi.org/10.1680/fcmh.66526.013

Chapter 3
The basics

3.1. Definitions

Since 1999, all FIDIC contracts begin with a list of definitions. All defined words or expressions commence with a capital letter, except 'day', 'month' and 'year'. FIDIC 2017 adopts the same practice.

The definitions in FIDIC 2017 are listed in alphabetical order, while in FIDIC 1999 they were grouped according to topics (i.e. The Contract; Parties and Persons; Dates, Tests, Periods and Completion; Money and Payments; Works and Goods; Other Definitions).

Most of the definitions are readily understandable but many are cross-referenced to one or more sub-clauses such that the user should study the referenced sub-clauses carefully to fully understand the definition.

Whereas RB 1999 contained 60 definitions, RB 2017 contains 88 definitions and YB 2017 contains 90. A complete list of definitions can be found in the glossary. However, some of the more significant additions/changes include the following

Sub-Clause 1.1.2	***'Advance Payment Certificate'*** *means a Payment Certificate issued by the Engineer for advance payment under Sub-Clause 14.2.2 [Advance Payment Certificate].*

Under RB 1999, YB 1999 and the Pink Book 2010 ('*Conditions of Contract for Construction for Building and Engineering Works designed by the Employer, Multilateral Development Bank Harmonised Edition*'), there was no differentiation between the first and subsequent Payment Certificates. All were referred to as 'Interim Payment Certificates' (except for the Final Payment Certificate). The need for an Interim Payment Certificate with respect to the advance payment was often overlooked. FIDIC 2017 emphasises the need for a specific Payment Certificate with respect to the advance payment.

Sub-Clause 1.1.6	***'Claim'*** *means a request or assertion by one Party to the other Party for an entitlement or relief under any Clause of these Conditions or otherwise in connection with, or arising out of, the Contract or the execution of the Works.*

'Claim' was not defined in earlier FIDIC Contracts. Note that the 2022 Reprint modifies this definition to exclude matters to be agreed or determined under Sub-Clause 3.7 (a) (see Chapter 19).

Sub-Clause 1.1.8	***'Compliance Verification System'*** *means the compliance verification system to be prepared and implemented by the Contractor for the Works in accordance with Sub-Clause 4.9.2 [Compliance Verification System].*
Sub-Clause 1.1.12	***'Contract Data'*** *means the pages, entitled contract data which constitute Part A of the Particular Conditions.*

'Contract Data' was the expression used in the Pink Book 2010, but in FIDIC 1999 contracts the equivalent expression was *'Appendix to Tender'*.

Sub-Clause 1.1.20	***'Cost Plus Profit'*** *means Cost plus the applicable percentage for profit stated in the Contract Data (if not stated, five percent (5%)).*

This is the same approach as used in the Pink Book 2010 and helps avoid long discussions about the rate of profit to be applied. The percentage is to be added to Cost, and such '*Cost Plus Profit*' is only to be added to the Contract Price where the Conditions state that the Contractor is entitled to payment of '*Cost Plus Profit*'.

Sub-Clause 1.1.22	***'DAAB'*** *or* ***'Dispute Avoidance/Adjudication Board'*** *means the sole member or three members (as the case may be) so named in the Contract, or appointed under Sub-Clause 21.1 [Constitution of the DAAB] or Sub-Clause 21.2 [Failure to Appoint DAAB Member(s)].*

This definition replaces '*DAB*' or '*Dispute Adjudication Board*' in FIDIC 1999 contracts and '*DB*' or '*Dispute Board*' in the Pink Book 2010. The change in name focuses attention on the dispute avoidance role which was sometimes overlooked in the FIDIC 1999 contracts.

Sub-Clause 1.1.24	***'Date of Completion'*** *means the date stated in the Taking-Over Certificate issued by the Engineer; or, if the last paragraph of Sub-Clause 10.1 [Taking Over the Works and Sections] applies, the date on which the Works or Section are deemed to have been completed in accordance with the Contract; or, if Sub-Clause 10.2 [Taking Over Parts] or Sub-Clause 10.3. [Interference with Tests on Completion] applies, the date on which the Works or Section or Part are deemed to have been taken over by the Employer.*

This definition has been modified in the 2022 Reprint to clarify the position with respect to taking over of Part of the Works (see Annex 2).

Sub-Clause 1.1.29	***'Dispute'*** *means any situation where:* *(a) one Party makes a claim against the other Party (which may be a Claim, as defined in these Conditions, or a matter to be determined by the Engineer under these Conditions, or otherwise);* *(b) the other Party (or the Engineer under Sub-Clause 3.7.2 [Engineer's Determination]) rejects the claim in whole or in part; and* *(c) the first Party does not acquiesce (by giving a NOD under Sub-Clause 3.7.5 [Dissatisfaction with Engineer's determination] or otherwise),* *provided however that a failure by the other Party (or the Engineer) to oppose or respond to the claim, in whole or in part, may constitute a rejection if, in the circumstances, the DAAB or the arbitrator(s), as the case may be, deem it reasonable for it to do so.*

Under earlier FIDIC contracts, questions often arose with respect to whether a matter was in dispute and, therefore, whether a DAB or DB had jurisdiction to deal with it. This definition seeks to avoid or answer this question. Note however that the 2022 Reprint modifies this definition (see Annex 2).

Sub-Clause 1.1.37	***'Exceptional Event'*** *means an event or circumstance as defined in Sub-Clause 18.1 [Exceptional Events].*

The expression replaces the expression 'Force Majeure' used in the FIDIC 1999 contracts and the Pink Book 2010. The 2022 Reprint inserts the word '*exceptional*' before '*event or circumstance*' (see Annex 2).

Sub-Clause 1.1.46	***'Joint Venture'*** *or* ***'JV'*** *means a joint venture, association, consortium or other unincorporated grouping of two or more persons, whether in the form of a partnership or otherwise.*
Sub-Clause 1.1.48	***'Key Personnel'*** *means the positions (if any) of the Contractor's Personnel, other than the Contractor's Representative, that are stated in the Specification.*
Sub-Clause 1.1.55	***'No-objection'*** *means that the Engineer has no objection to the Contractor's Documents, or other documents submitted by the Contractor under these Conditions, and such Contractor's Documents or other documents may be used for the Works.*

This term emphasises that under FIDIC contracts, the Engineer/Employer's Representative is almost never required to approve the Goods or Works provided or executed by the Contractor.

Sub-Clause 1.1.56	***'Notice'*** *means a written communication identified as a Notice and issued in accordance with Sub-Clause 1.3 [Notices and Other Communications].*

This definition is very important as the word 'Notice' appears 400 times in RB 2017 and even more in YB 2017.

Sub-Clause 1.1.58	***'Part'*** *means a part of the Works or part of a Section (as the case may be) which is used by the Employer and deemed to have been taken over under Sub-Clause 10.2 [Taking Over Parts].*

Although the mechanism existed in RB 1999 and YB 1999 (as well as in the Pink Book 2010) for deemed taking over of a part of the Works or Section used by the Employer, there was no definition of the relevant part which distinguished it from the remainder of the Works. The 2022 Reprint modifies this definition so that '*Part*' is not limited to a part of the Works or Section for which taking over is deemed to have occurred but may also refer to a part of the Works or of a Section, for which the Engineer/Employer's Representative issues a Taking-Over Certificate, at the discretion of the Employer (see Annex 2).

Sub-Clause 1.1.59	***'Particular Conditions'*** *means the document entitled particular conditions of contract included in the Contract, which consists of Part A – Contract Data and Part B – Special Provisions.*

Under FIDIC 1999, the expression '*Particular Conditions*' was not defined. However, it was stated that the Particular Conditions together with the General Conditions formed the Conditions of Contract. Similarly, the Pink Book 2010 provided no definition. However, it was stated that Particular Conditions ware composed of two parts, of which the Contract Data constituted Part A. In practice, Part B of the Particular Conditions under Pink Book contracts was often referred to as the '*Special Provisions*'. FIDIC 2017 follows this approach and Sub-Clause 1.1.75 provides a definition of '*Special Provisions*' (see below).

Sub-Clause 1.1.66	***'Programme'*** *means a detailed time programme prepared and submitted by the Contractor to which the Engineer has given (or is deemed to have given) a Notice of No-objection under Sub-Clause 8.3 [Programme].*

This definition serves to differentiate between programmes which have not yet been reviewed by the Engineer/Employer's Representative (or which have been subject to comments) and the programme to which there has been a Notice of No-objection, and the Contractor is allowed to use.

Sub-Clause 1.1.68	***'QM System'*** *means the Contractor's quality management system (as may be updated and/or revised from time to time) in accordance with Sub-Clause 4.9.1 [Quality Management System].*
Sub-Clause 1.1.70	***'Review'*** *means examination and consideration by the Engineer of a Contractor's submission in order to assess whether (and to what extent) it complies with the Contract and/or with the Contractor's obligations under or in connection with the Contract.*
Sub-Clause 1.1.75	***'Special Provisions'*** *means the document (if any), entitled special provisions which constitutes Part B of the Particular Conditions.*

The 'Special Provisions' add to, omit from or modify the General Conditions. In FIDIC 1999 contracts, these were referred to as the Particular Conditions.

3.2. Notices and Communications

Under FIDIC 1999 Contracts, questions often arose about whether a particular item of correspondence was a 'notice'. This was especially the case when the correspondence related to a Contractor's claim for an extension of time and/or for additional payment. A Contractor could lose its entitlement to more time or more money, if it was decided that the correspondence was not a valid notice sent within 28 days of when the Contractor became aware or should have become aware of the event or circumstance giving rise to the claim.

To avoid such questions, Sub-Clause 1.3 [*Notices and Other Communications*] of FIDIC 2017 seeks to clearly differentiate between Notices (a defined term) and other communications and sets out several formal requirements.

A Notice must:

- be in writing;
- state that it is a Notice.

It may be:

- a paper-original signed by the Contractor's Representative, the Engineer or the authorised representative of the Employer (as the case may be), which is either:
  - delivered by hand (against receipt), or
  - delivered by mail or courier (against receipt) at/to the address for the recipient's communications as stated in the Contract Data or other address for which the recipient has given Notice.

Alternatively, or in addition, the Notice may be:

- an electronic original generated by one of the systems of electronic transmission mentioned in the Contract Data, sent by the electronic address uniquely assigned to each of the authorised representatives; or
- a copy of the paper original transmitted electronically in the same manner.

If the system of electronic transmission is not stated in the Contract Data, a system acceptable to the Engineer may be used.

In short, a Notice must be in writing, must be clearly identified as a Notice and be sent by the correct person, to the correct person, at the correct address, and by the correct method.

If the communication is not a Notice, but another type of communication such as:

> *'acceptance, acknowledgement, advising, agreement, approval, certificate, Claim, consent, decision, determination, discharge, instruction, No-objection, record(s) of meeting, permission, proposal, record, reply, report, request, Review, Statement, statement, submission or any other similar type of communication'*

it must state the type of communication and must mention the provision(s) of the Contract which require(s) it to be issued.

In all other respects, the rules for such other communications are identical to those for a Notice.

Hint:

The above list of *'other communications'* includes *'No-objection'* and *'determination'*. However, every provision of RB 2017 which refers to 'No-objection' requires a Notice of No-objection to be sent and, when issuing a determination under Sub-Clause 3.7 [*Agreement or Determination*], the Engineer is required to do so by means of a *'Notice of Engineer's Determination'*.

In view of the potential consequences of not sending a Notice according to the rules set out under Sub-Clause 1.3, the Engineer/Employer's Representative should only inform the parties of its No-objection or its determination by a Notice, rather than by 'other communication'.

Hint:

Sub-Clause 1.3 [*Notices and Other Communications*] does not require the Notice to refer to the provision(s) of the Contract under which it is issued. This contrasts starkly with the other formalities and may be an oversight on the part of the drafters which might be corrected in the future. In the meantime, users are recommended to refer to the appropriate sub-clauses in their Notices to avoid any doubt about the purpose and validity of the Notice.

COMMUNICATIONS CHECKLIST			
Which Sub-Clause applies?			
What does the Sub-Clause require: a 'Notice' or 'other communication' such as a 'consent', a 'Statement', a 'Claim', etc.?			
NOTICE		**OTHER COMMUNICATIONS**	
What is the cut-off date for receipt (if any)?		What is the cut-off date for receipt (if any)?	
Who must the Notice be sent to?		Who must the communication be sent to?	
What address must be used?		What address must be used?	
Who must receive a copy?		Who must receive a copy?	
What address must be used?		What address must be used?	
How is the Notice to be sent?		How is the communication to be sent?	
Is this method authorised under Sub-Clause 1.3 and the Contract Data?		Is this method authorised under Sub-Clause 1.3 and the Contract Data?	
Is it signed by the correct person?		Is it signed by the correct person?	
Does it state that it is a Notice?		Does it state the type of communication: 'Statement', 'Claim', 'Request for consent', etc. mentioned in the Sub-Clause?	
Does it have a special title such as 'Notice of Dissatisfaction', 'Notice of Engineer's Determination', or 'Notice of Claim'?			
Have you mentioned the Sub-Clause?		Have you mentioned the Sub-Clause?	

Paper originals of Notices and other communications take effect when received or deemed to have been received at the recipient's address. Electronically transmitted Notices and other communications are deemed to be received on the day after transmission, provided that no notification of delivery failure is received by the sender.

Hint:

It is important to consider the rules with respect to date of receipt when Notices must be sent within a fixed period. Provisions such as Sub-Clause 20.2.1 [*Notice of Claim*] state that Notices of Claim must be sent within 28 days of the claiming Party becoming aware of the event or circumstance giving rise to the Claim or of when it should have become aware. Therefore, the rule with respect to the deemed receipt of electronic transmissions appears to be irrelevant. However, arguments are likely to arise about whether Notices of Claim transmitted electronically on day 28 are valid Notices. To avoid such arguments, users should ensure that Notices which are transmitted electronically are sent no later than day 27.

When a Notice or a Notice of Dissatisfaction (NOD) or a certificate is issued by a Party or the Engineer, the paper and/or electronic original must be sent to the intended recipient and a copy must be sent to the Engineer or the other Party, as the case may be.

Sub-Clause 1.3 goes on to state that all other communications must be copied to the Parties and/or the Engineer as stated under these Conditions or elsewhere in the Contract. This appears to address provisions which do not require both Parties and the Engineer to be informed about certain communications. For example: under Sub-Clause 4.3 [*Contractor's Representative*] unless the Contractor's Representative is named in the Contract, the Contractor must submit to the Engineer for consent the name and particulars of the person the Contractor proposes to appoint as Contractor's Representative. There is no requirement to copy this communication to the Employer.

Finally, Sub-Clause 1.3 states that all Notices and other types of communication must not be unreasonably withheld or delayed. Although this provision appears innocuous, it is often relied upon by Contractors when faced with an Engineer who delays the issue of an Interim Payment Certificate (IPC) or a Taking-Over Certificate (TOC) or who delays the processing of Claims by repeatedly requesting further particulars.

Example of a 'Notice':

BETTA CONTRACTORS LTD.

A.N. OTHER & Partners

Consulting Engineers

01 October 2022

For the attention of Mr. Other, the Engineer

Subject: **Notice of Unforeseeable Physical Conditions under Sub-Clause 4.12.1**

Notice of Claim under Sub-Clause 20.2.1

Dear Sir,

At approximately 8.15 pm yesterday, we encountered voids in the tunnel face at km 1.350 from which there was substantial flow of water (see attached photographs).

As you and your staff had already left the site and could not be contacted, we took the initiative to halt tunnel excavation and to begin grout injection of the ground ahead of the tunnel face. Grouting began at approximately 7.00 am today and we expect to continue until late tomorrow evening, at which stage we will attempt to recommence tunnelling. We trust that you approve of this action, but should you require any other measures, we would be pleased to receive your instructions.

These circumstances, which were not indicated in and could not be anticipated from the geological information available at the time of tender, are considered to be Unforeseeable. They will inevitably cause delay to the progress of the Works and additional costs. Although we are currently unable to estimate the extent, we hereby give notice under Sub-Clause 20.2.1 of our intention to claim an Extension of Time and reimbursement of the additional Costs under Sub-Clauses 4.12 [*Unforeseeable Physical Conditions*] and 8.5 [*Extension of Time*].

We will submit records of the manpower, plant and materials involved, for your agreement daily. Should you require any further records, please let us know.

Yours faithfully,

Contractor's Representative

cc Employer

Example of 'another communication':

> **BETTA CONTRACTORS LTD.**
>
> A.N. OTHER & Partners
>
> Consulting Engineers
>
> 15 September 2022
>
> **For the attention of Mr. Other, the Engineer**
>
> Subject: **Request for consent to the appointment of the Contractor's Representative under Sub-Clause 4.3**.
>
> Dear Sir,
>
> In accordance with Sub-Clause 4.3 [*Contractor's Representative*] we hereby request the Engineer's consent to the appointment of Mr. John Smith as Contractor's Representative for the project.
>
> A recent version of Mr. Smith's CV is attached.
>
> We would appreciate your early response to our request, but should you require any further information, we would be pleased to provide it.
>
> Yours faithfully,
>
> Contractor's Representative

3.3. Law and Language

Sub-Clause 1.4 [*Law and Language*] allows the Parties to choose the country (or jurisdiction) whose laws govern the interpretation of the Contract. If the Contract Data does not specify which country, the default position is that it is the Country where the Works (or most of the Works) are located.

It is unusual for a country or jurisdiction other than the Country where the Works are to be located to be chosen. This usually happens when the Works cross a border, such as a bridge over a river which represents the border between two countries. It can also happen when the project is privately financed with funds coming from another country and the financier does not agree that the local law should be the governing law.

The choice of another country as the source of the governing law with respect to interpretation of the Contract does not mean that the local laws do not apply to the execution of the Works. Sub-Clause 1.13 [*Compliance with Laws*] requires both the Contractor and the Employer to comply with all applicable Laws in performing the Contract. Sub-Clause 1.1.49 defines 'Laws' to mean:

> *'all national (or state or provincial) legislation, statutes, acts, decrees, rules, ordinances, orders, treaties, international law and other laws, and regulations and by-laws of any legally constituted public authority.'*

Thus, both Parties must comply with local regulations concerning permits, employment, safety, the environment, taxes and so forth.

Sub-Clause 1.4 [*Law and Language*] also permits the Parties to choose the ruling language with respect to interpretation of the Contract, as well as choosing the language for communications. The ruling language is to be stated in the Contract Data. If it is not, the ruling language is the language in which the Conditions of Contract are written. If there is more than one version of any part of the Contract, (for example: the Specifications or the Employer's Requirements), the version which is in the ruling language prevails.

If the language for communications is to be different from the ruling language, the language for communications must also be stated in the Contract Data. If the two languages are different, in the event of a dispute to be submitted for resolution by a third party such as a DAAB or an arbitral tribunal, it may be necessary to translate all communications into the ruling language, which may be very expensive and time-consuming.

3.4. Priority of Documents

FIDIC Contracts consist of several separate documents. Traditionally, a Red Book contract has at least eight separate documents or categories of document:

- the Contract Agreement;
- the Letter of Acceptance;
- the Letter of Tender (Bid) and the Appendix to Tender;
- the Particular Conditions;
- the General Conditions;
- the Specification;
- the Drawings; and
- the Schedules (such as BOQ, Dayworks Schedule, etc.) and any other documents forming part of the Contract.

In general, these documents must be read together, and they are taken as mutually explanatory. However, problems may arise if two or more documents contain conflicting provisions. In this case, reference must be made to the order of priority of the documents set out under Sub-Clause 1.5 [*Priority of Documents*]. The document at the top of the list has the highest priority and that at the bottom of the list, the lowest priority.

As a simple example, if the Drawings for a building project show wall-tiling in a room and the BOQ includes a unit rate for wall-tiling with tiles 15 cm × 15 cm, the two documents can be read together. The Contractor must supply and install tiles 15 cm × 15 cm and is to be paid at the rate stated in the BOQ. However, if the Drawings show wall-tiling with 20 cm × 20 cm tiles, and the Specification refers to wall-tiling with white tiles with 2 mm joints, whereas the BOQ contains only a rate for beige tiles 15 cm × 15 cm, the Parties and the Engineer must follow the order of priority. As a result, the Contractor must install white tiles (as mentioned in the Specification) which are 20 cm × 20 cm (as shown on the Drawings) with 2 mm joints (as mentioned in the Specification) at the unit rate stated in the BOQ. While this position was clear under RB 1999, it is less so under RB 2017 and it is further discussed in Chapter 12.2.

The list of documents set out above matches the list in RB 1999 except that there is no express mention under Sub-Clause 1.5 [*Priority of Documents*] of RB 1999 of the '*Appendix to Tender*' and there are no examples of the documents falling within the category referred to as '*the Schedules*'. YB 1999 and SB 1999 contain slightly different lists which reflect the nature of the contracts.

Each of the FIDIC 2017 contracts follows its predecessor except that the more recent lists separate the Particular Conditions into *Part A – Contract Data* (previously known as '*Appendix to Tender*') and *Part B – Special Provisions*. Moreover, if the Contractor is a Joint Venture (JV), under FIDIC 2017 the JV Undertaking is one of the documents which forms part of the Contract (although it is low in the order of priority).

Thus, the order of priority for RB 2017 is as follows:

(*a*) the Contract Agreement,
(*b*) the Letter of Acceptance,
(*c*) the Letter of Tender,
(*d*) the Particular Conditions Part A – Contract Data,
(*e*) the Particular Conditions Part B – Special Provisions,
(*f*) the General Conditions,
(*g*) the Specification,
(*h*) the Drawings,
(*i*) the Schedules,
(*j*) the JV Undertaking (if the Contractor is a JV), and
(*k*) any other documents forming part of the Contract.

In YB 2017, the order of priority is as follows:

(*a*) the Contract Agreement,
(*b*) the Letter of Acceptance,
(*c*) the Letter of Tender,
(*d*) the Particular Conditions Part A – Contract Data,
(*e*) the Particular Conditions Part B – Special Provisions,
(*f*) the General Conditions,
(*g*) the Employer's Requirements,
(*h*) the Schedules,
(*i*) the Contractor's Proposal,
(*j*) the JV Undertaking (if the Contractor is a JV), and
(*k*) any other documents forming part of the Contract.

In SB 2017, the order of priority is as follows:

(*a*) the Contract Agreement,
(*b*) the Particular Conditions Part A – Contract Data,
(*c*) the Particular Conditions Part B – Special Provisions,
(*d*) the General Conditions,
(*e*) the Employer's Requirements,
(*f*) the Schedules,
(*g*) the Tender,
(*h*) the JV Undertaking (if the Contractor is a JV), and
(*i*) any other documents forming part of the Contract.

The difference between RB 2017 and YB 2017 is that the Specification precisely defines the input to the Works which is in line with the Employer's responsibility under RB 2017 for design of the Works, whereas the Employer's Requirements under YB 2017 and SB 2017 define the output from the Works which the Contractor is expected to achieve from its design. It is therefore important to include, as a Contract document under YB 2017 or SB 2017, the *Contractor's Proposal* or its *Tender*, describing how it intends to satisfy the output criteria. In a two-stage bidding process the Contractor's Proposal is often called its *Technical Offer* or *Technical Bid*.

The differences between YB 2017 and SB 2017 reflect the nature of the EPC contract, under which the Contractor would normally have progressed its design to a greater degree than for its proposal for a 'design and build' project, and such design will have been discussed between the Parties and refined during a long tendering process.

The orders of priority listed above can be applied to many projects. However, it is not unusual for additional documents to form part of the Contract. Examples include questions and answers from the tendering/bidding phase; addenda to tender/bidding documents; and minutes of tender-clarification meetings.

The Special Provisions imposed by the World Bank and some other MDBs add three documents to the order of priority:	
Sub-Clause 1.5 ***Priority of Documents***	*The following documents are added in the list of Priority Documents after (e):* *'(f) the Particular Conditions Part C – Fraud and Corruption;* *(g) the Particular Conditions Part D – Environmental and Social (ES) Metrics for Progress Reports;* *(h) Particular Conditions Part E – Sexual Exploitation and Abuse (SEA) and/or Sexual Harassment Performance Declaration for Subcontractors';* *and the list renumbered accordingly.*

The position of these additional documents within the order of priority is clearly stated. However, problems can arise if additional documents are not correctly inserted in the order of priority.

In deciding where in the list to insert the additional documents, attention must be paid to legal principles which are generally applied to determine the order of priority in the absence of an expressly stated order. These are that specific documents take priority over general documents, and later documents take priority over earlier documents.

Thus, 'questions and answers' should be placed below the Letter of Tender but above the Particular Conditions as they are produced after the Particular Conditions but before the Letter of Tender. Addenda to tender documents should be placed immediately above the Specification/Employer's Requirements if they address solely technical matters, or above the Particular Conditions if they address administrative matters or a mixture of technical and administrative matters. Minutes of tender-clarification meetings should be inserted between the Letter of Tender and the Letter of Acceptance because the meetings were held after receipt of tenders but before acceptance of a tender.

A problem sometimes encountered with orders of priority is that the list includes documents which are included two or more times. This often happens when one or more documents is/are annexed to the Contract Agreement (which has highest priority) but is/are also included lower in the order of priority.

Another problem frequently encountered is where different versions of the same document appear at different places in the order of priority. Such a situation raises the question of whether the earlier version is totally replaced by the later version or whether both versions are valid.

3.5. Contract Agreement

Sub-Clause 1.6 [*Contract Agreement*] of RB 2017 and of YB 2017 contain identical wording. Both foresee that the Parties shall sign a Contract Agreement within 35 days of the Contractor receiving the Letter of Acceptance, unless they agree otherwise. This allows them to fix a different timescale or to agree that a Contract Agreement is unnecessary. The second possibility reflects the position in some jurisdictions, that the acceptance of an offer is sufficient to constitute a binding contract, without it being necessary to sign a Contract Agreement.

Under both contracts, the costs of any stamp duties or similar charges must be paid by the Employer.

If the Contractor is a JV, each member's authorised representative must sign the Contract Agreement.

Sub-Clause 1.6 [*Contract Agreement*] of SB 2017 contains different wording and a Contract Agreement must be signed before the Contract comes into full force. Thus, there is no provision for a Letter of Acceptance.

Under Sub-Clause 16.2 [*Termination by Contractor*] of all three contracts, the Contractor may give a Notice of termination of the Contract if the Employer fails to comply with Sub-Clause 1.6 [*Contract Agreement*], that is, if the Employer fails to sign a Contract Agreement within the appropriate timescale. There is no similar mention of a failure by the Contractor to comply with Sub-Clause 1.6 [*Contract Agreement*] among the circumstances listed under Sub-Clause 15.2 [*Termination for Contractor's Default*] which entitle the Employer to give Notice of termination for Contractor's default.

3.6. Joint Ventures

Under Sub-Clause 1.14 [*Joint and Several Liability*], if the Contractor is a JV, the members of the JV are jointly and severally liable to the Employer for the performance of the Contractor's obligations under the Contract. Each member of the JV can be pursued by the Employer to the full extent of the JV's liability towards the Employer. In practical terms, this means that if one member of the JV fails to perform in accordance with the Contract, the other members of the JV must stand in for the member that fails to perform and make good the failure. This situation arises most frequently when a member of the JV becomes bankrupt or insolvent.

The JV must be represented by a leader who has authority to bind the Contractor and each member of the JV.

The JV may not be modified either in its membership, its legal status or the distribution of scope among its members, without the prior consent of the Employer. However, such consent does not relieve the modified JV from any of its joint and several liabilities. A situation that is sometimes encountered is that a member of the JV is a 'sleeping partner' which lent its name and references to the JV but had no intention to contribute. This provision holds the other members of the JV liable but requires them to inform the Employer of (and obtain consent to) the redistribution of activities from that defined in the JV agreement. In such circumstances, and depending upon the legal jurisdiction, the Employer might be able to terminate the Contract.

It should be noted that under Sub-Clause 1.1.46 of RB 2017, 'Joint Venture' (JV) means '*a joint venture, association, consortium or other unincorporated grouping of two or more persons, whether in the form of a partnership or otherwise*'. It thus applies equally to the two most common forms of grouping which are the 'joint venture' in which all partners provide resources for the execution in common of the scope of works and the 'consortium' in which each partner takes responsibility for a specific part of the scope of works.

3.7. Limitation of Liability

Apart from the exceptions listed under Sub-Clause 1.15 [*Limitation of Liability*] and examined below, neither Party is liable to the other Party for loss of use of any Works, loss of profit, loss of any contract or for any indirect or consequential loss or damage which may be suffered by the other Party in connection with the Contract.

The exceptions are as follow:

(*a*) Sub-Clause 8.8 [*Delay Damages*]
Delay damages are usually considered as compensation for loss of use of all or part of the Works during the period of delay as well as costs which are a consequence of the delay, such as additional fees charged by the Engineer.

(*b*) Sub-paragraph (c) of Sub-Clause 13.3.1 [*Variation by Instruction*]
Under this sub-paragraph, the Contractor may request payment of the Cost incurred from any necessary modification to the Time for Completion (EOT) resulting from a Variation instruction. If the instruction is to omit work and the omitted work is to be executed by others, the Contractor is entitled to be paid the amount of any loss of profit and other losses and damages caused by the omission.

(*c*) Sub-Clause 15.7 [*Payment after Termination for Employer's Convenience*]
If the Employer terminates the Contract for its convenience (not due to the Contractor's default), the Contractor is entitled to be paid the amount of any loss of profit or other losses and damages caused by the termination. There is no requirement for a Notice of Claim under Sub-Clause 20.2. [*Claims for Payment and/or EOT*]

(*d*) Sub-Clause 16.4 [*Payment after Termination by Contractor*]
If the Contractor terminates the Contract due to the Employer's default, the Contractor is entitled to claim the amount of any loss of profit or other losses and damages caused by the termination. To do so, it must follow the procedure set out under Sub-Clause 20.2 [*Claims for Payment and/or EOT*].

(*e*) Sub-Clause 17.3 [*Intellectual and Industrial Property Rights*]
When one Party receives a claim from a third party for infringement of intellectual or industrial property rights and such infringement is the responsibility of the other Party, the responsible Party must indemnify and hold harmless the Party receiving the claim, including with respect to the legal

fees and expenses incurred consequent to the claim. Under some legal systems, including under English law, an indemnity given by one party in favour of another party may permit the receiving party to recover losses or damages that might otherwise be considered indirect or consequential.

(*f*) The first paragraph of Sub-Clause 17.4 [*Indemnities by Contractor*]
When the Employer receives a claim from a third party for personal injury or death or for property damage or loss, arising from or by reason of the Contractor's execution of the Works, the Contractor must indemnify and hold harmless the Employer, including with respect to the legal fees and expenses incurred consequent to the claim. See sub-paragraph (e) above with respect to the meaning of 'indemnity'.

(*g*) Sub-Clause 17.5 [*Indemnities by Employer*]
The Employer must indemnify and hold harmless the Contractor, including with respect to the legal fees and expenses incurred, when the Contractor receives from a third party a claim for personal injury or death or for property damage or loss, which is attributable to any negligence, wilful act or breach of the Contract by the Employer, the Employer's Personnel, or any of their respective agents, or a claim for property damage or loss arising out of any event described under Sub-Clause 17.2 [*Liability for Care of the Works*] as being the Employer's responsibility. See sub-paragraph (e) above with respect to the meaning of 'indemnity'.

The Contractor's total liability to the Employer under or in connection with the Contract is not to exceed the amount stated in the Contract Data. If no amount is stated, the total liability is not to exceed the Accepted Contract Amount.

However, the total liability may exceed such amount as follows:

(*a*) under Sub-Clause 2.6 [*Employer-Supplied Materials and Employer's Equipment*] if damage or loss occurs to Employer-Supplied Materials or to an item of Employer's Equipment while under the Contractor's control;

(*b*) under Sub-Clause 4.19 [*Temporary Utilities*] if the Employer is to provide, for the purposes of the Works, temporary utilities such as electricity or water, for which the Contractor is to pay at the rates stated in the Specification;

(*c*) under Sub-Clause 17 .3 [*Intellectual and Industrial Property Rights*] if the Employer receives a claim from a third party for infringement of intellectual or industrial property rights and such infringement is the responsibility of the Contractor, the Contractor must fully indemnify and hold harmless the Employer, including with respect to the legal fees and expenses incurred consequent to the claim; or

(*d*) under the first paragraph of Sub-Clause 17.4 [*Indemnities by Contractor*] if the Employer receives a claim from a third party for personal injury or death or for property damage or loss, arising from or by reason of the Contractor's execution of the Works, the Contractor must fully indemnify and hold harmless the Employer, including with respect to the legal fees and expenses incurred consequent to the claim.

If, by its fraud, gross negligence, deliberate default or reckless misconduct, one Party causes the other Party to incur loss or damage, the exclusions and limitations of liability set out under Sub-Clause 1.15 do not apply. An example of such a situation could be when a Contractor tries to force the Employer to agree to additional payment by intentionally delaying completion of the Works, with the knowledge that the delay will cause the Employer to incur claims from third parties, in excess of the amount sought by the Contractor.

Smith G
ISBN 978-0-7277-6652-6
https://doi.org/10.1680/fcmh.66526.029

Chapter 4
Basic obligations of the employer

4.1. Introduction

Clause 2 [*The Employer*] of FIDIC 2017 contracts sets out some obligations borne by the Employer. It can be considered to cover the items which the Employer must contribute to the project, such as the Site, finance, Site data, Employer's Personnel, Employer's other contractors, Employer-supplied materials, assistance with respect to procuring permits, and so forth.

4.2. Provision of the Site

Under Sub-Clause 2.1 [*Right of Access to the Site*], the Employer must give the Contractor the right of access to, and possession of all parts of the Site within the time or times stated in the Contract Data. However, the Employer is entitled to await receipt of the Performance Security before doing so.

Sub-Clause 1.1.74 defines the Site as:

> *'the places where the Permanent Works are to be executed and to which Plant and Materials are to be delivered, and any other places specified in the Contract as forming part of the Site'.*

This normally includes a reasonable working space outside the footprint of the Permanent Works.

The Special Provisions imposed by the World Bank and some other MDBs give a slightly different definition of the Site, which makes express reference to working space:

> *'"**Site**" means the places where the Permanent Works are to be executed, including storage and working area, and to which Plant and Materials are to be delivered, and any other places specified in the Contract as forming part of the Site.'*

Notwithstanding the contractual definition, the limits of the Site are often shown on a drawing or drawings which form part of the Contract. However, difficulties can arise if inconsistent terminology or undefined expressions or multiple limits are used. For example, if the drawings show the limit of the '*Right of Way*', does this entitle the Contractor to use the entire width of the corridor which is indicated? What if the drawing refers to '*Battery Limit*' rather than site boundary? What if the drawings refer to the '*Right of Way*' and also to the '*land acquisition limit*'?

To avoid misunderstandings, it is important to clearly define in the Contract the limits of the Site, either by a drawing or drawings or by topographical coordinates. If expressions such as *Right of Way*, *Battery Limits* or *Land Acquisition Boundary* are used, they should be explained, with reference to Sub-Clause 2.1.

Another difficulty which can arise, is that the Site area is insufficient to construct the Works, either because there is little or no working space outside the footprint of the Permanent Works or because the extent of the Permanent Works has been underestimated. The World Bank definition draws attention to the need for storage space and working space but does not define the amount which should be allowed.

Examples of each of these situations are given below.

Example 1:

A contract was awarded based on RB 1999 for construction of a large diameter sewer. The drawings of typical cross sections which were included within the Contract showed trenches with side slopes of 1:1.5. The Contractor was required to produce working drawings based on the typical cross sections. Some of the working drawings produced by the Contractor showed that if the typical cross section was to be followed, the trench excavation would be outside the site boundary. The Contractor proposed to reduce the trench width to stay within the site boundary, but the Engineer refused and argued that under Sub-Clause 4.13 [*Rights of Way and Facilities*] the Contractor was to bear all costs and charges for temporary rights of way which may be required for the purposes of the Works and was responsible for obtaining any additional facilities outside the Site required for the purposes of the Works. The Contractor objected, arguing that, as it was obliged to follow the typical cross section, trench excavation and back-filling were not Temporary Works, for which a temporary right of way was required but Permanent Works in relation to which, the Employer must provide the Site.

Example 2:

In relation to a road improvement project under RB 1999, the Contractor was provided with typical drawings, which the Contractor was to develop into working drawings. After producing the working drawings, the Contractor discovered that the toe of the new embankments often fell outside the Right of Way shown on the Contract drawings. The Contractor notified the Engineer when submitting the working drawings for review. The Engineer replied by rejecting the drawings on the basis that the Contractor must redesign or propose alternative construction methods to restrict activities to stay within the Right of Way.

The Employer's obligation is to provide: (a) the right of access and (b) possession of the Site. The right of access is not the same as the means of access which is the subject of Sub-Clause 4.15 [*Access Route*]. The right of access refers to the Contractor's entitlement to enter the Site. The means of access deals with the access routes to the Site. The Employer's obligation to provide the means of access (if any), is also mentioned in the second sentence of the first paragraph of Sub-Clause 2.1 [*Right of Access to the Site*]:

> *'If, under the Contract, the Employer is required to give (to the Contractor) ... means of access, the Employer shall do so in the time and manner stated in the Specification.'*

Thus, there is no general obligation under Sub-Clause 2.1 [*Right of Access to the Site*] for the Employer to provide the means of access to the Site. The Employer only carries such an obligation if it is expressly stated elsewhere in the Contract such as in the Specification or the Particular Conditions.

Possession of the Site relates to a temporary right to occupy the Site for the purpose of executing the Works. Such possession is expected to be 'undisturbed', 'unhindered' or 'quiet'. These expressions indicate that the Contractor's use of the Site must not be disturbed by matters such as legal proceedings brought by a third party claiming to have a legal title to the land. Circumstances such as this, which frequently arise in countries where the land registration system is not well developed, are the responsibility of the Employer.

Although possession of the Site is expected to be undisturbed, Sub-Clause 2.1 [*Right of Access to the Site*] expressly states that the right of access and possession may not be exclusive to the Contractor. This means that the Contractor cannot object to other contractors working on Site at the same time.

The Employer is not obliged to give the Contractor possession of the entire Site at a single date unless a single date is stated in the Contract Data. The Contract Data may provide dates for handover of the Site in phases. In such a case, the Employer must respect the stated dates.

If no dates are stated in the Contract Data, the Employer must provide the right of access to and possession of parts of the Site so that the Contractor can work in accordance with the Programme (or the initial programme submitted under Sub-Clause 8.3 [*Programme*]). Thus, if the Programme foresees that the Contractor will commence construction of a road at the southern end and work northwards, but the Contractor is obliged to commence at the northern end of the Site and work southwards because the southern part of the Site cannot be handed over by the Employer, free of claims from landowners, the Contractor may be entitled to reimbursement of any additional Cost Plus Profit and/or an EOT, subject to Sub-Clause 20.2 [*Claims for Payment and/or EOT*].

The Contractor's entitlement to additional payment and/or EOT related to late possession of the Site may be reduced to the extent that the late possession was caused by any error or delay by the Contractor, such as delayed submission of an applicable Contractor's Document.

Sub-Clause 8.3 [*Programme*] requires the Contractor to submit an initial programme within 28 days of receipt of the Engineer's Notice of Commencement Date. Thus, it could be difficult for the Employer to give possession of the first part of the Site in accordance with this initial programme unless the Employer has been given advance warning, prior to receipt of the initial programme, of which part of the Site is required for the Contractor to begin work.

Under the Special Provisions imposed by the World Bank and some other MDBs this requirement for advance warning of the Contractor's intentions is more pertinent.

Sub-Clause 8.1 [*Commencement of Work*] of these Special Provisions lists several conditions, all of which must be satisfied before the Engineer issues the Notice of the Commencement Date. This Notice must be issued not less than 14 days before the Commencement Date.

One of the listed conditions requires that the Contractor be given effective access to and possession of the Site as required for the commencement of the Works, before the Engineer issues the Notice of the Commencement Date. Thus, the Contractor must be given possession of the relevant part of the Site, and the

Notice of the Commencement Date, not less than 14 days before the Commencement Date, whereas the initial programme, which details the areas required for the initial activities, is only to be submitted within 28 days of receipt of the Notice of the Commencement Date.

4.3. Assistance with respect to permits, copies of Laws, and so forth

Under Sub-Clause 2.2 [*Assistance*], the Employer is required to provide reasonable assistance to the Contractor (if requested by the Contractor to do so) to obtain copies of the laws of the Country and any permits, licences, approvals and so forth required by the laws of the Country.

The assistance is to be provided promptly, that is, without delay. The Employer is not required to bear responsibility for obtaining the permits but only to provide the assistance which an average person would consider to be reasonable. This frequently involves issuing '*supporting letters*' or counter-signing importation documents and may also require the Employer to delegate someone to attend meetings with the customs authorities or other administrations to help resolve issues that arise.

Hint:

The Employer's obligation to provide assistance is subject to a request from the Contractor. If the Contractor does not ask for assistance, it cannot complain that none is provided.

As far as practicable, the Contractor's request should specify the assistance required, for example, provision of a supporting letter, attendance at a meeting and so forth.

The obligation to provide assistance consists of two parts. The first is to assist in obtaining copies of relevant Laws of the Country where the Works are located. These must be in the ruling language as stated in the Contract Data (or, if not stated, in the language of the Conditions).

The second part of the obligation is to assist in obtaining permits, possessions, licences or approvals required by the Laws of the Country which Sub-Clause 1.13 [*Compliance with Laws*] requires the Contractor to obtain for delivery of Goods including customs clearance and for the export of Contractor's Equipment when no longer required for the Works. '*Goods*' are defined in Sub-Clause 1.1.44 of RB 2017 to include the Materials and Plant for inclusion in the Works and the Contractor's Equipment and Temporary Works for the execution of the Works.

The second part of the obligation relates not only to the obtaining of permits, permissions, licences or approvals but also to the obtaining of information which the Contractor is required to submit in support of its application for the permits, permissions, licences or approvals. The requirement to assist in obtaining this particular information was not stated in FIDIC 1999 Contracts or the Pink Book.

4.4. Employer's Personnel and Other Contractors

Sub-Clause 1.1.33 of RB 2017 and YB 2017 defines '*Employer's Personnel*' to include the Engineer, the Engineer's Representative (if any) and assistants as well as the staff, labour and other employees of the Engineer and of the Employer responsible for fulfilment of the Employer's obligations under the Contract. It also includes others who are identified as Employer's Personnel by a Notice to the Contractor issued either by the Employer or the Engineer. The definition under SB 2017 is similar except that '*Engineer*' is replaced by '*Employer's Representative*'.

Under Sub-Clause 2.3 [*Employer's Personnel and Other Contractors*], the Employer must ensure that the Employer's Personnel cooperate with the Contractor with respect to the Contractor's obligations under Sub-Clause 4.6 [*Cooperation*] and that they comply with the same obligations as the Contractor with respect to health and safety and protection of the environment. The Employer must also ensure similar compliance by other contractors working for the Employer on or near the Site.

A new provision found in FIDIC 2017 which was not found in earlier FIDIC contracts, allows the Contractor to require the removal from Site of any member of the Employer's Personnel or of the Employer's other contractors who is found, based on reasonable evidence, to have engaged in corrupt, fraudulent, collusive or coercive acts. If the Contractor seeks to use this provision but the Employer disagrees with the finding of corrupt, fraudulent, collusive or coercive practice, the Contractor could consider that a Claim has arisen under Sub-Clause 20.1 [*Claims*] and seek a determination from the Engineer under Sub-Clause 3.7 [*Agreement or Determination*]. If no agreement is reached and the determination is contested by either Party, the dispute may be referred to the DAAB under Sub-Clause 21.4 [*Obtaining DAAB's Decision*].

4.5. Financial Arrangements

The Contract Data must provide detailed information concerning the Employer's financial arrangements to ensure prompt payment to the Contractor of due amounts. The Employer must immediately give Notice to the Contractor under Sub-Clause 2.4 [*Employer's Financial Arrangement*] with detailed particulars, if the Employer intends to make any change to these arrangements which would affect the Employer's ability to pay the remaining part of the Contract Price as estimated by the Engineer.

The Contractor may request that the Employer provide reasonable evidence that suitable financial arrangements are being maintained if the Contractor becomes aware of such a change in the Employer's financial arrangements for which no Notice has been received, or the Employer is late in making payment of any amount which has become due, or the Contractor is instructed to execute a Variation with a price which exceeds 10% of the Accepted Contract Amount, or a Variation which causes the accumulated total of Variations to exceed 30% of the Accepted Contract Amount.

There is a noticeable absence from the circumstances listed under FIDIC 2017 which permit the Contractor to seek evidence of the Employer's continued ability to pay the Contractor. This is in relation to the adjustment for changes in cost of labour and Goods under Sub-Clause 13.7 [*Adjustments for Changes in Cost*]. Often, the finance provided by the MDB does not cover the adjustments for changes in cost and these must be financed by the borrowing government. In times of high inflation such as has been witnessed worldwide since 2021, the cost adjustment under Sub-Clause 13.7 can be very significant and Contractors should be entitled to request reasonable evidence from the Employer that the necessary funds will be available to cover these adjustments when payment becomes due.

The Employer must provide the reasonable evidence within 28 days of receiving such a request from the Contractor. If the Employer fails to do so, the Contractor may issue a Notice under Sub-Clause 16.1 [*Suspension by Contractor*] indicating its intention to suspend work 21 days after giving the Notice unless and until the Employer has remedied the default. If the Contractor does so suspend or slow down work, it will be entitled to claim an EOT with respect to the delay incurred and/or payment of any additional Cost Plus Profit.

If, 42 days after issuing such a Notice under Sub-Clause 16.1, the Contractor has still not received the reasonable evidence of the financial arrangements, the Contractor will be entitled to give Notice under Sub-Clause 16.2.1 [*Termination by Contractor*] of its intention to terminate the Contract. Unless the Employer provides the reasonable evidence within 14 days of receiving such a Notice, the Contractor may, by giving a second notice (under Sub-Clause 16.2.2), immediately terminate the Contract.

Under FIDIC 1999, there was no obligation for the Employer to give evidence of its financial arrangements unless it received a request from the Contractor, or it intended to make a material change to the arrangements. Under the FIDIC Pink Book, the Employer was required to provide reasonable evidence of its financial arrangements prior to the issue of the Notice of Commencement Date.

Sub-Clause 2.4 of the Special Provisions imposed by the World Bank and some other MDBs when using FIDIC 2017 contracts follows the approach of the Pink Book:

> *'The Employer shall submit, before the Commencement Date, reasonable evidence that financial arrangements have been made for financing the Employer's obligations under the Contract.'*

Under Sub-Clause 8.1 of the Special Provisions, delivery of the reasonable evidence is a condition that must be satisfied before the Engineer can issue the Notice of the Commencement Date.

Sub-Clause 2.4 of the Special Provisions also follows the FIDIC Pink Book with respect to suspension by the Bank of disbursements under the loan agreement. Such a suspension might occur if the Borrower is in serious breach of the loan agreement. In such circumstances:

> *'if the Bank has notified to the Borrower that the Bank has suspended disbursements under its loan, which finances in whole or in part the execution of the Works, the Employer shall give notice of such suspension to the Contractor with detailed particulars, including the date of such notification, with a copy to the Engineer, within 7 days of the Borrower having received the suspension notification from the Bank. If alternative funds will be available in appropriate currencies to the Employer to continue making payments to the Contractor beyond a date 60 days after the date of Bank notification of the suspension, the Employer shall provide reasonable evidence in its notice of the extent to which such funds will be available.'*

Under Sub-Clause 16.1 [*Suspension by the Contractor*] of the Special Provisions, if the Contractor is not provided with reasonable evidence that alternative funds will be available, the Contractor may by notice suspend work or reduce the rate of work at any time, but not less than 7 days after the Borrower received the suspension notification from the Bank. There is no requirement to wait 21 days after giving Notice under Sub-Clause 16.1 before suspending or slowing down work.

Under Sub-Clause 16.2.2 [*Termination*] of the Special Provisions, if the Bank suspends the loan and the Contractor has not received payment under Sub-Clause 14.7 [*Payment*] within 14 days of submission of a Statement of the amount shown on the Statement, the Contractor may either suspend work or reduce the rate of work under Sub-Clause 16.1 or terminate the Contract by giving Notice to the Employer, with a copy to the Engineer. The termination takes effect 14 days after the giving of the Notice. In these circumstances, it seems that the Contractor must choose between suspension or termination and cannot suspend initially and then terminate later.

4.6. Site Data

Under FIDIC 1999, the subject of site data was addressed by Sub-Clause 4.10 [*Site Data*].

The initial responsibility for providing site data lies with the Employer, whereas responsibility for interpreting the data lies with the Contractor. Therefore, under FIDIC 2017, the subject is addressed by Sub-Clause 2.5 [*Site Data and Items of Reference*] as well as Sub-Clause 4.10 [*Use of Site Data*].

Similarly, under FIDIC 1999, the subject of original points, lines and levels of reference was addressed by Sub-Clause 4.7 [*Setting Out*]. Again, the initial responsibility for providing the relevant information lay with the Employer, whereas responsibility for checking the information lay with the Contractor. Therefore, under FIDIC 2017, the subject is addressed under Sub-Clause 2.5 [*Site Data and Items of Reference*] as well as Sub-Clause 4.7 [*Setting Out*].

Prior to the Base Date (that is prior to the date 28 days before the latest date for submission of the Tender), the Employer is to have provided to the Contractor all relevant data in the Employer's possession on the topography of the Site and on the sub-surface, hydrological, climatic and environmental conditions at the Site. However, the disclosure obligation is ongoing, and the Employer must also promptly make available all relevant data that comes into its possession after the Base Date.

Problems sometimes arise with respect to the language of the data. The data provided by the Employer after award of the Contract must be in the ruling language of the Contract or the language for communications, if different. However, the position with respect to the data provided prior to the Base Date is less clear as the Contract does not yet exist and there is, therefore, no contractual obligation with respect to the language.

The position is sometimes made clearer by the Instructions to Tenderers/Bidders which may specify the language to be used for the Tender/Bid.

In this respect the Instructions to Bidders contained in the Standard Bidding Documents used by many of the MDBs and other International Financing Institutions state:

> *'The Bid, as well as all correspondence and documents relating to the Bid exchanged by the Bidder and the Employer, shall be written in the language specified in the BDS [Bid Data Sheet]. Supporting documents and printed literature that are part of the Bid may be in another language provided they are accompanied by an accurate translation of the relevant passages in the language specified in the BDS, in which case, for purposes of interpretation of the Bid, such translation shall govern.'*

Although the Employer is under an obligation to provide the Contractor with all data in its possession, the data is given for information only, and it is the Contractor's responsibility under Sub-Clause 4.10 [*Use of Site Data*] to interpret the data.

As stated above, as well as providing site data, the Employer must provide details concerning the original survey control points and reference lines and levels. The details must be shown on the Drawings and/or given in the Specifications or issued by a Notice from the Engineer. Under Sub-Clause 4.7 [*Setting Out*], the Contractor must verify the accuracy of the reference information before using it.

The Contract Data may indicate the period during which the Contractor must complete the verification. If no period is stated and the reference information is contained within the Specification or shown on Drawings, the period is 28 days after the Commencement Date. If the information is issued by a Notice from the Engineer, the Contractor must complete the verification as soon as practicable.

If the Contractor discovers an error in the information, it must give Notice to the Engineer under Sub-Clause 4.7.2 describing the error. The Engineer must then follow the procedure set out under Sub-Clause 4.7.3 [*Agreement or Determination of rectification measures, delay and/or Cost*] and Sub-Clause 3.7 [*Agreement or Determination*] to agree or determine whether there is an error in the items of reference and, if so, the consequences of that error.

Sub-Clause 4.7.3 [*Agreement or Determination of rectification measures, delay and/or Cost*] of RB 2017 begins:

> *'After receiving a Notice from the Contractor under Sub-Clause 4.7.2 [Errors] the Engineer shall proceed under Sub-Clause 3.7 [Agreement or Determination] to agree or determine:*
>
> (*a*) *whether or not there is an error in the items of reference;*
> (*b*) *whether or not (taking account of cost and time) an experienced contractor exercising due care would have discovered such an error*
>
> - *when examining the Site, the Drawings and the Specification before submitting the Tender; or*
> - *if the items of reference are specified on the Drawings and/or in the Specification and the Contractor's Notice is given after the expiry of the period stated in sub-paragraph (a) of Sub-Clause 4.7.2; and*
>
> (*c*) *what measures (if any) the Contractor is required to take to rectify the error (and, for the purpose of Sub-Clause 3.7.3 [Time limits], the date the Engineer receives the Contractor's Notice under Sub-Clause 4.7.2 [Errors] shall be the date of commencement of the time limit for agreement under Sub-Clause 3.7.3).'*

The Errata published by FIDIC soon after publication of RB 2017 modify the above sub-paragraph (b) so that it reads:

> *'(b) whether or not (taking account of cost and time) an experienced contractor exercising due care would have discovered such an error*
>
> - *when examining the Site, the Drawings and the Specification before submitting the Tender; or*
> - *when examining the items of reference within the period stated in sub-paragraph (a) of Sub-Clause 4.7.2, if the items of reference are specified on the Drawings and/or in the Specification; and'*

This correction does not clarify the procedure because if the Engineer is reacting to a valid Notice from the Contractor under Sub-Clause 4.7.2 [*Errors*] it should not be necessary to determine whether an experienced contractor would have discovered the error within the stated period. By sending the Notice, the Contractor has demonstrated that it did discover the error!

The remainder of Sub-Clause 4.7.3 states:

> *'If, under sub-paragraph (b) above, an experienced contractor would not have discovered the error:*
>
> (i) *Sub-Clause 13.3.1 [Variation by Instruction] shall apply to the measures that the Contractor is required to take (if any); and*
> (ii) *if the Contractor suffers delay and/or incurs Cost as a result of the error, the Contractor shall be entitled subject to Sub-Clause 20.2 [Claims for Payment and/or EOT] to EOT and/or payment of such Cost Plus Profit.'*

Thus, although not expressly stated, if the Contractor finds an error in the reference information contained in the Specifications or shown in the Drawings after the period stated in the Contract Data (if any) or more than 28 days from the Commencement Date if no period is stated, it seems that it is not entitled to an EOT and/or additional payment because there is no express requirement for the Engineer to seek to agree or to determine the matter.

If the reference information was contained in the Specification or shown on the Drawings and an experienced contractor, exercising due care (taking account of cost and time) would have discovered the error before submitting its Tender, the Contractor is not entitled to an EOT and/or additional payment.

If the reference information was contained in the Specification or shown on the Drawings and an experienced contractor, exercising due care, would not have discovered the error before submitting its Tender but would have discovered the error during the period stated in the Contract Data (if any) or within 28 days of the Commencement Date (if no period is stated) the Contractor is not entitled to an EOT and/or additional payment. This appears to be a mistake. If the Contractor discovered an error during the verification period which would not have been found by an experienced contractor before submitting its Tender, the Contractor should be entitled to an EOT and/or additional payment.

If the reference information was contained in the Specification or shown on the Drawings and an experienced contractor exercising due care would not have discovered the error during the period stated in the Contract Data (if any) or within 28 days of the Commencement Date (if no period is stated), and the Contractor gave Notice of the error within the period, it is entitled to an EOT and/or additional payment. This is clearly unworkable. If an experienced contractor would not have discovered the error during the period stated in the Contract Data or within 28 days of the Commencement Date, the Contractor could not submit Notice of the error within the said time to protect its rights to compensation.

Sub-Clause 4.7.3 does not state that the Contractor is entitled to an EOT and/or additional payment if the reference information provided by Notice from the Engineer is found to contain an error. This oversight has been corrected by FIDIC in the 2022 Reprint (see Chapter 19 below).

Under Sub-Clause 2.5 [*Site Data and Items of Reference*] of SB 2017, the Employer has no responsibility for the accuracy, sufficiency or completeness of site data and/or reference information. The only exceptions to this general rule are listed under Sub-Clause 5.1 [*General Design Obligations*].

Figure 4.1 Errors in site data or items of reference

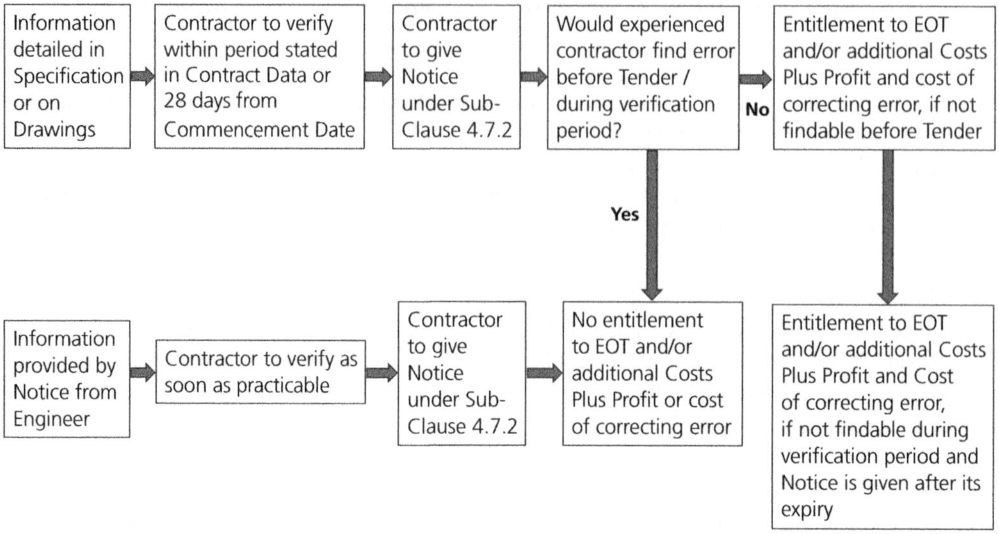

Smith G
ISBN 978-0-7277-6652-6
https://doi.org/10.1680/fcmh.66526.039

Chapter 5
The role of the engineer/employer's representative

5.1. The Person

Under Sub-Clause 3.1 [*The Engineer*], the Employer must appoint the Engineer under RB 2017 and YB 2017 or the Employer's Representative under SB 2017. (Under SB 1999, the Employer was not under an obligation to appoint an Employer's Representative.)

Sub-Clause 1.1.35 of RB 2017 defines the '*Engineer*' as the person named in the Contract Data appointed by the Employer to act as the Engineer for the purposes of the Contract, or any replacement appointed under Sub-Clause 3.6 [*Replacement of the Engineer*]. Sub-Clause 1.1.31 of SB 2017 contains a similar definition with respect to the 'Employer's Representative'.

In view of these definitions, the Engineer under RB 2017 or YB 2017 and the Employer's Representative under SB 2017, must be appointed before the Tender documents are issued to contractors so that the appointed person can be named in the Contract Data.

If no Engineer is appointed, or its appointment expires, it will be impossible to operate an RB or YB contract because the duties of the Engineer are extensive, and many cannot be undertaken by the Employer. This situation arises most frequently because the Works Contract is awarded before the Engineer is selected or because completion of the Works is delayed and the Employer fails to extend the Engineer's contract, which is usually time based. In such cases, the Contractor sometimes terminates the Works Contract because the absence of the Engineer is seen as a material breach of an Employer's obligation. The situation can also arise when the Employer fails to pay the Engineer and the Engineer suspends its services or terminates its agreement.

If a company is appointed as the Engineer or Employer's Representative, the company must give Notice to the Parties of the natural person who is appointed to act on its behalf. In the case of RB 2017 and YB 2017, that person must be a professional engineer with suitable qualifications, experience and competence to fulfil the role of the Engineer under the Contract and be fluent in the ruling language. Under SB 2017, the Employer's Representative, or the natural person appointed to act on its behalf, must act as a skilled professional, be competent to fulfil the role and be fluent in the ruling language.

If the Employer wishes to replace the Engineer, it must give the Contractor a Notice under Sub-Clause 3.6 [*Replacement of the Engineer*] of the name, address and relevant experience of the intended replacement. The Notice must be issued not less than 42 days before the intended date of replacement. From the date of receipt of the Employer's Notice, the Contractor has 14 days in which to object, with reasons, by Notice to the Employer. If no objection is raised, the Contractor is deemed to have accepted the replacement.

If the Contractor does object within the 14-day period, and the objection is reasonable, the Employer cannot replace the Engineer with the proposed replacement.

The Employer may replace the Engineer immediately if the Engineer is unable to act because of death, illness, disability or resignation. The Employer may also replace the Engineer immediately if the Engineer is a company and becomes unable or unwilling to carry out any of its duties, other than for a cause attributable to the Employer. An example of such a circumstance could be the imposition of sanctions against the Country which prevent the company from continuing its involvement in the project. Such immediate replacement is subject to Notice from the Employer to the Contractor, informing the Contractor of the reasons for the replacement and the name, address, and relevant experience of the replacement. Such immediate replacement is considered to be temporary until another replacement is appointed or unless the Contractor accepts that the person appointed temporarily should continue as a permanent replacement.

The Employer is not permitted to appoint a replacement Engineer because the Engineer has suspended its services due to late payment by the Employer, as such a suspension is considered to be attributable to the Employer.

Under SB 2017, if the Employer wishes to replace the Employer's Representative, the procedure is the same as under RB 2017 and YB 2017, but the Notice must be given to the Contractor not less than 14 days before the intended date of replacement.

5.2. Authority

Under Sub-Clause 3.1 [*The Engineer*] of RB 2017 and YB 2017, the person appointed as the Engineer must have all the authority necessary to act as the Engineer under the Contract. Similarly, the Employer's Representative under SB 2017 must have and is deemed to have (unless and until the Contractor is notified otherwise) full authority to represent the Employer under the Contract except that the Employer's Representative does not have authority to terminate the Contract under Sub-Clause 15 [*Termination by Employer*]. The authority of the natural person takes effect upon receipt by both Parties of the Notice informing the Parties of that person's appointment.

Under Sub-Clause 3.2 [*Engineer's Duties and Authority*] of RB 2017 and YB 2017, the Engineer must act as a skilled professional and is deemed to act for the Employer when carrying out duties or exercising any authority specified or implied in the Contract (except under Sub-Clause 3.7 [*Agreement or Determination*]).

If the Engineer is required to obtain the consent of the Employer before exercising its authority, such requirements must be stated in the Particular Conditions and the Employer may not impose additional constraints, except with the agreement of the Contractor. However, whenever the Engineer exercises a specified authority for which the Employer's consent is required, such consent is deemed to have been given. Thus, it is not necessary for the Contractor to check that the Employer was consulted and gave its consent, as the Employer is bound towards the Contractor as if it had been consulted and had consented. However, in such circumstances, the Employer might be able to hold the Engineer liable for breaching the Engineer's contract by exercising authority without the Employer's consent.

The position with respect to constraints on the Engineer's authority was largely similar under FIDIC 1999. However, under the Pink Book, there was an express requirement for the Engineer to obtain the prior approval of the Employer before:

(*a*) *agreeing or determining an extension of time and/or additional cost under Sub-Clause 4.12 [Unforeseeable Physical Conditions].*
(*b*) *instructing a Variation under Sub-Clause 13.1 [Right to Vary], except:*
   (i) *in an emergency as determined by the Engineer, or*
   (ii) *if such a Variation would increase the Accepted Contract Amount by less than the percentage specified in the Contract Data.*
(*c*) *approving under Sub-Clause 13.3 [Variation Procedure], a proposal for Variation submitted by the Contractor in accordance with Sub-Clause 13.1 [Right to Vary] or Sub-Clause 13.2 [Value Engineering].*
(*d*) *specifying under Sub-Clause 13.3 [Variation Procedure], the amount payable in each of the applicable currencies.*

Moreover, under the Pink Book, the Employer was able to change the authority of the Engineer simply by informing the Contractor.

In practice, within the Particular Conditions under FIDIC 1999 and the Pink Book, Employers have often inserted additional constraints on the Engineer's authority, particularly with respect to determinations related to claims for EOT and/or additional payment and for instructing or approving Variations.

To counter this practice, RB 2017 and YB 2017 include an express statement that there is no requirement for the Engineer to obtain the Employer's consent before the Engineer exercises its authority under Sub-Clause 3.7 [*Agreement or Determination*] and the Employer may not impose such a constraint after award of the Contract. However, it is too soon to know whether this will be sufficient to dissuade Employers from adding such a constraint.

Under Sub-Clause 3.2 [*Engineer's Duties and Authority*] of RB 2017 and YB 2017, the Engineer has no authority to amend the Contract. Nor does the Engineer have the power to relieve either Party of any duty, obligation or responsibility, except under certain limited circumstances stated in the Conditions. An example of such a limited power can be found under Sub-Clause 9.4 [*Failure to Pass Tests on Completion*] which permits the Engineer to issue a TOC despite the Works having repeatedly failed Tests on Completion, but the Engineer can only do so if requested by the Employer and subject to the Employer's right to claim a reduction in the Contract Price.

Under SB 2017, as the Employer's Representative has full power to represent the Employer except with respect to termination, it does have power to amend the Contract with the agreement of the Contractor.

5.3. Delegation

Under Sub-Clause 3.3 [*The Engineer's Representative*] of RB 2017 and YB 2017, the Engineer may appoint an Engineer's Representative who has all the authority necessary to act on behalf of the Engineer at the Site. However, the Engineer's Representative has no power to appoint its own replacement.

The Engineer's Representative, who must be a qualified engineer with the necessary qualifications, experience and competence as well as being fluent in the ruling language, must be based at the Site for the whole time that the Works are being executed. If the Engineer's Representative is to be temporarily absent from the Site, an equivalent replacement must be appointed by the Engineer by giving Notice to the Contractor.

Given that the Engineer is not obliged to appoint an Engineer's Representative, but if he does, the Engineer's Representative or his temporary replacement must always be based at the Site, it must be implied that if the Engineer does not appoint an Engineer's Representative, the Engineer must always be based at the Site.

Under Sub-Clause 3.4 [*Delegation by the Engineer*] of RB 2017 and YB 2017, the Engineer may appoint assistants to whom it assigns duties and delegates authority. The assistants must be suitably qualified natural persons with the necessary experience and competence as well as being fluent in the language for communications. The Engineer may assign the duties and delegate authority by sending Notice to the Parties, describing the assigned duties and the delegated authority of each assistant. The assignment and delegation take effect upon receipt of the Notice by both Parties and may be revoked in the same manner.

The Engineer may not delegate authority with respect to Sub-Clause 3.7 [*Agreement or Determination*] or Sub-Clause 15.1 [*Notice to Correct*]. Under FIDIC 1999 and the Pink Book, the Engineer could not delegate authority with respect to determinations but delegation of authority with respect to notices to correct was not forbidden.

Any act by an assistant, within the limits of the assignment or delegation mentioned in the Notice from the Engineer, has the same effect as if the act had been by the Engineer. However, the assistant is not authorised to issue instructions which exceed the authority granted by the Notice. An example of such exceeding of authority could be the signing of Daywork Records by an assistant who is not authorised to do so.

Thus, when a Contractor receives an instruction from an Engineer's assistant or when an assistant signs an inspection form, a non-compliance notice or similar, the Contractor should always consider whether the assistant has the necessary authority. If the Contractor questions whether the assistant has the necessary authority, the Contractor may, by giving a Notice, refer the matter to the Engineer. If the Engineer does not respond within 7 days of receiving the Contractor's Notice, the instruction, communication or Notice is deemed to be confirmed. A period of 7 days is excessive if the Contractor wishes to receive confirmation before complying with the instruction.

Under the last paragraph of Sub-Clause 3.2 [*Engineer's Duties and Authority*] of RB 2017 and YB 2017, any acceptance, agreement, approval, check, certificate, comment, consent, disapproval, examination, inspection, instruction, Notice, No-objection, record(s) of meeting, permission, proposal, record, reply, report, request, Review, test, valuation or similar act (including the absence of any such act) by the Engineer, the Engineer's Representative or any assistant does not relieve the Contractor from any duty, obligation or responsibility.

Under Sub-Clause 3.3 [*Delegated Persons*] of SB 2017, any such acceptance, agreement, approval and so forth by a person to whom the relevant authority has been delegated by Notice to the Contractor has the same effect as if the said act had been performed by the Employer. However, unless stated otherwise in the communication from the delegated person, the act does not relieve the Contractor from any duty, obligation, or responsibility. Moreover, the Contractor may challenge any instruction, communication or Notice from the delegated person by giving Notice to the Employer. If the Employer does not respond within 7 days of receiving the Contractor's Notice, the instruction, communication, or Notice is deemed to be confirmed. As stated above, the period of 7 days is excessive if the Contractor awaits confirmation before complying with the instruction.

5.4. Instructions

One of the most significant sub-clauses in FIDIC contracts deals with the power of the Engineer to issue instructions to the Contractor. In RB 2017 and YB 2017, the relevant sub-clause is Sub-Clause 3.5 [*Engineer's Instructions*]. In SB 2017, a similar provision with respect to the Employer/Employer's Representative is contained within Sub-Clause 3.4 [*Delegation by the Engineer*], the only difference being that the instruction must state the obligation(s) and sub-clause to which it relates.

The Engineer, the Engineer's Representative or an assistant to whom the necessary authority has been delegated (or the Employer's Representative or an assistant under SB 2017) may give an instruction to the Contractor at any time and the Contractor must comply. The only limitation on this power (except under very limited circumstances which are explained below) is that the instruction must be necessary for the execution of the Works. Clearly, this is subjective and not very restrictive.

The FIDIC 2017 Contracts Guide [2022] suggests that the Contractor must even comply with an instruction from the Engineer that is not validly given (such as an instruction to complete the Works within the Time for Completion, despite delays having arisen from causes which entitle it to an EOT), but advises the Contractor to serve Notice on the Engineer/Employer, stating that the instruction is considered to be a Variation under Sub-Clause 3.5 (a) of RB 2017 or YB 2017, Sub-Clause 3.4 (a) of SB 2017.

If the instruction under the FIDIC 2017 contracts states that it constitutes a Variation, Sub-Clause 13.3.1. [*Variation by Instruction*] applies. If the instruction does not mention that it is a Variation but the Contractor believes that it is (or that it involves work that is a part of an existing Variation) or if the Contractor considers that the instruction does not comply with applicable Laws or will reduce the safety of the Works or is technically impossible, the Contractor must not comply with the instruction but must immediately give Notice to the Engineer (or the Employer under SB 2017) with reasons. The instruction will be considered revoked unless the Engineer (or the Employer under SB 2017) responds within 7 days, by giving a Notice which confirms, reverses or varies the instruction.

The Contractor must comply with this response. This means that even if the Contractor considers that the instruction does not comply with applicable Laws or that it will reduce the safety of the Works, if the Engineer (Employer under SB 2017) confirms the instruction, the Contract requires the Contractor to comply.

> **Hint:**
>
> This contradicts the Contractor's obligation under Sub-Clause 1.13 [*Compliance with Laws*] to comply with all applicable Laws. Moreover, in many legal jurisdictions, the Contractor will be liable if it complies with the instruction, knowing that the instruction does not comply with the Laws or that it will have a negative impact on safety. Therefore, the Contractor should seek legal advice before complying.

To some extent, Sub-Clause 3.5 [*Engineer's Instructions*] of RB 2017 and YB 2017 (Sub-Clause 3.4 [*Instructions*] of SB 2017) is also in conflict with Sub-Clause 13.1 [*Right to Vary*], which entitles the Contractor to object to a Variation instruction on more grounds than those listed under Sub-Clause 3.5 [*Engineer's Instructions*]. Under Sub-Clause 13.1 of RB 2017, the Contractor is not bound by a Variation instruction if the Contractor promptly gives Notice to the Engineer (with supporting particulars) that:

(*a*) *the varied work was Unforeseeable having regard to the scope and nature of the Works described in the Specification;*
(*b*) *the Contractor cannot readily obtain the Goods required for the Variation; or*
(*c*) *it will adversely affect the Contractor's ability to comply with Sub-Clause 4.8 [Health and Safety Obligations] and/or Sub-Clause 4.18 [Protection of the Environment].*

Sub-Clause 13.1 [*Right to Vary*] of YB 2017 and SB 2017 contains two more possible reasons for refusing to immediately comply, which are:

(*d*) *it will have an adverse impact on the achievement of the Schedule of Performance Guarantees; or*
(*e*) *it may adversely affect the Contractor's obligation to complete the Works so that they shall be fit for the purpose(s) for which they are intended under Sub-Clause 4.1 [Contractor's General Obligations].*

Thus, if the Contractor receives an instruction from the Engineer which does not state that it is a Variation, but the Contractor considers that it does constitute a Variation, the Contractor should immediately send Notice to the Engineer under Sub-Clause 3.5 [*Engineer's Instructions*] and should list any of the applicable reasons listed under Sub-Clause 13.1 [*Right to Vary*] which entitle the Contractor to refuse to comply.

It should be noted however, that even after receiving Notice from the Contractor under Sub-Clause 13.1 [*Right to Vary*], the Engineer may confirm or vary its instruction, in which case, the Contractor must comply.

Under RB 1999 and the Pink Book, it was possible for the Engineer or a delegated assistant to issue an instruction orally and for this instruction to be confirmed in writing by the Contractor within two working days. This confirmation constituted the written instruction unless it was contested by the Engineer or delegated assistant within two more working days. Such oral instructions are not foreseen under FIDIC 2017 contracts, but the Notes on the Preparation of Special Provisions propose wording similar to that contained within RB 1999 for insertion in the Special Provisions if the Parties wish to allow oral instructions.

Figure 5.1 Engineer's instructions

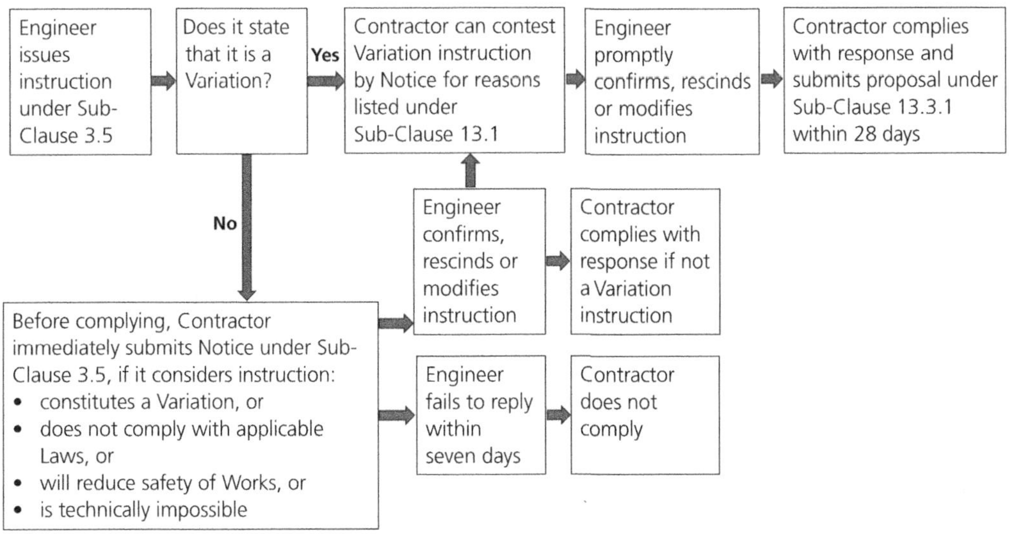

5.5. Determinations

In the FIDIC 1999 contracts and the Pink Book, Sub-Clause 3.5 [*Determinations*] consisted of a little more than 100 words. It was a relatively simple process in two stages: the Engineer was to consult the Parties to try to obtain an agreement, but if unsuccessful, the Engineer issued a determination (with reasons) which was binding on the Parties unless and until revised by the Dispute Adjudication Board.

Under FIDIC 2017 contracts (Sub-Clause 3.7 in RB 2017 and YB 2017, Sub-Clause 3.5 in SB 2017), the procedure is much more complex and extends to two and a half pages.

The following table lists sub-clauses under RB 1999 and RB 2017 which provide for determinations by the Engineer. RB 1999 provided for determinations issued in accordance with Sub-Clause 3.5 [*Determinations*] but also provided, in three instances, for determinations by the Engineer without reference to Sub-Clause 3.5. RB 2017 has eliminated this dichotomy and all determinations are to be issued in accordance with the procedure set out under Sub-Clause 3.7 [*Agreement or Determination*].

At first sight, the scope for Engineer's determinations appears to be less extensive under RB 2017 than under RB 1999. However, this is not the case. RB 2017 makes less reference to Engineer's determinations only because Sub-Clauses 20.1 and 20.2 provide for determinations in relation to all circumstances that might give rise to a Claim, whereas under RB 1999, several sub-clauses which address a potential source of claim, make reference to Engineer's determinations. However, under RB 1999 the approach is inconsistent. For example, Sub-Clause 2.1 [*Right of Access to the Site*] contains an express reference to Sub-Clause 3.5 [*Determinations*] when a claim arises but Sub-Clause 8.7 [*Delay Damages*] does not. RB 2017 is much more consistent in this respect.

RB 1999		RB 2017
Sub-Clauses providing for Sub-Clause 3.5 determination	Other determinations by Engineer	Sub-Clauses providing for Sub-Clause 3.7 determination
1.9 Delayed Drawings or Instructions		
2.1 Right of Access to the Site		
2.5 Employer's Claims		
4.7 Setting Out		4.7 Setting Out
4.12 Unforeseeable Physical Conditions		4.12 Unforeseeable Physical Conditions
4.19 Electricity, Water and Gas		
4.20 Employer's Equipment and Free Issue Materials		
4.24 Fossils		
7.4 Testing		
8.4 Extension of Time for Completion		8.5 Extension of Time for Completion
9.4 Consequences of Suspension		
10.2 Taking Over of Parts of the Works		10.2 Taking Over of Parts of the Works
10.3 Interference with Tests on Completion		
		11.2 Cost of Remedying Defects
11.4 Failure to Remedy Defects		
11.8 Contractor to Search		
		12.1 Works to be Measured
12.3 Evaluation		12.3 Valuation of the Works
12.4 Omissions		
13.2 Value Engineering		
		13.3 Variation procedure
		13.5 Daywork
13.7 Adjustments for Changes in Legislation		
	13.8 Adjustments for Changes in Cost	
14.4 Schedule of Payments		14.4 Schedule of Payments

RB 1999				**RB 2017**	
Sub-Clauses providing for Sub-Clause 3.5 determination		Other determinations by Engineer		Sub-Clauses providing for Sub-Clause 3.7 determination	
		14.5	Plant and Materials intended for the Works	14.5	Plant and Materials intended for the Works
				14.6	Issue of IPC
15.3	Valuation at date of Termination			15.3	Valuation after Termination for Contractor's Default
				15.6	Valuation after Termination for Employer's Convenience
16.1	Contractor's Entitlement to Suspend Work				
17.4	Consequences of Employer's Risks				
19.4	Consequences of Force Majeure				
		19.6	Optional Termination, Payment and Release	18.5	Optional Termination
20.1	Contractor's Claims			20.1	Claims
				20.2	Claims for Payment and/or EOT

Sub-Clause 3.7 [*Agreement or Determination*] of RB 2017 and YB 2017 begins with the statement that the Engineer must act neutrally when carrying out its duties under the sub-clause and is not deemed to act for the Employer. This statement should be read in conjunction with the provision under Sub-Clause 3.2 [*Engineer's Duties and Authority*], that there shall be no requirement for the Engineer to obtain the Employer's consent before exercising the Engineer's authority under Sub-Clause 3.7 [*Agreement or Determination*]. There is a clear intention therefore that the Engineer will work in an unbiased manner in seeking an agreement or in determining the issue. In this respect the Notes on the Preparation of Tender Documents state:

> *'By these statements it is intended that, although the Engineer is appointed by the Employer and acts for the Employer in most other respects under the Contract, when acting under this Sub-Clause the Engineer treats both Parties even-handedly, in a fair-minded and unbiased manner.'*

Under Sub-Clause 3.5 [*Agreement or Determination*] of SB 2017 there is no requirement for the Employer's Representative to act neutrally but nevertheless, the Employer's Representative is not deemed to act for the Employer.

> **Hint:**
>
> Given that there is no provision (like that contained in Sub-Clause 3.2 of RB 2017 and YB 2017) that there shall be no requirement for the Employer's Representative to obtain the Employer's consent before exercising authority under Sub-Clause 3.5 of SB 2017, it is difficult to imagine how the Employer's Representative can be considered not to act for the Employer in this situation.
>
> Indeed, the Notes on the Preparation of Special Provisions suggest that the second leg of the Sub-Clause 3.5 procedure might be replaced by a referral to the DAAB for a decision rather than requesting a determination of the issue by the Employer's Representative.

As in the FIDIC 1999 contracts, the procedure under FIDIC 2017 consists of two stages: a first stage during which the Engineer/Employer's Representative is to seek to obtain an agreement between the Parties and, if this is not possible, a second stage during which the Engineer/Employer's Representative issues a determination with respect to the issue.

5.5.1 Consultations to reach agreement

The Engineer has a period of 42 days under Sub-Clause 3.7.1 [*Consultation to reach agreement*] in which to try to obtain an agreement between the Parties through joint or separate consultations with the Parties. To maximise the chances of success within this period, which is known as the '*time limit for agreement*', the Engineer must start the consultations as soon as possible.

The Engineer is to encourage discussion between the Parties and must provide the Parties with a record of the consultations. This implies that the consultations will consist of more than a simple exchange of correspondence, which was the approach often taken by some Engineers under the FIDIC 1999 contracts and the Pink Book.

The Engineer may propose a different period for attempting agreement, but unless both Parties agree, the period of 42 days is fixed. The start date of the period depends on the type of issue being addressed (Sub-Clause 3.7.3 [*Time limits*]).

For a matter to be agreed or determined which is not a Claim, the start date is the date stated in the relevant sub-clause.

An example is found under Sub-Clause 11.2 [*Cost of Remedying Defects*] and relates to disagreements about the cause of defects discovered during the DNP. If the Contractor considers that the repair work is attributable to a cause for which it is not responsible, the Contractor must give a Notice to the Engineer promptly and the date of this Notice is the start date of the time limit for agreement under Sub-Clause 3.7.3 [*Time limits*].

If the matter which is to be agreed or determined is a Claim under sub-paragraph (c) of Sub-Clause 20.1 [*Claims*], the start date is the date at which the Engineer receives a Notice from the claiming Party under Sub-Clause 20.1 [*Claims*]. Sub-paragraph (c) covers claims for entitlements other than additional payment and/or extensions of the DNP or EOT. The Notice is to be given as soon as practicable after the claiming Party becomes aware of a disagreement with the Engineer or the other Party with respect to the alleged entitlement (or there has been no response within a reasonable period).

If the matter to be agreed or determined is a Claim under Sub-Clause 20.1 [*Claims*], that is sub-paragraph (a) (a claim from the Employer for additional payment and/or an extension of the DNP) or sub-paragraph (b) (a claim from the Contractor for additional payment and/or an EOT), then the start date is:

- the date at which the Engineer receives a fully detailed Claim under Sub-Clause 20.2.4 [*Fully detailed Claim*]; or
- if the Claim is of continuing effect under Sub-Clause 20.2.6 [*Claims of continuing effect*], the date at which the Engineer receives an interim Claim or a final fully detailed Claim (as the case may be).

Given that Sub-Clause 20.2.6 [*Claims of continuing effect*] requires interim Claims to be submitted at monthly intervals, in such circumstances there will be overlapping consultation periods (see below). In consequence, it is highly unlikely that agreement will be reached before submission of the final fully detailed claim.

If the Parties reach an agreement within the time limit for agreement, they must sign a document recording the agreement and this is to be attached to a '*Notice of the Parties' Agreement*' issued by the Engineer.

An example of a Notice of Parties' Agreement is as follows:

A.N. OTHER & Partners
Consulting Engineers

XANADU ROADS DEPARTMENT
Attention: Mr. B. Ware, Chief Engineer

BETTA CONTRACTORS LTD.
Attention: Mr. Alright, Contractor's Representative

31 December 2020

Subject: **NOTICE OF THE PARTIES' AGREEMENT**

Dear Sirs,

In accordance with Sub-Clause 3.7.1 [*Consultation to reach agreement*], I am pleased to formally provide you with signed originals of the agreement which was reached following consultations with respect to Contractor's Claim N°1 which arose from the COVID 19 pandemic.

I remind you that in accordance with Sub-Clause 3.7.4 [*Effect of the agreement or determination*], the agreement is binding on both Parties.

Yours faithfully,

The Engineer
encl. Copy of the signed agreement related to Contractor's Claim No. 1

Figure 5.2 Agreement of a matter which is not a Claim

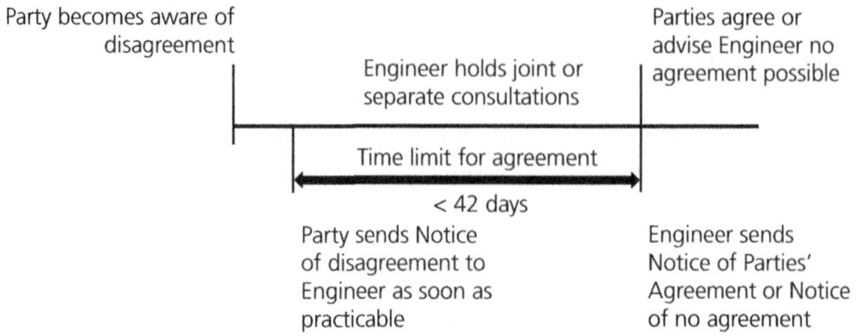

Figure 5.3 Agreement of a matter which is a Claim

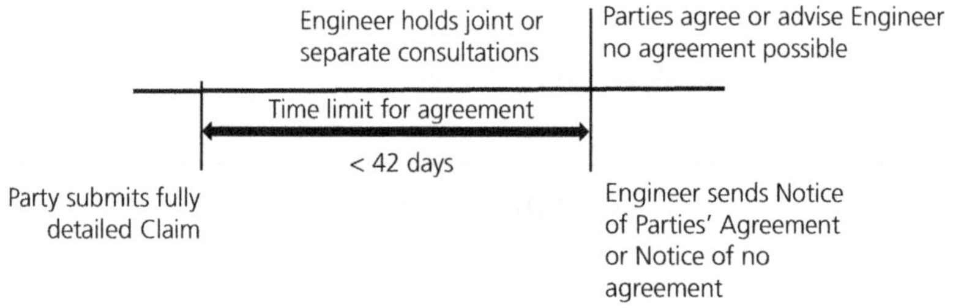

Figure 5.4 Agreement of a matter which is a Claim of continuing effect

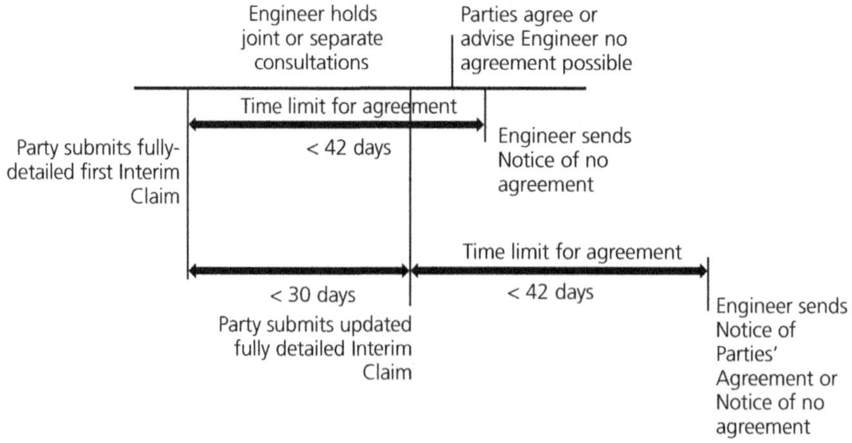

If the Parties fail to reach an agreement within the time limit for agreement, or they jointly inform the Engineer that no agreement is possible within the time limit, the Engineer must issue a Notice to the Parties under Sub-Clause 3.7.1 [*Consultation to Reach Agreement*] recording this fact and immediately proceed to the determination stage.

5.5.2 Engineer's Determination

Within 42 days of the issue of the Engineer's Notice of no agreement under Sub-Clause 3.7.1 [*Consultation to reach agreement*] (or within such other period as may be proposed by the Engineer and agreed by both Parties) the Engineer must issue to both Parties a Notice of the Engineer's Determination under Sub-Clause 3.7.2 [*Engineer's Determination*] which describes the determination in detail, with reasons and detailed supporting particulars. This time limit is known as '*the time limit for determination*'.

The determination must be fair and must be in accordance with the Contract, taking due regard of all relevant circumstances. In other words, it must be an unbiased interpretation of the Contract and its application to the facts.

Sub-Clause 3.7.3 [*Time limits*] ends with the statement that if the Engineer fails to issue the Notice of the Parties' Agreement or the Notice of the Engineer's Determination within the relevant time limit:

(*a*) in the case of a Claim, the Engineer is deemed to have given a determination rejecting the Claim and a Party must serve a NOD within 28 days if it wishes to refer the Dispute to the DAAB (see below); or

(*b*) in the case of a matter to be agreed or determined which is not a Claim, the matter is deemed to be a Dispute and may be referred by either Party to the DAAB for its decision under Sub-Clause 21.4 [*Obtaining DAAB's Decision*]. In such circumstances there is no requirement to submit a NOD (and Sub-Clause 3.7.5 [*Dissatisfaction with Engineer's determination*] and sub-paragraph (a) of Sub-Clause 21.4.1 [*Reference of a Dispute to the DAAB*] do not apply).

Taken literally, this means that if the Parties are unable to reach an agreement within the time limit for agreement, the Engineer is deemed to have given a determination rejecting the Claim. This was an error which has been corrected in the 2022 Reprint (see Chapter 19).

Figure 5.5 Determination of a matter which is not a Claim. Note: This is the position prior to the 2022 Reprint

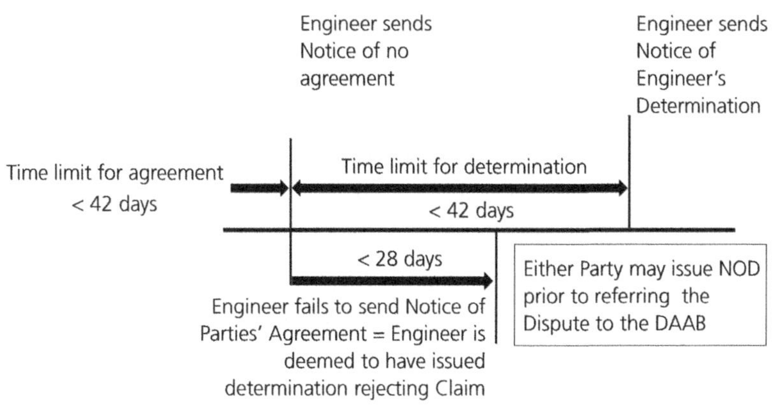

Figure 5.6 Determination of a matter which is a Claim. Note: This is the position prior to the 2022 Reprint

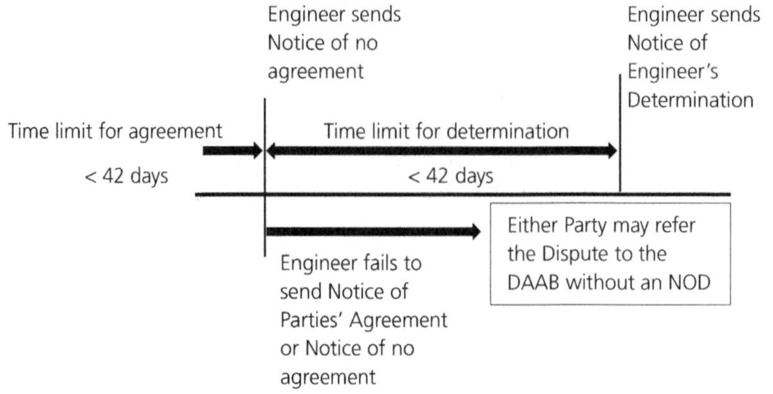

An example of a Notice of Engineer's Determination is as follows:

A.N. OTHER & Partners
Consulting Engineers

XANADU ROADS DEPARTMENT
Attention: Mr. B. Ware, Chief Engineer

BETTA CONTRACTORS LTD.
Attention: Mr. Alright, Contractor's Representative

31 December 2020

Subject: **NOTICE OF THE ENGINEER'S DETERMINATION**

Dear Sirs,

In accordance with Sub-Clause 20.2.5 [*Agreement or Determination of the Claim*] and Sub-Clause 3.7.2 [*Engineer's Determination*], please find below the Engineer's Determination with respect to Contractor's Claim N°1 which arose from the COVID 19 pandemic.

The Contractor's Claim was primarily based on Sub-Clause 18.4 [*Consequences of an Exceptional Event*]. The Exceptional Event was alleged to be the declaration by the World Health Organisation of the COVID 19 pandemic in March 2020 and the restrictions on travel imposed by numerous countries in response. To be entitled to an EOT and/or payment of Costs, the Contractor is required to have served Notice within 14 days of when it became aware or should have become aware of the Exceptional Event. The Notice was to specify the obligations which the Contractor was prevented by the Exceptional Event from performing. Moreover, a further Notice was to be submitted under Sub-Clause 20.2 [*Claims for Payment and/or EOT*] within 28 days of when the Contractor became aware or should have become aware of the Exceptional Event.

> Although the said Notices were received on time, you have never been able to explain which obligations were prevented by the declaration of the pandemic. Instead, you have sought to ascribe to the pandemic the travel restrictions imposed by various governments. In reality, the travel restrictions result from changes in legislation. As such, I do not agree that Sub-Clause 18.4 [*Consequences of an Exceptional Event*] applies. However, I do accept your alternative argument that Sub-Clause 8.5 (d) entitles you to an EOT to the extent that you suffered Unforeseeable shortages in the availability of personnel caused by the pandemic or governmental actions.
>
> The detailed analysis attached to this determination shows that the restrictions imposed by the Government of the Republic of Xanadu were of limited impact. Accordingly your entitlement to EOT is determined to be 62 days and the Time for Completion is hereby extended by this amount.
>
> With respect to your claim for reimbursement of additional Costs Plus Profit, you are reminded that Sub-Clause 8.5 [*Extension of Time for Completion*] gives no entitlement to additional payment. Accordingly, your claim for reimbursement is rejected.
>
> Yours faithfully,
>
> The Engineer
> encl. Engineer's detailed analysis of delays and additional Costs.

If, within 14 days of the Notice of the Parties' Agreement or the Notice of the Engineer's Determination being issued by the Engineer or received by the Parties, a typographical, clerical or arithmetical error is found:

(*a*) by the Engineer: then the Engineer must immediately inform the Parties; or
(*b*) by a Party: then that Party must give a Notice to the Engineer, stating that it is given under Sub-Clause 3.7.4 [*Effect of the agreement or determination*] and clearly identifying the error.

If the Engineer does not agree that there was an error, it must immediately advise the Parties accordingly.

Figure 5.7 Error in agreement

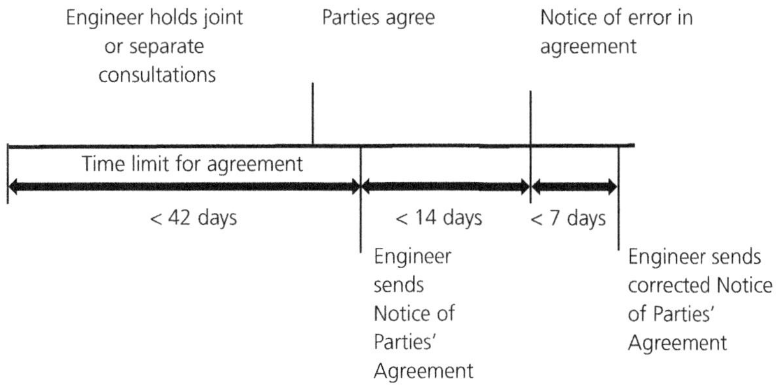

Figure 5.8 Error in determination

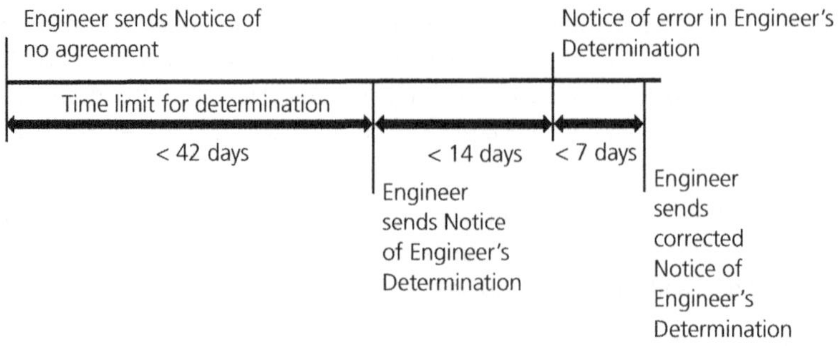

If the Engineer agrees that there was an error, it must within 7 days of finding the error or of receiving Notice of the error, give a Notice to both Parties of the corrected agreement or determination. The corrected agreement or determination is to be treated as the agreement or determination for the purpose of the Conditions and replaces the original agreement or determination.

5.5.3 Dissatisfaction with Engineer's determination

If either Party is dissatisfied with a determination of the Engineer, it may give a NOD to the other Party under Sub-Clause 3.7.5 [*Dissatisfaction with Engineer's determination*], with a copy to the Engineer, stating that it is a Notice of Dissatisfaction with the Engineer's Determination and setting out the reasons for dissatisfaction.

The NOD must be given within 28 days of the Notice of the Engineer's Determination being received or, if applicable, after receiving the Notice of the corrected determination or, in the case of a deemed determination rejecting the Claim, within 28 days of the expiry of the time limit for determination.

After the issue of the NOD, either Party may refer the dispute to the DAAB for a decision under Sub-Clause 21.4 [*Obtaining DAAB's Decision*].

If no NOD is given by either Party within the period of 28 days, the determination of the Engineer is deemed to have been accepted by both Parties and becomes final and binding on them.

If the dissatisfied Party is dissatisfied with only part of the Engineer's determination, that part must be clearly identified in the NOD and that part, along with any other parts of the determination that are affected by or rely on the contested part for completeness, must be severed from the remainder of the determination. The remainder of the determination then becomes final and binding on both Parties as if the NOD had not been given.

An example of a NOD is as follows:

> **BETTA CONTRACTORS LTD.**
>
> A.N. OTHER & Partners
> Consulting Engineers
>
> 27 January 2021
>
> **For the attention of Mr. Other, the Engineer**
>
> Subject: **Notice of Dissatisfaction with Engineer's Determination**
>
> Dear Sir,
>
> In accordance with Sub-Clause 3.7.5 [*Dissatisfaction with Engineer's determination*], we hereby give Notice of our dissatisfaction with the Engineer's Determination dated 31 December 2020 related to our Claim arising from the COVID 19 pandemic.
>
> The reason for our dissatisfaction is that you have taken no account of our alternative argument that the travel restrictions give rise, under Sub-Clause 13.6 [*Adjustments for Changes in Laws*], to an entitlement to an EOT and/or reimbursement of additional Costs.
>
> Yours faithfully,
>
> Contractor's Representative
>
> cc The Employer

5.5.4 Compliance with the agreement or determination

Under Sub-Clause 3.7.4 [*Effect of the agreement or determination*], the agreement or determination is binding on both Parties and must be complied with by both Parties and the Engineer unless and until corrected due to an error or, in the case of a determination, it is revised under Clause 21 [*Disputes and Arbitration*].

If an agreement or determination concerns the payment of an amount from one Party to the other Party, the Contractor must include the amount in its next Statement and the Engineer must include the amount in the next Payment Certificate.

If either Party fails to comply with an agreement of the Parties or a final and binding determination of the Engineer (that is, a determination for which no NOD has been served), the other Party may, without prejudice to any other rights it may have, refer the failure itself directly to arbitration under Sub-Clause 21.6 [*Arbitration*]. In such a case, the first and the third paragraphs of Sub-Clause 21.7 [*Failure to Comply with DAAB's Decision*] apply to the reference in the same way as these paragraphs apply to a final and binding decision of the DAAB.

Thus, if a Party serves a NOD with respect to an Engineer's determination but does not comply, the other Party may not refer the failure to comply to arbitration as the determination will not be final and binding. The other Party must await a favourable decision from the DAAB before being able to enforce compliance or request the DAAB to order compliance as an interim measure under Rule 5.1 (j) of the DAAB Procedural Rules.

Figure 5.9 Failure to comply with agreement

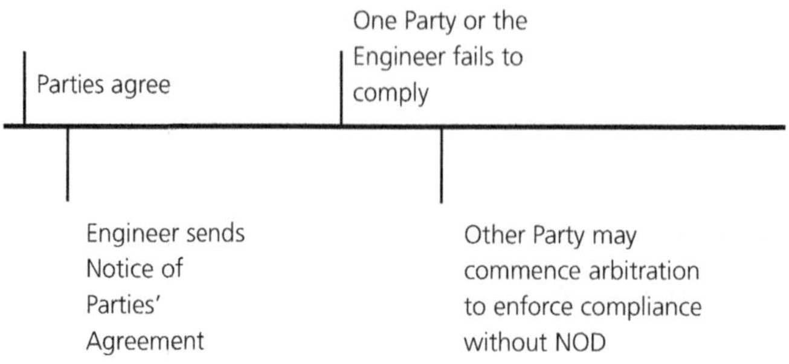

Figure 5.10 Failure to comply with final and binding determination

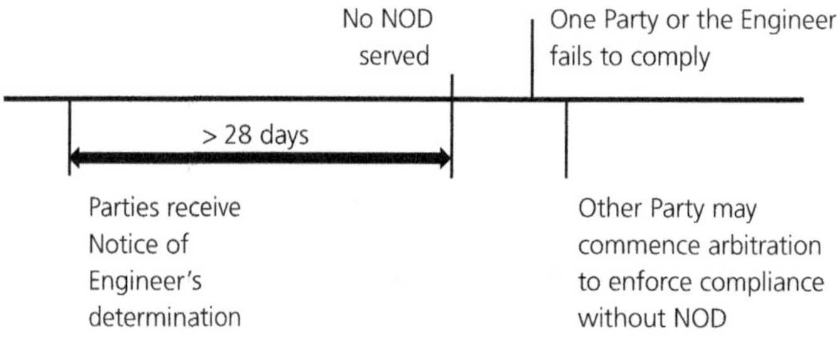

Figure 5.11 Failure to comply with binding but not final determination

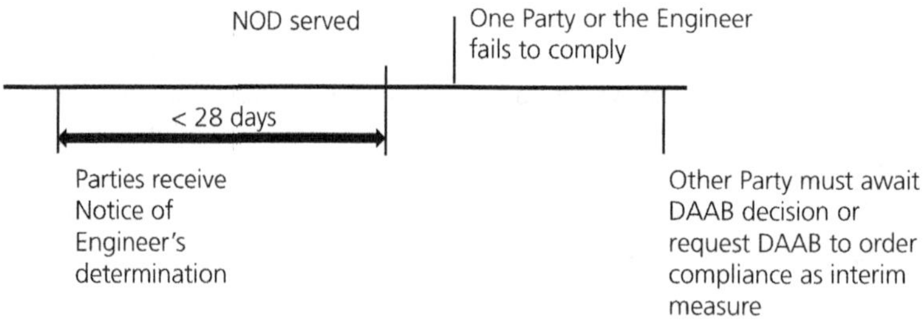

Smith G
ISBN 978-0-7277-6652-6
https://doi.org/10.1680/fcmh.66526.057

Chapter 6
Basic obligations of the contractor

6.1. Introduction

Clauses 4, 5 and 6 of FIDIC 2017 contracts deal with the Contractor's input to the Works: the resources that it will employ either directly or through subcontractors, its management and supervision, health and safety measures, protection of the environment, the methods of working, means of access and working areas, and coordination with others employed on or near the Site, as well as the documents that it is to provide and the financial securities to be provided to the Employer.

The obligations and procedures with respect to 'Contractor's Documents' are addressed in Chapter 7 below, and the obligations of the Parties with respect to the financial securities are examined in Chapter 8 as part of the start-up procedures.

6.2. General Obligations

The first paragraph of Sub-Clause 4.1 [*Contractor's General Obligations*] of RB 2017 states that the Works are to be executed in accordance with the Contract and that the Contractor undertakes that the completed Works will be in accordance with the Contract documents as altered or modified by Variations. This is slightly different from the corresponding paragraph in RB 1999.

Under RB 1999, the Contractor was to design (to the extent specified in the Contract), execute and complete the Works in accordance with the Contract and with the Engineer's instructions, and was to remedy any defects in the Works.

The new text is clearer in that if the Contractor executes the Works in accordance with the Contract documents as modified by Variations, its obligations with respect to design (if any) will be met, as will be its obligations to comply with Engineer's instructions and to remedy defects for which it is liable, both before Taking Over of the Works and during the Defects Notification Period. It is not, therefore, necessary to detail these aspects within the paragraph.

However, in some countries, users of FIDIC contracts believe that the Contractor's basic obligation is to work for the duration of the Time for Completion. They believe that once the Time for Completion has expired, the Contractor is no longer obliged to work, and the Employer may no longer be able or permitted to process payments. To keep the Contract 'alive', the Time for Completion is extended regardless of whether the Contractor has a contractual entitlement to such an extension. The wording of the first paragraph of Sub-Clause 4.1 of RB 1999 was helpful in avoiding this misunderstanding in that it expressly referred to the obligation to complete the Works. The omission from the corresponding paragraph of RB 2017 of the requirement to complete the Works may lead to the continued mistaken belief that the contract is for a fixed duration.

Under YB 2017 and SB 2017, not only must the Contractor ensure that the Works are executed in accordance with the Contract. It must also ensure that, when completed, the Works (or a Section or Part or a major item of Plant) shall be fit for their intended purpose(s). Such purpose must be defined and described in the Employer's Requirements but in the absence of such a definition or description, the Works, Section, Part or Plant must be fit for their ordinary purpose.

6.3. Resources

Among its general obligations, the Contractor must provide the Personnel, Goods, consumables and other things and services, whether of a temporary or permanent nature, required to fulfil the Contractor's obligations under the Contract including Plant (and spare parts if any) and the Contractor's Documents specified in the Contract.

YB 2017 and SB 2017 go further than RB 2017 in that the Contractor is required to do any work necessary to satisfy the Employer's Requirements, Contractor's Proposal and Schedules, or any work that is **implied by the Contract**, as well as all works which are necessary for the **stability**, or for the **completion**, or for the **safe and proper operation** of the Works even if not mentioned in the Contract.

Similar wording is found in YB 1999 and SB 1999, but the scope of '*work ... implied by the Contract*' and '*works ... necessary .... for the completion*' and '*work ... necessary for proper operation of the Works*' is subjective and open to interpretation.

Example:

Under a contract for expansion of an existing power station project based on SB 1999, for which the Contractor was to provide spare parts for two years, the Employer argued incorrectly that it was implicit that the Contractor also had to build an air-conditioned warehouse in which to store the spare parts.

6.4. Methods of working

The Contractor is responsible for the adequacy and safety of its operations, its methods of construction and its Temporary Works. Whenever required by the Engineer, the Contractor must submit details of the arrangements and methods which it proposes to adopt for the execution of the Works (often called 'Method Statements'). The Contractor must not make any significant alteration to these arrangements and methods without first submitting details to the Engineer.

Although there is no stated requirement under Sub-Clause 4.1 [*Contractor's General Obligations*] for these Method Statements to be approved or to be subject to a Notice of No-objection before being implemented, if the said documents are described in the Specification as '*Contractor's Documents*' to be submitted for review by the Engineer under Sub-Clause 4.4 [*Contractor's Documents*] of RB 2017 (or are specified as such in the Employer's Requirements under Sub-Clause 5.2 [*Contractor's Documents*] of YB 2017 or SB 2017), then the Engineer/Employer's Representative has a period of 21 days to review the said Contractor's Document. Under YB 2017 and SB 2017, Contractor's Documents which are submitted for review are not to be used before they have been the subject of a Notice of No-objection issued under Sub-Clause 5.2 or deemed to have been issued. The position is similar under Sub-Clause 4.1 [*Contractor's General Obligations*] of RB 2017 with respect to any part of the Works for which the Specification requires the Contractor to provide the design.

Under Sub-Clause 4.1 [*Contractor's General Obligations*] of RB 2017, unless stated elsewhere in the Contract, the Contractor is responsible for all Contractor's Documents, Temporary Works and the design of each item of Plant and Materials. This is required for the item to be in accordance with the Contract. But the Contractor is not otherwise responsible for the design or specification of the Permanent Works.

This general position can be modified if the Contract specifies that the Contractor shall design a part of the Works (see Chapter 7 below).

6.5. Contractor's Representative

Under Sub-Clause 4.3 [*Contractor's Representative*], the Contractor must appoint someone to represent it on Site. That person must have all necessary authority to act on behalf of the Contractor, except to appoint its own replacement.

The Contractor's Representative must be based at the Site for the whole time that the Works are being executed, and their whole time must be given to overseeing performance of the Contract. If the Contractor's Representative is to be temporarily absent, a suitable replacement must be temporarily appointed, subject to the Engineer's prior consent.

Note:

It is surprising how often these three apparently simple requirements are not met.

On occasions, the appointed Contractor's Representative is a manager based in the Contractor's head office or is a Regional Manager based in a neighbouring country. In both cases, the lack of full-time involvement and infrequent visits lead to breakdowns in communication and deteriorating working relationships.

The possibility of the Contractor's Representative being temporarily replaced is also open to abuse, with 'temporarily' sometimes being interpreted to cover periods of many months. Synonyms for 'temporarily' include 'momentarily', 'briefly' and 'provisionally', none of which could be applied reasonably to a period of several months on a project which is scheduled to be completed within, say, two or three years.

If the Contractor's Representative is not named in the Contract, the Contractor must submit to the Engineer for consent, the name and particulars of the person the Contractor proposes to appoint. This must be done before the Commencement Date. If the Engineer's consent is withheld or subsequently withdrawn, or if the appointed person fails to act as Contractor's Representative, the Contractor must similarly submit the name and particulars of a replacement.

The Contractor's Representative must be qualified, experienced and competent in the '*main engineering discipline applicable to the Works*'. The requirement with respect to the engineering discipline was not found in the FIDIC 1999 contracts. The Notes on the Preparation of Special Provisions suggest that the '*main engineering discipline applicable to the Works*' should be the engineering discipline which is of highest value proportionate to the value of the Works.

If the Engineer does not respond to the Contractor within 28 days of receiving the Contractor's submission, by giving a Notice objecting to the proposed person (or any replacement), the Engineer is deemed to have given their consent.

The Contractor must not replace the Contractor's Representative without the Engineer's prior consent (unless the Contractor's Representative is unable to act as a result of death, illness, disability or resignation). In such a case, the appointment of a replacement is treated as temporary until the Engineer gives consent to this or another replacement.

The Contractor's Representative acts for and on behalf of the Contractor throughout the performance of the Contract. They issue and receive all Notices and other communications under Sub-Clause 1.3 [*Notices and Other Communications*] and receive instructions under Sub-Clause 3.5 [*Engineer's Instructions*].

Hint:

Authority for such issuing and receiving of Notices, instructions and other communications cannot be delegated. Thus, if Notices or other communications are issued or received in the name of the Contractor by someone other than the Contractor's Representative, they are invalid. Similarly, an Engineer's instruction which is received by someone other than the Contractor's Representative will not have been validly transmitted.

Except for such authority to issue or receive Notices, other communications and Engineer's instructions, the Contractor's Representative may delegate their powers, functions and authority to a suitably competent and experienced person and may also revoke the delegation by Notice to the Engineer.

The Contractor's Representative and any person receiving delegated authority must be fluent in the language for communications defined in Sub-Clause 1.4 [*Law and Language*]. However, the Notes on the Preparation of Special Provisions offer additional text to be added to Sub-Clause 4.3 [*Contractor's Representative*] if it is acceptable for the Contractor's Representative and/or delegates not to be fluent in the language for communications. This text foresees the availability of a sufficient number of competent interpreters for the Contractor's Representative and/or delegates to be able to fulfil their duties.

6.6. Key Personnel

FIDIC 2017 contracts include provisions not found in the FIDIC 1999 contracts, which relate to 'Key Personnel', defined under Sub-Clause 1.1.48 of RB 2017 as the positions (if any) of the Contractor's personnel other than the Contractor's Representative, that are stated in the Specification. If the Specification does not identify any Key Personnel, the provisions set out under Sub-Clause 6.12 [*Key Personnel*] do not apply.

The Contractor's bid must name each person to be appointed to one of the positions listed in the Specification as being Key Personnel. If no-one is named in the bid, or if the appointed person fails to act in the relevant position, the Contractor must submit, for the Engineer's consent, the name and particulars of another person to fill the position. If the Engineer's consent is withheld or subsequently revoked, the Contractor must similarly submit the name and particulars of a suitable replacement.

The Standard Procurement Documents used by the World Bank and some other MDBs in conjunction with FIDIC 2017 contracts require the person named in the bid as the Contractor's Representative and each person named as Key Personnel to sign Form PER-2 which contains a declaration that the information with respect to their qualifications and experience given in the bid is correct and that they will be available for the duration stated in the bid. Any misrepresentation or omission can be taken into account during the evaluation of the bid or can result in the person's dismissal from the project.

If the Engineer does not respond to the Contractor within 14 days of receiving the Contractor's submission to give a Notice objecting to the proposed person (or any replacement) with reasons, the Engineer is deemed to have given their consent.

Note:

Under Sub-Clause 6.12 [*Key Personnel*], the Engineer is required to give reasons for withholding consent to the appointment of Key Personnel, whereas under Sub-Clause 4.3 [*Contractor's Representative*], there is no such requirement when withholding consent to the appointment of the Contractor's Representative. This is likely to be an oversight on the part of the drafters of the FIDIC 2017 contracts, which will be corrected in due course.

The Contractor must not replace any of the Key Personnel without the Engineer's prior consent (unless the appointed person is unable to act because of death, illness, disability or resignation). In such a case, the appointment of a replacement is treated as temporary until the Engineer gives consent to this or another replacement.

If any of the Key Personnel are to be temporarily absent during execution of the Works, a suitable replacement must be temporarily appointed, subject as always, to the Engineer's prior consent.

All Key Personnel must be based at the location where Works are being executed, whether on or off Site and must be fluent in the language for communications defined in Sub-Clause 1.4 [*Law and Language*]. However, the Notes on the Preparation of Special Provisions offer additional text to be added to Sub-Clause 6.12 [*Key Personnel*] if it is acceptable for any member of the Key Personnel not to be fluent in the language for communications. This text foresees the availability of a sufficient number of competent interpreters for that member to be able to fulfil their duties.

6.7. Contractor's Personnel

Under Sub-Clause 1.1.17, '*Contractor's Personnel*' is defined to mean:

> *'the Contractor's Representative and all personnel whom the Contractor utilises on Site or other places where the Works are being carried out, including the staff, labour and other employees of the Contractor and of each Subcontractor; and any other personnel assisting the Contractor in the execution of the Works'.*

Under Sub-Clause 6.9 [*Contractor's Personnel*], the Contractor's Personnel (including Key Personnel, if any) must be appropriately qualified, skilled, experienced and competent in their respective trades or occupations. In specific circumstances, the Engineer is empowered to require the Contractor to remove and replace any person employed on the Site or Works.

Under the Special Provisions imposed by the World Bank and some other MDBs for use with FIDIC 2017 contracts, the Engineer may also require the removal of any member of the Contractor's Personnel who engages in behaviour which breaches the Code of Conduct for Contractor's Personnel (see Chapter 6.15 below). In addition, the Special Provisions oblige the Contractor to act, if any of the listed circumstances arise, even if the Engineer does not request removal.

In practice, some Engineers request the removal of a member of the Contractor's Personnel for a reason which does not fall within the circumstances listed under Sub-Clause 6.9; for example, because of a personality clash. In such a situation, the Contractor might oppose the Engineer's request, but resistance is likely to be unproductive.

6.8. Contractor's Superintendence

Throughout the duration of the Works and completion of all remedial work with respect to defects arising during the DNP, the Contractor is required by Sub-Clause 6.8 [*Contractor's Superintendence*] to provide enough supervisors:

(*a*) who are fluent in or have adequate knowledge of the language for communications (defined in Sub-Clause 1.4 [*Law and Language*]); and
(*b*) who have adequate knowledge of the operations to be carried out (including the methods and techniques required, the hazards likely to be encountered and methods of preventing accidents) for the satisfactory and safe execution of the Works.

As is the case for the Contractor's Representative and Key Personnel, the Notes on the Preparation of Special Provisions offer additional wording with respect to language ability.

6.9. Contractor's Equipment

Under Sub-Clause 1.1.16, 'Contractor's Equipment' is defined to mean 'all apparatus, equipment, machinery, construction plant, vehicles and other items required by the Contractor for the execution of the Works'. It does not include Temporary Works. Nor does it include Plant, Materials and other things which form or will form part of the Permanent Works.

Under Sub-Clause 4.17 [*Contractor's Equipment*], the Contractor is responsible for all such Contractor's Equipment. When an item of Contractor's Equipment is delivered to Site, the Contractor must give a Notice to the Engineer confirming the delivery date and providing details of ownership. The Notice must be given within 7 days of delivery. Thereafter, the item of Contractor's Equipment must not be removed from Site without the Engineer's consent. However, consent is not required for minor items or for vehicles transporting Goods or transporting Contractor's Personnel off Site.

6.10. Goods

Under Sub-Clause 1.1.44 '*Goods*' are defined to mean Contractor's Equipment, Materials, Plant and/or Temporary Works.

Under Sub-Clause 4.17 [*Transport of Goods*], the Contractor must give a Notice to the Engineer not less than 21 days before the date on which any Plant or a major item of other Goods (as stated in the Specification) will be delivered to the Site.

The Contractor is responsible for packing, handling, transporting, storing and protecting all Goods, including related customs clearances, import permits, fees and charges, and all obligations necessary for their delivery to the Site.

In this respect, it should be remembered that under Sub-Clause 2.2 [*Assistance*], the Contractor may request the Employer to provide reasonable assistance to allow the Contractor to obtain necessary permits, permissions and so forth. for the delivery of Goods, including customs clearance and for the export of Contractor's Equipment when it is no longer required.

The Contractor must indemnify and hold the Employer harmless against all damage, losses and expenses resulting from the import, transport and handling of all Goods, and must negotiate and pay all third-party claims arising from their import, transport and handling.

Under the Standard Procurement Documents imposed by the World Bank and some other MDBs, for use with FIDIC 2017 contracts, the Employer must choose one of the following alternatives as an addition to Sub-Clause 14.1 [*Contract Price*] when preparing the bidding documents:

[Alternative 1]

'Notwithstanding the provisions of subparagraph (b), Contractor's Equipment, including essential spare parts therefor, imported by the Contractor for the sole purpose of executing the Contract shall be exempt from the payment of import duties and taxes upon importation.'

[Alternative 2]

'Notwithstanding the provisions of subparagraph (b), Contractor's Equipment, including essential spare parts therefore, imported by the Contractor for the sole purpose of executing the Contract shall be temporarily exempt from the payment of import duties and taxes upon initial importation, provided the Contractor shall post with the customs authorities at the port of entry an approved export bond or bank guarantee, valid until the Time for Completion plus six months, in an amount equal to the full import duties and taxes which would be payable on the assessed imported value of such Contractor's Equipment and spare parts, and callable in the event the Contractor's Equipment is not exported from the Country on completion of the Contract. A copy of the bond or bank guarantee endorsed by the customs authorities shall be provided by the Contractor to the Employer upon the importation of individual items of Contractor's Equipment and spare parts. Upon export of individual items of Contractor's Equipment or spare parts, or upon the completion of the Contract, the Contractor shall prepare, for approval by the customs authorities, an assessment of the residual value of the Contractor's Equipment and spare parts to be exported, based on the depreciation scales and other criteria used by the customs authorities for such purposes under the provisions of the applicable Laws. Import duties and taxes shall be due and payable to the customs authorities by the Contractor on (a) the difference between the initial imported value and the residual value of the Contractor's Equipment

and spare parts to exported; and (b) on the initial imported value of the Contractor's Equipment and spare parts remaining in the Country after completion of the Contract. Upon payment of such dues within 28 days of being invoiced, the bond or bank guarantee shall be reduced or released accordingly; otherwise the security shall be called in the full amount remaining.'

The second alternative is a substantial change from PB 2010, which provided a full exemption from import duties and taxes on Contractor's Equipment and related spare parts without a requirement for a bond or bank guarantee.

Hint:

In practice it is not unusual for Employers to include such mention of tax exemptions in the Contract but to fail to take the necessary steps to establish the exemptions. In many countries, the Customs Authorities will not allow imported Goods to leave the port unless and until they have received payment of import duties and taxes or have received written instructions from the Ministry of Finance not to apply the duties and taxes. This situation invariably delays mobilisation by the Contractor and subsequent execution of the Works, leading to claims for EOT and additional payment.

6.11. Contractor's Records and Progress Reports

Sub-Clause 6.10 [*Contractor's Records*] requires the Contractor to compile a daily record of the Contractor's Personnel and Goods used (unless otherwise agreed with the Engineer). The said records are to be included in the monthly progress report submitted under Sub-Clause 4.20 [*Progress Reports*].

For each work activity shown in the Programme, at each work location and for each day of work, the progress report must include records of:

(*a*) occupations and actual working hours of each class of the Contractor's Personnel;
(*b*) the type and actual working hours of each of the Contractor's Equipment;
(*c*) the types of Temporary Works used;
(*d*) the types of Plant installed in the Permanent Works; and
(*e*) the quantities and types of Materials used.

The Special Provisions imposed by the World Bank and some other MDBs for use with FIDIC 2017 contracts contain additional requirements under Sub-Clause 6.23 [*Employment Record of Workers*]:

'The Contractor shall keep complete and accurate records of the employment of labour at the Site. The records shall include the names, ages, genders, hours worked, and wages paid to all workers. These records shall be summarized on a monthly basis and submitted to the Engineer. These records shall be included in the details to be submitted by the Contractor under Sub-Clause 6.10 [Records of Contractor's Personnel and Equipment].'

Sub-Clause 4.20 [*Progress Reports*] requires the Contractor to submit monthly progress reports to the Engineer.

The submission of the progress report is also required under Sub-Clause 14.3 [*Application for Interim Payment*] as part of the supporting documents which must accompany the Contractor's Statement. The period for the Engineer to issue the related IPC under Sub-Clause 14.6 [*Issue of IPC*] only commences upon receipt of the Contractor's Statement and supporting documents, including the progress report.

Each progress report must be in the format stated in the Specification or in a format acceptable to the Engineer and must be submitted in one paper original and one electronic copy with additional paper copies (if any) as stated in the Contract Data.

The requirements for such reports are more detailed in the FIDIC 2017 contracts than FIDIC 1999.

The first report must cover the period up to the end of the first month following the Commencement Date. Subsequent reports must be submitted at monthly intervals, each within 7 days of the last day of the month to which it relates, until the Date of Completion[2] or the date at which any outstanding work listed in the Taking-Over Certificate is completed.

Unless otherwise stated in the Specification, in addition to the records of Contractor's Personnel required under Sub-Clause 6.10 [*Contractor's Records*], each progress report must include:

- charts, diagrams and detailed descriptions of progress of all phases of the Works, including Contractor's Documents, procurement, manufacture, delivery to Site, construction, erection and testing;
- photographs and/or video recordings showing the status of manufacture and of progress on and off Site;
- for the manufacture of each main item of Plant and Materials, full details of the manufacturer, location, percentage progress, and the actual or expected dates of commencement of manufacture, inspections, tests, shipment and arrival at Site;
- copies of quality management documents, inspection reports, test results, and compliance verification documentation (including certificates of Materials);
- a list of Variations, and any Notices given (by either Party) under Sub-Clause 20.2.1 [*Notice of Claim*];
- health and safety statistics, including details of any hazardous incidents and activities relating to environmental aspects and public relations; and
- comparisons of actual and planned progress, with details of any events or circumstances which may adversely affect the completion of the Works in accordance with the Programme and the Time for Completion, and the measures being (or to be) adopted to overcome delays.

The last of the items listed places the Contractor under an obligation to collate information concerning actual start and finish dates of all activities in the Programme and to identify each month the current critical path and likely delays to completion. Having identified possible delays to completion, the

[2] Sub-Clause 1.1.24 defines 'Date of Completion' as the date stated in the Taking-Over Certificate issued by the Engineer; or the date on which the Works or Section are deemed to have been completed under Sub-Clause 10.1 [*Taking Over the Works and Sections*]; or the date on which the Works or Section or Part are deemed to have been taken over by the Employer under Sub-Clause 10.2 [*Taking Over Parts*]; or Sub-Clause 10.3 [*Interference with Tests on Completion*]. This definition was slightly modified by the 2022 Reprint (see Chapter 19).

Contractor must take measures to minimise these delays. This obligation to minimise delays is often referred to as an obligation to 'mitigate'. Whether the Contractor is required to bear additional Costs to mitigate the impact of any delays will depend on the governing law of the Contract.

The Special Provisions imposed by the World Bank and some other MDBs for use with FIDIC 2017 contracts replace 'health and safety statistics, including details of any hazardous incidents and activities relating to environmental aspects and public relations' by 'the Environmental and Social (ES) metrics set out in Particular Conditions – Part D'. The said ES metrics give extensive details of the information that is to be provided with respect to ES aspects of the project.

In addition, under the Special Provisions the World Bank and some other MDBs also impose another requirement with respect to progress reports:

> *'In addition to the reporting requirement of this sub-paragraph (g) of Sub-Clause 4.20 [Progress Reports] the Contractor shall inform the Engineer immediately of any allegation, incident or accident, which has or is likely to have a significant adverse effect on the environment, the affected communities, the public, Employer's Personnel or Contractor's Personnel. This includes, but is not limited to, any incident or accident causing fatality or serious injury; significant adverse effects or damage to private property; or any allegation of SEA*[3] *and/or SH. In case of SEA and/or SH, while maintaining confidentiality as appropriate, the type of allegation (sexual exploitation, sexual abuse or sexual harassment), gender and age of the person who experienced the alleged incident should be included in the information.*
>
> *The Contractor, upon becoming aware of the allegation, incident or accident, shall also immediately inform the Engineer of any such incident or accident on the Subcontractors' or suppliers' premises relating to the Works which has or is likely to have a significant adverse effect on the environment, the affected communities, the public, Employer's Personnel or Contractor's, its Subcontractors' and suppliers' personnel. The notification shall provide sufficient detail regarding such incidents or accidents. The Contractor shall provide full details of such incidents or accidents to the Engineer within the timeframe agreed with the Engineer.*
>
> *The Contractor shall require its Subcontractors and suppliers (other than Subcontractors) to immediately notify the Contractor of any incidents or accidents referred to in this Subclause.'*

Note:

Nothing stated in any progress report constitutes a Notice under a sub-clause of the FIDIC 2017 Conditions. This statement in the last paragraph of Sub-Clause 4.20 [*Progress Reports*] was not included in earlier FIDIC contracts and is aimed at ending discussion about what constitutes a Notice.

[3] SEA stands for Sexual Exploitation and Abuse. SH stands for Sexual Harassment.

6.12. Cooperation and coordination

Under Sub-Clause 4.6 [*Co-operation*] of FIDIC 1999 contracts, the Contractor was obliged to allow appropriate opportunities for work by the Employer's Personnel, other contractors employed by the Employer and the personnel of any legally constituted public authorities. Such work could be on or near the Site.

The appropriate opportunities were to be specified in the Contract and therefore deemed to be included in the Contract Price. If not specified in the Contract, they could be instructed by the Engineer. If the instruction from the Engineer caused the Contractor to incur delays and/or Unforeseeable Cost, the instruction constituted a Variation entitling the Contractor to an extension of time and/or additional payment. '*Unforeseeable*' was stated under Sub-Clause 1.1.6.8 of RB 1999 to mean '*not reasonably foreseeable by an experienced contractor by the Base Date*'. The '*Base Date*' was stated under Sub-Clause 1.1.3.1 to mean '*the date 28 days prior to the latest date for submission of the Tender*'.

Sub-Clause 4.6 [*Co-operation*] of FIDIC 2017 contracts is broadly similar. However, the more recent contracts not only require the Contractor to allow appropriate opportunities for work by others but also require the Contractor to cooperate with the others and to use '*all reasonable endeavours*' to coordinate the Contractor's activities with those of the others but only to the extent (if any) stated in the Specification or as instructed by the Engineer.

The meaning of '*all reasonable endeavours*' varies from one legal jurisdiction to another and can be onerous. However, the extent of the obligation is limited to that stated in the Specification or, if not stated, to that fixed by the instruction from the Engineer.

If the Contractor suffers delay and/or incurs Cost as a result of such an instruction from the Engineer, the Contractor shall be entitled subject to Sub-Clause 20.2 [*Claims for Payment and/or EOT*] to an EOT and/or payment of such Cost Plus Profit to the extent that cooperation, allowance of opportunities and coordination was Unforeseeable having regard to that stated in the Specification.

Under FIDIC 1999, there was no requirement for the Contractor to give notice of a claim with respect to an instruction from the Engineer related to cooperation with others. Any such instruction was a Variation if it caused delays or Unforeseeable Cost. Under FIDIC 2017, the Contractor is required to submit a Notice under Sub-Clause 20.2.1 within 28 days of receipt of the instruction. Moreover, it is not Unforeseeable Cost which triggers the right to compensation but the extent that cooperation, allowance of opportunities and coordination was Unforeseeable when compared to the description in the Specification. The meaning of this requirement is unclear. If the Specification states that the Contractor must coordinate its work with the activities of another contractor, is a claim for an extension of time due to delay caused by the other contractor excluded on the basis that the coordination (and its consequences) was not Unforeseeable?

Hint:

To avoid future disputes, during contract negotiations, the Parties should consider amending Sub-Clause 4.6 [*Co-operation*] or the Specification to fix precise limits with respect to the Contractor's obligation to cooperate.

6.13. Health and safety

The Contractor's health and safety obligations fall mainly under three sub-clauses: Sub-Clause 4.8 [*Health and Safety Obligations*], Sub-Clause 4.21 [*Security of the Site*] and Sub-Clause 6.7 [*Health and Safety of Personnel*]. The MDBs have substantially expanded these obligations.

Under Sub-Clause 4.8 [*Health and Safety Obligations*] of FIDIC 2017, the Contractor must:

(*a*) comply with all applicable health and safety regulations and Laws;
(*b*) comply with all applicable health and safety obligations specified in the Contract;
(*c*) comply with all directives issued by the Contractor's health and safety officer (appointed under Sub-Clause 6.7 [*Health and Safety of Personnel*]) (see below);
(*d*) take care of the health and safety of all persons entitled to be on the Site as well as any other places where the Works are being executed;
(*e*) keep the Site and the Works (and any other places where the Works are being executed) clear of unnecessary obstruction so as to avoid danger to these persons;
(*f*) provide fencing, lighting, safe access, guarding and watching of:

  (i) the Works, until the Works are taken over under Clause 10 [*Employer's Taking Over*]; and
  (ii) any part of the Works where the Contractor is executing outstanding works or remedying any defects during the DNP; and

(*g*) provide any Temporary Works (including roadways, footways, guards and fences) which may be necessary, because of the execution of the Works, for the use and protection of the public and of owners and occupiers of adjacent land and property.

The Contractor must also submit to the Engineer 'for information' a health and safety manual which has been specifically prepared for the Works, the Site and other places (if any) where the Contractor intends to execute the Works. The manual must be provided within 21 days of the Commencement Date and before commencing any construction on the Site.

The health and safety manual (which is additional to any other similar document required under applicable health and safety regulations and Laws) must address all the health and safety requirements:

(i) that are stated in the Specification;
(ii) that comply with all the Contractor's health and safety obligations under the Contract; and
(iii) that are necessary for a healthy and safe working environment for all persons entitled to be on the Site and any other places where the Works are being executed.

This manual must be revised as necessary by the Contractor or the health and safety officer, or when reasonably requested by the Engineer. Each revision of the manual must be promptly submitted to the Engineer.

The World Bank's Special Provisions require the health and safety manual and any updates to be submitted to the Engineer '*for review*' in accordance with the procedures set out under Sub-Clause 4.4.1 [*Preparation and Review*]. The said sub-clause relates to Contractor's Documents and is examined in Chapter 7.

With respect to the content of the manual, the World Bank has again added requirements:

'The health and safety manual shall set out all the health and safety requirements under the Contract,

(*a*) *which shall include at a minimum:*
  - (i) *the procedures to establish and maintain a safe working environment without risk to health at all workplaces, machinery, equipment and processes under the control of the Contractor, including control measures for chemical, physical and biological substances and agents;*
  - (ii) *details of the training to be provided, records to be kept;*
  - (iii) *the procedures for prevention, preparedness and response activities to be implemented in the case of an emergency event (i.e. an unanticipated incident, arising from both natural and manmade hazards, typically in the form of fire, explosions, leaks or spills, which may occur for a variety of different reasons including failure to implement operating procedures that are designed to prevent their occurrence, extreme weather or lack of early warning);*
  - (iv) *the measures to be taken to avoid or minimize the potential for community exposure to water-borne, water-based, water-related, and vector-borne diseases,*
  - (v) *the measures to be implemented to avoid or minimize the spread of communicable diseases (including transfer of Sexually Transmitted Diseases or Infections (STDs), such as HIV virus) and non-communicable diseases associated with the execution of the Works, taking into consideration differentiated exposure to and higher sensitivity of vulnerable groups. This includes taking measures to avoid or minimize the transmission of communicable diseases that may be associated with the influx of temporary or permanent Contract-related labour;*
  - (vi) *the policies and procedures on the management and quality of accommodation and welfare facilities if such accommodation and welfare facilities are provided by the Contractor in accordance with Sub-Clause 6.6; and*

(*b*) *any other requirements stated in the Specification.'*

The Special Provisions imposed by the World Bank and some other MDBs for use in relation to FIDIC 2017 contracts add the following at the end of Sub-Clause 4.8 [*Health and Safety Obligations*]:

(*h*) *provide health and safety training of Contractor's Personnel as appropriate and maintain training records;*

(*i*) *actively engage the Contractor's Personnel in promoting understanding, and methods for, implementation of health and safety requirements, as well as in providing information to Contractor's Personnel, and provision of personal protective equipment without expense to the Contractor's Personnel;*

(*j*) *put in place workplace processes for Contractor's Personnel to report work situations that they believe are not safe or healthy, and to remove themselves from a work situation which they have reasonable justification to believe presents an imminent and serious danger to their life or health;*

> (*k*) *Contractor's Personnel who remove themselves from such work situations shall not be required to return to work until necessary remedial action to correct the situation has been taken. Contractor's Personnel shall not be retaliated against or otherwise subject to reprisal or negative action for such reporting or removal;*
>
> (*l*) *subject to Sub-Clause 4.6, collaborate with the entities and Personnel under paragraph (a), (b) and (c) of Sub-Clause 4.6, in applying the health and safety requirements. This is without prejudice to the responsibility of the relevant entities for the health and safety of their own personnel; and*
>
> (*m*) *establish and implement a system for regular (not less than six-monthly) review of health and safety performance and the working environment.'*

Under Sub-Clause 4.21 [*Security of the Site*], the Contractor is responsible for keeping unauthorised persons off the Site. In this context, '*authorised persons*' means the Contractor's Personnel, the Employer's Personnel (including the Engineer and delegated assistants) and other personnel such as the Employer's other contractors, who are identified as '*authorised*' in a Notice to the Contractor from either the Employer or the Engineer.

> **In this respect, the Special Provisions imposed by the World Bank** and some other MDBs for use in relation to FIDIC 2017 contracts have again added significantly more requirements.
>
> *'Subject to Sub-Clause 4.1, the Contractor shall submit for the Engineer's No-objection a security management plan that sets out the security arrangements for the Site.*
>
> *The Contractor shall (i) conduct appropriate background checks on any personnel retained to provide security; (ii) train the security personnel adequately (or determine that they are properly trained) in the use of force (and, where applicable, firearms) and appropriate conduct towards Contractor's Personnel, Employer's Personnel and affected communities; and (iii) require the security personnel to act within the applicable Laws and any requirements set out in the Specification.*
>
> *The Contractor shall not permit any use of force by security personnel in providing security except when used for preventive and defensive purposes in proportion to the nature and extent of the threat.*
>
> *In making security arrangements, the Contractor shall also comply with any additional requirements stated in the Specification.'*

Under Sub-Clause 6.7 [*Health and Safety of Personnel*] of FIDIC 2017, the Contractor must take all necessary precautions to maintain the health and safety of the Contractor's Personnel. The Contractor must ensure that:

(*a*) medical staff, first aid facilities, sick bay, ambulance services and any other medical services stated in the Specification ('Employer's Requirements' in YB 2017 and SB 2017) are available at all times at the Site and at any accommodation for Contractor's and Employer's Personnel; and

(*b*) suitable arrangements are made for all necessary welfare and hygiene requirements and for the prevention of epidemics.

This must be done in collaboration with local health authorities.

The Contractor must appoint a health and safety officer to the Site, who is to:

(i) be duly qualified, experienced and competent for this responsibility; and
(ii) have full authority to issue instructions and to take protective measures to prevent accidents and to maintain the health and safety of all personnel authorised to enter and/or work on the Site.

Given the definition of '*authorised personnel*' under Sub-Clause 4.21 [*Security of the Site*], some confusion could arise over the authority of the Contractor's health and safety officer with respect to the personnel of other contractors working on the Site.

The Contractor must provide everything which is required by the health and safety officer to exercise this responsibility and authority throughout the execution of the Works.

In addition to the reporting requirement of sub-paragraph (*g*) of Sub-Clause 4.20 [*Progress Reports*], the Contractor must provide the Engineer with details of any accident as soon as practicable after its occurrence and, in the case of an accident which causes serious injury or death, must inform the Engineer immediately.

The World Bank has deleted this requirement and instead adds a reference to the text added to Sub-Clause 4.20 [*Progress Reports*] (See Chapter 6.11 above).

Finally, under FIDIC 2017, Sub-Clause 4.8 [*Health and Safety Obligations*], the Contractor must, as stated in the Specification or as reasonably required by the Engineer, maintain records and make reports (in compliance with the applicable health and safety regulations and Laws) concerning the health and safety of persons and any damage to property.

6.14. Environment

Under Sub-Clause 4.18 [*Protection of the Environment*] of FIDIC 2017, the Contractor must take all necessary measures to:

- protect the environment (both on and off the Site);
- comply with the environmental impact statement for the Works (if any); and
- limit damage and nuisance to people and property resulting from pollution, noise and other results of the Contractor's operations and/or activities.

The Contractor must ensure that emissions, surface discharges, effluent and any other pollutants from the Contractor's activities do not exceed the values indicated in the Specification/Employer's Requirements, or those prescribed by applicable Laws.

This is largely similar to FIDIC 1999 except for the mention of the environmental impact statement.

The Special Provisions used by the World Bank replace the standard wording of Sub-Clause 4.18 [*Protection of the Environment*] of FIDIC 2017 by the following:

> *'The Contractor shall take all necessary measures to:*
> *(a) protect the environment (both on and off the Site); and*
> *(b) limit damage and nuisance to people and property resulting from pollution, noise and other results of the Contractor's operations and/or activities.*
>
> *The Contractor shall ensure that emissions, surface discharges, effluent and any other pollutants from the Contractor's activities shall exceed neither the values indicated in the Specification, nor those prescribed by applicable Laws.*
>
> *In the event of damage to the environment, property and/or nuisance to people, on or off Site as a result of the Contractor's operations, the Contractor shall agree with the Engineer the appropriate actions and time scale to remedy, as practicable, the damaged environment to its former condition. The Contractor shall implement such remedies at its cost to the satisfaction of the Engineer.'*

The obligation to restore the environment to its former condition, with the risk of a call against the Contractor's Environmental and Social Performance Security in the event of failure to do so, is a powerful incentive to ensure that no damage is caused.

6.15. Code of Conduct

The Special Provisions imposed by the World Bank and some other MDBs for use with FIDIC 2017 seek to adapt the FIDIC 2017 contracts to include many of the aspects which were included in the Pink Book but not found in RB 1999, particularly those aspects which addressed 'social issues' under Clause 6 [*Staff and Labour*] such as the treatment of workers and the local community. Moreover, further measures have been introduced with the aim of improving the protection of the local community. These include the introduction of a Code of Conduct that is to be acknowledged by all members of the Contractor's Personnel, which, by definition, includes workers employed by subcontractors.

The Standard Procurement Documents include a model of the Code of Conduct (see Annex 3) which all bidders are to use as a template for the Code of Conduct to be submitted with the bid and to be enforced during execution of the Works. The model sets out the minimum content to which bidders may add.

Sub-Clause 4.25 [*Code of Conduct*] of the World Bank's Special Provisions deals with the use of the Code of Conduct during the execution of the Works. The Contractor must take all necessary measures to ensure that each member of the Contractor's Personnel (including workers employed by subcontractors) is made aware of the Code of Conduct including specific prohibited behaviour and understands the consequences of engaging in such behaviour.

These measures include providing instructions and documentation that can be understood by each member of the Contractor's Personnel and seeking to obtain the person's signature acknowledging receipt of the instructions and/or documentation.

The Contractor must also ensure that the Code of Conduct is visibly displayed in multiple locations on the Site and any other place where the Works will be carried out, as well as in areas outside the Site accessible to the local community and people affected by the project. The posted Code of Conduct must be in languages which are understandable to Contractor's Personnel, Employer's Personnel and the local community.

In addition, the Contractor's Management Strategy and Implementation Plans (see Chapter 6.16 below) must detail procedures for the Contractor to verify compliance with these obligations.

6.16. Management Strategies and Implementation Plans

In addition to the Code of Conduct, the World Bank and some other MDBs require bidders to include with their bids 'Management Strategies and Implementation Plans' (MSIPs) to manage key environmental and social (ES) risks mentioned by the Employer in the Instructions to Bidders.

These strategies and plans must describe in detail the actions, materials, equipment, management processes and so forth that will be implemented by the Contractor and its subcontractors with regard to the ES provisions of the Contract.

The 'Mobilization Schedule' which each bidder includes with its bid must take into account that no mobilisation to Site is to take place before the Engineer is satisfied that appropriate measures are in place to address the ES risks, including (as a minimum) the application of the MSIPs and the Code of Conduct.

This latter point is covered by additional text added by the Special Provisions to Sub-Clause 4.1 [*Contractor's General Obligations*]. The relevant part of the Special Provisions states:

> *'The Contractor shall not carry out mobilization to Site (e.g. limited clearance for haul roads, site accesses and work site establishment, geotechnical investigations or investigations to select ancillary features such as quarries and borrow pits) unless the Engineer gives a Notice of No-objection to the Contractor, a Notice that shall not be unreasonably delayed, to the measures the Contractor proposes to manage the environmental and social risks and impacts, which at a minimum shall include applying the Management Strategies and Implementation Plans (MSIPs) and Code of Conduct for Contractor's Personnel submitted as part of the Bid and agreed as part of the Contract.*
>
> *The Contractor shall submit to the Engineer for Review any additional MSIPs as are necessary to manage the ES risks and impacts of ongoing Works (e.g. excavation, earth works, bridge and structure works, stream and road diversions, quarrying or extraction of materials, concrete batching and asphalt manufacture). These MSIPs collectively comprise the Contractor's Environmental and Social Management Plan (C-ESMP). The Contractor shall review the C-ESMP, periodically (but not less than every six (6) months), and update it as required to ensure that it contains measures appropriate to the Works. The updated C-ESMP shall be submitted to the Engineer for Review.*
>
> *The C-ESMP shall be part of the Contractor's Documents. The procedures for Review of the C-ESMP and its updates shall be as described in Sub-Clause 4.4.1 [Preparation and Review].'*

The following is added as (g) of Sub-Clause 4.1; sub-paragraphs (g) and (h) of the Sub-Clause are then renumbered as (h) and (i) respectively.

(g) 'if so stated in the Specification, the Contractor shall:

(i) *design structural elements of the Works taking into account climate change considerations;*
(ii) *apply the concept of universal access (the concept of universal access means unimpeded access for people of all ages and abilities in different situations and under various circumstances;*
(iii) *consider the incremental risks of the public's potential exposure to operational accidents or natural hazards, including extreme weather events; and*
(iv) *any other requirement stated in the Specification.'*

The following is also added at the end of Sub-Clause 4.1 [*Contractor's General Obligations*]:

'The Contractor shall provide relevant contract-related information, as the Employer and/or Engineer may reasonably request to conduct Stakeholder engagements. 'Stakeholder' refers to individuals or groups who:

(i) *are affected or likely to be affected by the Contract; and*
(ii) *may have an interest in the Contract.*

The Contractor shall also directly participate in Stakeholder engagements, as the Employer and/or Engineer may reasonably request.'

6.17. Subcontractors, Nominated Subcontractors and Suppliers

6.17.1 Subcontractors

In RB 2017, the topic of subcontracting is addressed under Sub-Clause 5.1 [*Subcontractors*] and Sub-Clause 5.2 [*Nominated Subcontractors*]. In YB 2017 and SB 2017, the corresponding sub-clauses are Sub-Clause 4.4 and Sub-Clause 4.5.

In RB 2017, Sub-Clause 1.1.78 defines '*Subcontractor*' to be:

'any person named in the Contract as a subcontractor, or any person appointed by the Contractor as a subcontractor or designer, for a part of the Works; and the legal successors in title to each of these persons.'

Sub-Clause 1.1.80 of YB 2017 and Sub-Clause 1.1.70 of SB 2017 give the same definition.

Under FIDIC 1999 contracts, the Contractor was not permitted to subcontract the whole of the Works. Under FIDIC 2017, the Contractor is not permitted to subcontract works with a total accumulated value greater than the percentage of the Accepted Contract Amount stated in the Contract Data. If no such percentage is stated, the position is the same as under FIDIC 1999, that is: the Contractor may not subcontract the whole of the Works.

In practice, the only contractors who subcontract 'the whole of the Works' are those that 'sell' the contract to another company or lend their name and references for bidding purposes. It is entirely understandable that such practices must be prohibited.

Some other contractors award subcontract packages which together represent a large proportion of the physical activity, but they maintain managerial control (planning, quality, coordination, safety, insurance, financial management, contracts management and general management, etc.). This is not necessarily detrimental to the project and may be beneficial.

Imposing a ceiling on the amount of work that may be subcontracted, could entail the exclusion of a very competent Subcontractor only because the value of the subcontract package takes the accumulated value of subcontracted work beyond the ceiling set in the Contract Data.

Moreover, the enforcement of the limit may be difficult. For example, the Notes on the Preparation of Special Provisions for RB 2017 suggest wording for use when the Employer decides that it is unnecessary for the Contractor to seek consent before awarding a subcontract package that amounts to less than 1% of the Accepted Contract Amount. If this wording is adopted, the Contractor could award, say, 20 small packages that together represent almost 20% of the Accepted Contract Amount, without the Engineer/Employer's Representative being informed. In such circumstances, how can the ceiling be enforced? Enforcement requires the Engineer/Employer's Representative to be informed of every subcontract package, no matter how small, including professional consultancy agreements with law firms, accountants, auditors and so forth.

As well as the possible ceiling on the proportion of the Works that the Contractor is permitted to subcontract, Sub-Clause 5.1 [*Subcontractors*] permits the Employer to identify in the Contract Data, specific parts of the Works that the Contractor is not permitted to subcontract.

Unless the proposed Subcontractor is merely a supplier of Materials or is named in the Contract, the Contractor is required to obtain the prior consent of the Engineer/Employer's Representative before appointing the Subcontractor.

To obtain such consent the Contractor must provide the following information:

(*a*) Name
(*b*) Address
(*c*) Work to be subcontracted
(*d*) Detailed particulars and relevant experience of intended Subcontractor
(*e*) Further information reasonably required.

An example of such a request for consent follows:

BETTA CONTRACTORS LTD.

A.N. OTHER & Partners
Consulting Engineers

15 September 2022

For the attention of Mr. Other, the Engineer

Subject: **Request for consent to the appointment of the Earthworks Subcontractor under Sub-Clause 5.1.**

Dear Sir,

In accordance with Sub-Clause 5.1 [*Subcontractors*], we hereby request the Engineer's consent to the appointment of ACE Contracting as the subcontractor for all earthworks on the project.

We attach for your information:

a) details of the subcontractor's corporate structure, resources and experience;

b) details of the work to be subcontracted;

c) a draft of the proposed subcontract (excluding prices).

We would appreciate your early response to our request, but should you require any further information, we would be pleased to provide it.

Yours faithfully,

Contractor's Representative

The World Bank and some other MDBs add to the information to be supplied by the Contractor when seeking the Engineer's consent:

'The Contractor's submission to the Engineer shall also include the Subcontractor's declaration in accordance with the Particular Conditions – Part E – Sexual Exploitation and Abuse (SEA) and/or Sexual Harassment Performance Declaration for Subcontractors.'

Moreover, the MDB add to Sub-Clause 5.1 [*Subcontractors*]:

'All subcontracts relating to the Works shall include provisions which entitle the Employer to require the subcontract to be assigned to the Employer under sub-paragraph (a) of Sub-Clause 15.2.3 [After Termination].'

This means that when seeking consent for the appointment of the intended Subcontractor, the Contractor must also enclose a copy of the draft Subcontract Agreement so that the Engineer/Employer's

Representative can verify the presence of this provision. However, in doing so, there is no obligation on the Contractor to provide details of the Subcontractor's rates or prices.

Finally, the MDB insert under Sub-Clause 5.1:

'Where practicable, the Contractor shall give fair and reasonable opportunity for contractors from the Country to be appointed as Subcontractors.'

If no Notice of objection is received within 14 days of receipt of the Contractor's request (or receipt of further information, if requested), consent is deemed to have been given.

Prior to the start of the Subcontractor's work, the Contractor is required to give Notice to the Engineer/Employer's Representative at least 28 days prior to the intended start date. Notice is also to be given at least 28 days prior to the commencement of the Subcontractor's work on Site.

The Contractor is responsible for the work of the Subcontractor, for managing and coordinating the work of all Subcontractors, and for the acts or defaults of any Subcontractor, any Subcontractor's agents, or employees, as if they were the acts or defaults of the Contractor.

The World Bank and some other MDBs add to this provision:

'The Contractor shall require that its Subcontractors execute the Works in accordance with the Contract, including complying with the relevant ES requirements and the obligations set out in Sub-Clause 4.25 above.'

Difficulties may arise if a Subcontractor was named in the accepted bid but, after award of the Contract, the Contractor seeks to replace the named company. If the named company has become insolvent or bankrupt, both the Contractor and the Employer have little choice but to select and consent to a replacement Subcontractor. However, if acceptance of the Contractor's bid was dependent on the appointment of the named Subcontractor and the Contractor is unable or unwilling to proceed with the appointment, the Employer might be entitled to terminate the Contract (if permitted by the governing Laws).

Hint:

Contractors should carefully select potential subcontractors before bidding (preferably allowing themselves flexibility to appoint alternatives) and should 'lock in' key subcontractors so that they cannot increase their prices after acceptance of the Contractor's bid.

6.17.2 Nominated Subcontractors

Sub-Clause 5.2 [*Nominated Subcontractors*] of RB 2017 (and Sub-Clause 4.5 of YB 2017 and SB 2017) define 'nominated Subcontractor' to be a Subcontractor named as a nominated Subcontractor in the

Specification/Employer's Requirements or a Subcontractor which the Contractor under Sub-Clause 13.4 [*Provisional Sums*], is instructed to employ as a Subcontractor. In other words, it is a Subcontractor chosen by the Employer, not the Contractor.

> **Note:**
>
> There is frequent misunderstanding that a subcontractor which a bidder decides to mention in its bid is a '*nominated Subcontractor*'. It is not. Such a subcontractor should be described as a 'named subcontractor' because it was named in the bid but was not imposed by the Employer.

If the nominated Subcontractor was named as such in the Contract, the Contractor cannot object to the nomination. However, if the Engineer (Employer's Representative under SB 2017) instructs the Contractor to employ a nominated Subcontractor, the Contractor can object by giving a Notice within 14 days of receiving the instruction. The Notice must provide detailed particulars in support of the objection.

If the objection is reasonable, the Contractor will not be under an obligation to employ the nominated Subcontractor, unless the Employer agrees to indemnify the Contractor against the consequences. An objection is deemed reasonable if it arises from (among other things) any of the following:

(*a*) *'there are reasons to believe that the Subcontractor does not have sufficient competence, resources or financial strength;*

(*b*) *the subcontract does not specify that the nominated Subcontractor shall indemnify the Contractor against and from any negligence or misuse of Goods by the nominated Subcontractor, the nominated Subcontractor's agents and employees; or*

(*c*) *the subcontract does not specify that, for the subcontracted work (including design, if any), the nominated Subcontractor shall:*

- (i) *undertake to the Contractor such obligations and liabilities as will enable the Contractor to discharge the Contractor's corresponding obligations and liabilities under the Contract, and*
- (ii) *indemnify the Contractor against and from all obligations and liabilities arising under or in connection with the Contract and from the consequences of any failure by the Subcontractor to perform these obligations or to fulfil these liabilities'.*

> **The World Bank** and some other MDBs add a third point to the list of circumstances which are deemed to constitute a reasonable objection to the appointment of a nominated Subcontractor:
>
> *'(iii) be paid only if and when the Contractor has received from the Employer payments for sums due under the Subcontract referred to under Sub-Clause 5.2.3 [Payment to nominated Subcontractors].'*
>
> This provision is illegal in some jurisdictions where the government has passed legislation to ensure that small subcontractors are paid even if the Contractor is not paid.

Hint:

If the objection is not accepted by the Engineer/Employer's Representative as reasonable, the Contractor is obliged to appoint the nominated company. In doing so, and if the Contractor is convinced that it has reasonable grounds for contesting the appointment, it should reserve its rights to continue to contest the nomination.

Unless the Contractor's objection is accepted as reasonable, the Contractor is obliged to accept the nomination and enter the subcontract with the nominated Subcontractor. By doing so, he accepts liability for any underperformance of the nominated Subcontractor, its failures and those of its workers.

The remainder of Sub-Clause 5.2 [*Nominated Subcontractors*] deals with payment.

Under Sub-Clause 5.2.3 [*Payments to nominated Subcontractors*], the Contractor must pay to the nominated Subcontractor all amounts due in accordance with the subcontract, which must be included in the Contract Price together with related charges in accordance with sub-paragraph (b) of Sub-Clause 13.4 [*Provisional Sums*], except as stated in Sub-Clause 5.2.4 [*Evidence of Payments*] (see below).

The reference to '*related charges*' concerns a mark-up for overhead charges and profit, calculated as a percentage of the amounts paid (or to be paid) to the nominated Subcontractor in accordance with sub-paragraph (b) of Sub-Clause 13.4 [*Provisional Sums*].

Under Sub-Clause 5.2.4 [*Evidence of Payments*], before issuing a Payment Certificate which includes an amount payable to a nominated Subcontractor, the Engineer may request the Contractor to supply reasonable evidence that the nominated Subcontractor has received all amounts due in accordance with the previous Payment Certificates, less applicable deductions for retention.

The Employer may pay directly to the nominated Subcontractor, part or all of such amounts previously certified (less applicable deductions) as are due to the nominated Subcontractor and for which the Contractor has failed to:

(*a*) submit to the Engineer the requested reasonable evidence of payment; or
(*b*) satisfy the Engineer in writing that the Contractor is reasonably entitled to withhold or refuse to pay these amounts and submit to the Engineer reasonable evidence that the nominated Subcontractor has been notified of such Contractor's entitlement.

In such circumstances, the Engineer must give a Notice to the Contractor stating the amount paid directly to the nominated Subcontractor by the Employer and, in the next IPC after this Notice, must include the paid amount as a deduction under sub-paragraph (b) of Sub-Clause 14.6.1 [*The IPC*].

6.17.3 Suppliers

FIDIC 2017 excludes suppliers from most provisions dealing with Subcontractors, as was the case for FIDIC 1999. Indeed, unlike '*Subcontractor*', there is no definition for '*supplier*', and '*suppliers*' are expressly excluded wholly or partly from the scope of several sub-clauses such as Sub-Clause 5.1 [*Subcontractors*] of RB 2017.

This is not so for the World Bank and some other MDBs who impose specific requirements through an additional sub-clause (Sub-Clause 4.24) which forces the Contractor to assume responsibility for any suppliers' behaviour which falls short of the standards expected by the MDB:

*'**4.24.1 Forced Labour***
The Contractor shall take measures to require its suppliers (other than Subcontractors) not to employ or engage forced labour including trafficked persons as described in Sub-Clause 6.21. If forced labour/trafficking cases are identified, the Contractor shall take measures to require the suppliers to take appropriate steps to remedy them. Where the supplier does not remedy the situation, the Contractor shall within a reasonable period substitute the supplier with a supplier that is able to manage such risks.

4.24.2 Child labour
The Contractor shall take measures to require its suppliers (other than Subcontractors) not to employ or engage child labour as described in Sub-Clause 6.22. If child labour cases are identified, the Contractor shall take measures to require the suppliers to take appropriate steps to remedy them. Where the supplier does not remedy the situation, the Contractor shall within a reasonable period substitute the supplier with a supplier that is able to manage such risks.

4.24.3 Serious Safety Issues
The Contractor, including its Subcontractors, shall comply with all applicable safety obligations, including as stated in Sub-Clauses 4.8, 5.1 and 6.7. The Contractor shall also take measures to require its suppliers (other than Subcontractors) to adopt procedures and mitigation measures adequate to address safety issues related to their personnel. If serious safety issues are identified, the Contractor shall take measures to require the suppliers to take appropriate steps to remedy them. Where the supplier does not remedy the situation, the Contractor shall within a reasonable period substitute the supplier with a supplier that is able to manage such risks.

4.24.4 Obtaining natural resource materials in relation to supplier
The Contractor shall obtain natural resource materials from suppliers that can demonstrate, through compliance with the applicable verification and/or certification requirements, that obtaining such materials is not contributing to the risk of significant conversion or significant degradation of natural or critical habitats such as unsustainably harvested wood products, gravel or sand extraction from riverbeds or beaches.

If a supplier cannot continue to demonstrate that obtaining such materials is not contributing to the risk of significant conversion or significant degradation of natural or critical habitats, the Contractor shall within a reasonable period substitute the supplier with a supplier that is able to demonstrate that they are not significantly adversely impacting the habitats.'

Smith G
ISBN 978-0-7277-6652-6
https://doi.org/10.1680/fcmh.66526.081

Chapter 7
Design

7.1. Employer's design

The full title of RB 2017 is 'FIDIC Conditions of Contract for Construction for Building and Engineering Works designed by the Employer'. Under Sub-Clause 4.1 of RB 2017 [*Contractor's General Obligations*], the Contractor is to execute the Works in accordance with the Contract such that the completed Works will be in accordance with the documents forming the Contract, as altered or modified by Variations.

The documents forming the Contract are listed under Sub-Clause 1.1.10:

Sub-Clause 1.1.10	***'Contract'*** *means the Contract Agreement, the Letter of Acceptance, the Letter of Tender, any addenda referred to in the Contract Agreement, these Conditions, the Specification, the Drawings, the Schedules, the Contractor's Proposal, the JV Undertaking (if applicable) and the further documents (if any) which are listed in the Contract Agreement or in the Letter of Acceptance.*

However, the Errata published by FIDIC clarify that the reference to '*the Contractor's Proposal*' is to be deleted. Thus, with respect to the Employer's design, the relevant Contract documents are the Specification, the Drawings and any further documents listed in the Contract Agreement or in the Letter of Acceptance and to a lesser extent, the Schedules.

Other relevant definitions are as follow:

Sub-Clause 1.1.30	***'Drawings'*** *means the drawings included in the Contract, and any additional and modified drawings issued by (or on behalf of) the Employer in accordance with the Contract.*
Sub-Clause 1.1.76	***'Specification'*** *means the document entitled specification included in the Contract, and any additions and modifications to the specification in accordance with the Contract. Such document specifies the Works.*

Thus, the Contractor commits to construct the Works in accordance with documents to which additions or modifications may be made after award of the Contract, in accordance with the Contract. It is necessary to maintain the ability to add to or modify the Drawings and/or Specification to overcome errors and omissions in the Employer's design and/or changes in the Employer's requirements.

The provisions of the Contract which address such additions and modifications are found under Sub-Clause 1.9 [*Delayed Drawings or Instructions*], Sub-Clause 3.5 [*Engineer's Instructions*] and Sub-Clause 13.3.1 [*Variation by Instruction*].

Under Sub-Clause 3.5 [*Engineer's Instructions*], the Engineer is permitted to issue instructions to the Contractor at any time, which may be necessary for the execution of the Works. Under Sub-Clause 13.1 [*Right to Vary*], the Contractor must not make any alteration to and/or modification of the Permanent Works, unless and until the Engineer instructs a Variation under Sub-Clause 13.3.1 [*Variation by Instruction*]. However, even if the Engineer does not acknowledge that an instruction is a Variation, the Contractor must comply with all instructions from the Engineer under Sub-Clause 3.5 [*Engineer's Instructions*].

In such circumstances, if the Contractor considers that the instruction constitutes a Variation it should immediately, and before commencing any work related to the instruction, give a Notice to the Engineer with reasons. If the Engineer does not respond within 7 days of receiving this Notice, by giving a Notice confirming, reversing or varying the instruction, the Engineer is deemed to have revoked the instruction. Otherwise, the Contractor must comply with the Engineer's response.

Sub-Clause 1.9 [*Delayed Drawings or Instructions*] of RB 2017 deals with delay by the Engineer in issuing any additional drawing or instruction that the Contractor considers to be necessary.

Note:

Sub-Clause 1.9 [*Delayed Drawings or Instructions*] contains three parts, the first of which is often overlooked by Contractors. The first part requires the Contractor to give a Notice to the Engineer whenever the Works **are likely to be delayed or disrupted if** any necessary drawing or instruction is not issued to the Contractor within a particular time. The Notice, which must give a reasonable period of warning to the Engineer, **must include details of the necessary drawing or instruction, and why and by when it is needed, together with details of the nature and amount of the delay or disruption likely to be suffered if it is late.**

This requirement already existed in RB 1999 but few Contractors complied. This could be understandable given the amount of information that is required. If a Contractor finds a 'clash' on a reinforcement drawing when fixing the reinforcement and requests an instruction, is it reasonable to expect them to provide a forecast of the disruption that they are likely to suffer if the instruction does not come within minutes or hours? The excessive demands of the sub-clause give the impression that it has been designed to protect the Engineer who is late in issuing a drawing or instruction.

Example of Notice of likely delay if drawing or instruction is late:

BETTA CONTRACTORS LTD.

A.N. OTHER & Partners
Consulting Engineers

01 November 2022

For the attention of Mr. Other, the Engineer

Subject: **Notice under Sub-Clause 1.9 [*Delayed Drawings or Instructions*]**

Dear Sir,

Please be informed that a conflict exists between the Site limit as shown on Drawing N°KRA/GS/DR 002 and the position of the northern toe of the embankment between Km 101+350 and Km 102+765, based on the typical cross section applied to the vertical profile of the road as shown on Drawing N°KRA/GS/DR0042. If the typical cross section is followed, the toe of the embankment falls outside the Site limit by a maximum of 1.610 m, which is likely to lead to difficulties with adjacent landowners. We request your instruction in this respect.

Unless your instruction is received by close of business on 04 November 2022, our site clearance team will be unable to work in the area from 07h00 on 06 November 2022.

In such case, we will be obliged to claim an EOT due to the delay incurred together with reimbursement of the additional Costs Plus Profit, which will include standby costs and/or relocation costs together with additional time-related costs due to the delay.

Yours faithfully,

Contractor's Representative
cc The Employer

The second part of the sub-clause states that if the Contractor suffers delay and/or incurs Cost as a result of a failure of the Engineer to issue the requested drawing or instruction within the period specified in the Notice, the Contractor is entitled, subject to Sub-Clause 20.2 [*Claims for Payment and/or EOT*] to an EOT and/or payment of such Cost Plus Profit.

The third part of the sub-clause states that if the Engineer's failure was caused by an error or delay by the Contractor, the Contractor is not entitled to such EOT and/or Cost Plus Profit.

7.2. Employer's Requirements

Under YB 2017 and SB 2017, the 'Employer's Requirements' are the equivalent document to the 'Specification' under RB 2017.

YB 2017 **Sub-Clause 1.1.33** **SB 2017** **Sub-Clause 1.1.31**	***'Employer's Requirements'*** *means the document entitled employer's requirements, as included in the Contract, and any additions and modifications to such document in accordance with the Contract. Such document describes the purpose(s) for which the Works are intended, and specifies Key Personnel (if any), the scope, and/or design and/or other performance, technical and evaluation criteria, for the Works.*

In reality, under YB 2017, the content of the Employer's Requirements is often much more extensive than this definition suggests and includes a significant amount of design by the Employer. Indeed, one of the largest sources of disputes under FIDIC Yellow Book contracts is the interface between the design conceived by or on behalf of the Employer and the development of this design by the Contractor.

In this respect, the Notes on the Preparation of the Special Provisions recommends that if the Employer's Requirements include an outline design, tenderers should be advised of the extent to which the Employer's outline design is a suggestion or a requirement.

Under Sub-Clause 5.1 [*General Design Obligations*] of YB 2017, promptly after receiving a Notice under Sub-Clause 8.1 [*Commencement of Works*], the Contractor must scrutinise the Employer's Requirements (including design criteria and calculations, if any). If this scrutiny reveals any error, fault or other defect in the Employer's Requirements, Sub-Clause 1.9 [*Errors in the Employer's Requirements*] applies (unless it is an error in the items of reference specified in the Employer's Requirements, in which case Sub-Clause 4.7 [*Setting Out*] applies).

Sub-Clause 1.9 [*Errors in the Employer's Requirements*] of YB 2017 states that if the Contractor discovers an error, fault or defect in the Employer's Requirements as a result of its scrutiny under Sub-Clause 5.1 [*General Design Obligations*], the Contractor must give a Notice to the Engineer within the period stated in the Contract Data (if no period is stated, the period is 42 days) calculated from the Commencement Date.

If, at a later date (after expiry of this scrutiny period), the Contractor finds such an error, fault or defect in the Employer's Requirements, the Contractor must also give a Notice to the Engineer describing the error, fault or defect.

The Engineer must then proceed in accordance with Sub-Clause 3.7 [*Agreement or Determination*] to agree or determine:

(*a*) whether or not there is an error, fault or defect in the Employer's Requirements;
(*b*) whether or not (taking account of cost and time) an experienced contractor exercising due care would have discovered the error, fault or other defect:

- when examining the Site and the Employer's Requirements before submitting the Tender; or
- if the Contractor's Notice is given after the expiry of the scrutiny period stated in the Sub-Clause 1.9 [*Errors in the Employer's Requirements*], when scrutinising the Employer's Requirements under Sub-Clause 5.1 [*General Design Obligations*]; and

(*c*) what measures (if any) the Contractor is required to take to rectify the error, fault or defect.

The time limit for agreement under Sub-Clause 3.7.3 [*Time limits*] begins at the date the Engineer receives the Contractor's Notice under Sub-Clause 1.9 [*Errors in the Employer's Requirements*].

If there is an error, fault or defect which an experienced contractor would not have discovered before submitting its Tender the measures that the Contractor is instructed to take to rectify the error, fault or defect constitute a Variation under Sub-Clause 13.3.1 [*Variation by Instruction*]. In addition, if the Contractor suffers delay and/or incurs Cost as a result of the error, fault or defect, the Contractor is entitled subject to Sub-Clause 20.2 [*Claims for Payment and/or EOT*] to EOT and/or payment of such Cost Plus Profit.

It is implied that if the error, fault or defect would have been discovered before submitting its Tender, by an experienced contractor exercising due care, the measures to rectify the error, fault or defect would not constitute a Variation and the Contractor would not be entitled to an EOT and/or additional payment.

Similarly, if the error, fault or defect could have been discovered during the scrutiny period by an experienced Contractor exercising due care, the necessary rectification measures would not constitute a Variation.

In reality, the sub-clause aims to address three different periods:

(i) the pre-bid period;
(ii) the scrutiny period;
(iii) the post-scrutiny period.

In line with these aims, following receipt of the Contractor's Notice, the Engineer must consider whether the defect was discoverable by an experienced contractor exercising due care (taking account of cost and time) during the relevant period. If an experienced Contractor could not have detected the defect in the preceding period, it is entitled to compensation, subject always to compliance with Sub-Clause 20.2 [*Claims for Payment and/or EOT*].

SB 2017 does not contain provisions similar to those found under Sub-Clause 1.9 [*Errors in the Employer's Requirements*] of YB 2017. This is because, under Sub-Clause 5.1 [*General Design Obligations*] of SB 2017, the Contractor is deemed to have scrutinised, prior to the Base Date, the Employer's Requirements (including design criteria and calculations, if any) and upon signing the Contract, accepts that the Employer shall not be responsible for any error, inaccuracy or omission of any kind in the Employer's Requirements as originally included in the Contract and shall not be deemed to have given any representation of accuracy or completeness of any data or information (except as noted below). Any data or information received by the Contractor, from the Employer or otherwise, does not relieve the Contractor from the Contractor's responsibility for the execution of the Works.

As a result, the Employer bears no responsibility for any error, inaccuracy or omission of any kind in the Employer's Requirements and is not deemed to have given any representation of accuracy or completeness of any data or information, but for a few exceptions as follow:

(*a*) portions, data and information which are stated in the Contract as being immutable or the responsibility of the Employer;
(*b*) definitions of intended purposes of the Works or any parts thereof;

(*c*) criteria for the testing and performance of the completed Works; and

(*d*) portions, data and information which cannot be verified by the Contractor, except as otherwise stated in the Contract.

This allocation of risk was the same under SB 1999. Unfortunately, it was not uncommon for Employers to modify item (d) above so that the Contractor was held responsible even for portions, data and information which could not be verified by the Contractor. Such a practice is totally contrary to FIDIC's Golden Principle GP3:

> *'The Particular Conditions must not change the balance of risk/reward allocation provided for in the General Conditions.'*

> **Note:**
>
> Given that upon signing the Contract, the Contractor waives all recourse against the Employer (except in very limited circumstances) for errors, inaccuracies or omissions in the data provided to bidders, it is necessary to allow bidders sufficient time to analyse and verify the data provided. For this reason, the bidding periods under SB 2017 should be substantially greater than those for projects to be executed under RB 2017 or YB 2017 contracts.

7.3. Design by Contractor

Theoretically RB 2017 is to be used for projects for which the Employer provides the design. The full title of RB 2017 is: '*FIDIC Conditions of Contract for Construction* ***for Building and Engineering Works designed by the Employer***' (emphasis added). The Notes at the start of the FIDIC document state:

> *'the FIDIC 1999 Red Book has been in widespread use for nearly two decades. In particular, it has been recognised for, among other things, its principles of balanced risk sharing between the Employer and the Contractor in* ***projects where the Contractor constructs the works in accordance with a design provided by the Employer.****'*

However, the Notes go on to state that:

> *'the works may include some elements of Contractor-designed civil, mechanical, electrical and/or construction works'.*

Thus, there is a possibility for the Contractor to be allocated responsibility for the design of part of the Works.

This possibility is again evident under Sub-Clause 1.2 [*Interpretation*] paragraph (j), where it is clarified that *"execute the Works' or 'execution of the Works' means the construction and completion of the Works and the remedying of any defects* ***(and shall be deemed to include design to the extent, if any, specified in the Contract)****'* (emphasis added).

In this respect, the fifth paragraph of Sub-Clause 4.1 [*Contractor's General Obligations*] begins:

> *'If the Contract specifies that the Contractor shall design any part of the Permanent Works, then unless otherwise stated in the Particular Conditions:*
>
> (*a*) *the Contractor shall prepare, and submit to the Engineer for Review, the Contractor's Documents for this part (and any other documents necessary to complete and implement the design during the execution of the Works and to instruct the Contractor's Personnel)...'*

Thus, it is confirmed that under RB 2017 (as it was under RB 1999) the Contractor may be responsible for the design of part of the Works. If so, the part to be designed by the Contractor must be clearly defined in the Contract (usually within the Specification).

Note:

Being given responsibility for designing a part of the Works is not the same as being given responsibility for a part of the design for all the Works, that is: being responsible for a design stage for all the Works. Thus, the Contractor might be given total responsibility for the design of a specific part of the Works such as deep foundations, but not given responsibility for detailed design of all the Works.

Under Sub-Clause 5.1 [*General Design Obligations*] of YB 2017 and SB 2017, the Contractor is to carry out, and be responsible for, the design of the Works.

Under Sub-Clause 5.3 [*Contractor's Undertaking*] of YB 2017 and SB 2017, the Contractor undertakes that the design, the Contractor's Documents, the execution of the Works and the completed Works will be in accordance with:

(*a*) the Laws of the Country; and
(*b*) the documents forming the Contract, as altered or modified by Variations.

In addition, under Sub-Clause 4.1 [*Contractor's General Obligations*], the Works (or any Section or any major item of Plant), when completed, must be fit for the intended purpose(s) as defined and described in the Employer's Requirements. If no purpose is defined or described, the Works, Section or major item of Plant must be fit for their ordinary purpose(s).

It should be noted that if the Contractor is required to design part of the Works under RB 2017, the Contractor is under a similar obligation to ensure that the part he designs is fit for the intended purpose(s) as specified in the Contract or fit for its ordinary purposes(s) if no such specification is given.

This is higher than the standard which FIDIC sets with respect to its own members under the FIDIC '*White Book*'[4] which is generally to 'exercise reasonable skill, care and diligence', that is, to do as most consulting engineers would do under the circumstances.

[4] Client/Consultant Model Services Agreement

The obligation under Sub-Clause 5.3 [*Contractor's Undertaking*] is reinforced by Sub-Clause 5.4 [*Technical Standards and Regulations*] of YB 2017 and SB 2017 which requires that the execution of the Works (including the design) and the completed Works (including defects remedied by the Contractor) must comply with the Country's technical standards, building, construction and environmental Laws, Laws applicable to the product being produced from the Works, and other standards specified in the Employer's Requirements, applicable to the Works or defined by applicable Laws.

Accordingly, the Contractor must comply with all the applicable technical or other standards and Laws in force when the Works or Section or Part are taken over under Clause 10 [*Employer's Taking Over*]. This is expressly stated in the second paragraph of the sub-clause.

Yet in the third paragraph of the sub-clause, it is stated that if changed or new applicable standards come into force in the Country after the Base Date, the Contractor shall promptly give a Notice to the Engineer.

Following receipt of the Notice, the Engineer may request a proposal from the Contractor with respect to compliance with the new or changed standard. If the Engineer considers that compliance is required and such compliance requires change(s) to the execution of the Works which constitute a Variation, then the Engineer is to instruct a Variation under Clause 13 [*Variations and Adjustments*]. This suggests that the Engineer might not require the Contractor to comply with the changed or new standard, which contradicts the second paragraph.

Hint:

Given the contradiction between the second and third paragraphs of Sub-Clause 5.4 [*Technical Standards and Regulations*], the Contractor should be particularly attentive with respect to changed or new standards, and should ensure that Notice is given promptly and that the Engineer/Employer clearly states whether compliance is required or not.

Note:

Under Sub-Clause 13.6 [*Adjustments for Changes in Legislation*], if a change in the Laws of the Country occurs after the Base Date, which causes the Contractor to suffer delay and/or additional Costs, the Contractor shall be entitled, subject to Sub-Clause 20.2 [*Claims for Payment and/or EOT*], to EOT and/or payment of such Cost.

Under Sub-Clause 5.8 [*Design Error*] of YB 2017 and SB 2017, the Contractor is responsible for the correction, at its own cost, of any errors, omissions and so forth in the Contractor's design and/or the Contractor's Documents, and the impact on the Works. The relevant Contractor's Documents will be considered as having been the subject of a Notice under sub-paragraph (b) of Sub-Clause 5.2.2 [*Review by Engineer*], that is, they failed to comply with the Employer's Requirements and/or the Contract.

In other words, if there is an error or omission in the Contractor's design, it will be considered that it was found by the Engineer during the review, even if it was not.

7.4. Preparation and processing of Contractor's Documents

If the Contractor is obliged to design a part of the Works under RB 2017, the above quote from Sub-Clause 4.1. [*Contractor's General Obligations*] requires it to prepare and submit for Review, the related Contractor's Documents.

RB 2017 **Sub-Clause 1.1.15**	***'Contractor's Documents'*** means *'the documents prepared by the Contractor as described in Sub-Clause 4.4 [Contractor's Documents], including calculations, digital files, computer programs and other software, drawings, manuals, models, specifications and other documents of a technical nature.'*

Sub-Clause 1.1.14 of YB 2017 and Sub-Clause 1.1.12 of SB 2017 give similar definitions.

Sub-Clause 4.4 [*Contractor's Documents*] of RB 2017 states that the Contractor's Documents shall comprise the documents:

(*a*) *'stated in the Specification;*

(*b*) *required to satisfy all permits, permissions, licences and other regulatory approvals which are the Contractor's responsibility under Sub-Clause 1.13 [Compliance with Laws];*

(*c*) *described in Sub-Clause 4.4.2 [As-Built Records] and Sub-Clause 4.4.3 [Operation and Maintenance Manuals], where applicable; and*

(*d*) *required under sub-paragraph (a) of Sub-Clause 4.1 [Contractor's General Obligations], where applicable.'*

Sub-paragraph (a) of Sub-Clause 4.1 [*Contractor's General Obligations*], states that for any part of the Works which the Contractor is required by the Contract to design, the Contractor shall prepare and submit to the Engineer for Review the Contractor's Documents (and any other documents necessary to complete and implement the design during the execution of the Works and to instruct the Contractor's Personnel).

Thus, the Specification must contain a list or descriptions of the documents which together will constitute the Contractor's Documents. (Sub-Clause 5.2 of YB 2017 and SB 2017 contain similar explanations with respect to the composition of the Contractor's Documents except that there is no provision similar to sub-paragraph (d).)

In this respect, the Notes on the Preparation of Special Provisions emphasise:

> *'It is important that the Specification should clearly specify which Contractor's Documents the Employer requires the Contractor to prepare, which may not necessarily include (for example) all the technical documents which the Contractor's Personnel will need in order to execute the Works.*

Further, if there are Contractor's Documents that the Contractor must submit to the Engineer for review, it is important that they are clearly identified in the Specification.'

The reference in Sub-Clause 4.1 (a) to 'any other documents necessary to complete and implement the design during the execution of the Works and to instruct the Contractor's Personnel' includes documents that are often called 'Shop Drawings' or 'Working Drawings'. These are not design documents in that they do not define that which is to be constructed but are documents which explain to the workers how to implement the design such as bar-bending schedules or formwork drawings.

That the Contractor is required to prepare such other documents is entirely understandable, but it is less easy to understand why the Contractor must submit such other documents for Review. In effect, this requirement means that the '*other documents*' are no different from the Contractor's Documents, which is contrary to the Notes on the Preparation of Special Provisions which indicate that Contract Documents do not necessarily include such other documents.

It is also relevant that, whereas Sub-Clause 5.2.1 [*Preparation by Contractor*] of YB 2017 and SB 2017 also refers to '*any other documents necessary to complete and implement the design during execution of the Works and to instruct the Contractor's Personnel*', there is no requirement for these other documents to be submitted for Review.

Moreover, the procedure described under Sub-Clause 4.4 [*Contractor's Documents*] of RB 2017 for the submission and Review of Contractor's Documents, makes no mention of such '*other documents*'.

It appears therefore, that the requirement under Sub-Clause 4.1 [*Contractor's General Obligations*] of RB 2017 for submission of '*other documents*' for Review by the Engineer is a mistake by the drafters.

Hint:

Include the following Special Provision to address the apparent error in Sub-Clause 4.1 [*Contractor's General Obligations*] of RB 2017:

'Replace sub-paragraph (a) of the fifth paragraph by: 'the Contractor shall prepare, and submit to the Engineer for Review, the Contractor's Documents for this part (and shall also prepare any other documents necessary to complete and implement the design during the execution of the Works and to instruct the Contractor's Personnel).''

Sub-Clause 1.1.70 **RB 2017** **Sub-Clause 1.1.71** **YB 2017** **Sub-Clause 1.1.61** **SB 2017**	***'Review'*** *means examination and consideration by the Engineer of a Contractor's submission in order to assess whether (and to what extent) it complies with the Contract and/or with the Contractor's obligations under or in connection with the Contract.*

Sub-Clause 4.4 [*Contractor's Documents*] of RB 2017, states that if the Specification or the Conditions of Contract specify that a Contractor's Document is to be submitted to the Engineer for Review, it must be submitted together with a Notice from the Contractor stating that the Contractor's Document is ready for Review and that it complies with the Contract.

After receiving the Contractor's Document and this Notice from the Contractor, the Engineer has 21 days within which to respond to the Contractor:

(*a*) by a Notice of No-objection (which may include comments concerning minor matters which will not substantially affect the Works); or
(*b*) by a Notice that the Contractor's Document fails to comply with the Contract to the extent stated, with reasons.

If the Engineer fails to respond within the period of 21 days, the Engineer is deemed to have given a Notice of No-objection to the Contractor's Document.

If the Engineer gives a Notice that the Contractor's Document fails to comply with the Contract, the Contractor must revise the Contractor's Document and resubmit it to the Engineer for a further Review. The period of 21 days for Review begins at the date that the Engineer receives the revised document.

YB 2017 and SB 2017 contain largely similar provisions with respect to Contractor's Documents. However, there are some noticeable differences.

Firstly, with respect to Review by the Engineer, the term '*Contractor's Document*' excludes any of the Contractor's Documents which are not specified in the Employer's Requirements, or the Conditions of Contract, as being required to be submitted for Review but includes all documents on which a specified Contractor's Document relies for completeness.

Secondly, the Notice from the Contractor which must accompany the Contractor's Document submitted for Review is referred to as the 'Contractor's Notice'. As well as stating that the Contractor's Document is considered by the Contractor to be ready for Review and that it complies with the Contract, the Contractor's Notice must state that the Contractor's Document is ready for use, and that it complies with the Employer's Requirements and the Conditions, or the extent to which it does not do so.

Thirdly, if the Engineer gives no Notice within the Review Period, the Engineer shall be deemed to have given a Notice of No-objection to the Contractor's Document (but only if all other Contractor's Documents on which that Contractor's Document relies have received, or are deemed to have received, a Notice of No-objection.

Fourthly, if the Engineer instructs that further Contractor's Documents are reasonably required to demonstrate that the Contractor's design complies with the Contract, the Contractor must promptly prepare and submit the additional Contractor's Documents at the Contractor's cost.

Fifthly, if the Employer incurs additional costs as a result of the resubmission of a Contractor's Document and its subsequent Review, the Employer is entitled, subject to Sub-Clause 20.2 [*Claims for Payment and/or EOT*], to recover from the Contractor the costs reasonably incurred. Such a possibility existed under YB 1999 and SB 1999 but it was less apparent and rarely used. Of course, to successfully claim from the Contractor the additional costs incurred by the additional Review by the Engineer, the Employer must provide evidence that the Engineer is entitled to be paid by the Employer for the additional Review and the basis for calculating the amount. This would probably require the Engineer to keep a record of the time spent on the Review, which must be reasonable.

Finally, under Sub-Clause 5.2.3 [*Construction*], construction of any part of the Works is not to begin until a Notice of No-objection has been given (or is deemed to have been given) for all relevant Contractor's Documents. The Contractor may subsequently modify any design or Contractor's Documents which have previously been submitted for Review, by giving a Notice to the Engineer with reasons, but in this case, the Contractor must suspend work on the affected part of the Works and must not resume until a Notice of No-objection is given (or is deemed to have been given) by the Engineer for the revised documents.

Note:

Under Sub-Clause 5.2 [*Contractor's Documents*] of YB 1999, the Employer's Requirements could define the Contractor's Documents that were to be submitted for review only and/or those that were to be submitted for review and approval. YB 2017 does not foresee 'approval' by the Engineer of Contractor's Documents.

The Notes on the Preparation of Tender Documents attached to YB 2017 recognise that Review by the Engineer might be insufficient to satisfy the applicable law which might impose mandatory review/checking of certain elements of design by an authorised professional or other legally recognised individual and/or verification that such design is in accordance with the applicable law.

FIDIC considers that it is essential to draw the attention of tenderers to such mandatory requirements in the Instruction to Tenderers and to amend Sub-Clause 5.2 [*Contractor's Documents*] to unambiguously address the following:

(a) *'the mandatory review/checking and/or verification process(es) required by the applicable law, and details of the submission procedure(s) associated with such process(es);*

(b) *which element(s) of design, and which type(s) of Contractor's Documents associated with such element(s), shall be subject to the mandatory review/checking and/or verification process(es);*

(c) *whether, and to what extent, the mandatory review/checking and/or verification process(es) of an element of design (and the Contractor's Documents associated with such element) shall replace the Engineer's review under this Sub-Clause;*

(d) *a statement that any Notice of No-objection (or deemed Notice of No-objection) from the Engineer with respect to any Contractor's Document shall not replace the mandatory review/checking and/or verification of the design (or a revised design).'*

7.5. As-Built Records and Operation and Maintenance Manuals

Under RB 2017, '*as-built*' records and operation and maintenance manuals are dealt with under Sub-Clause 4.4.2 [*As-Built Records*] and Sub-Clause 4.4.3 [*Operation and Maintenance Manuals*]. In both cases, the sub-clauses begin with a statement that the sub-clause does not apply if the Specification does not state that the relevant documents are to be prepared by the Contractor. Therefore, before looking further at the two sub-clauses, it is necessary to verify that the Specification does place the Contractor under an obligation to prepare and submit the relevant documents.

Note:

Under Sub-Clause 4.1 [*Contractor's General Obligations*] of RB 1999 (fifth paragraph) and the Pink Book 2010 (sixth paragraph), the Contractor was only required to prepare and submit 'as-built' records for those parts of the Permanent Works which it designed (if any). However, many Employers overlooked this and wrongly assumed that the Contractor must produce 'as-built' records for the whole of the Permanent Works.

If the Specification does allocate responsibility for as-built records to the Contractor, the Contractor must prepare, and keep up-to-date, a complete set of as-built records of the execution of the Works, showing the exact as-built locations, sizes and details of the work as executed by the Contractor.

The Specification must define the format, referencing system, system of electronic storage and other relevant details which the Contractor must follow. If not, the Contractor must propose such details to the Engineer, who must decide whether the proposed details are acceptable.

The as-built records, which must be kept on Site, must be submitted to the Engineer for Review at the appropriate time. The Works are not to be considered completed for the purposes of taking over under Sub-Clause 10.1 [*Taking Over the Works and Sections*] until the Engineer has given (or is deemed to have given) a Notice of No-objection to the as-built records under sub-paragraph (i) of Sub-Clause 4.4.1 [*Preparation and Review*].

Hint:

Too many contractors complete construction and testing of the Works without preparing the 'as-built' records and are then surprised that the TOC is refused. It is recommended (and it is an obligation) that the Contractor complete the 'as-built' records in line with the progress of the Works and submit them for Review as soon as possible.

Under Sub-Clause 4.4.3 [*Operation and Maintenance Manuals*], the Contractor has similar obligations to those under Sub-Clause 4.4.2 [*As-Built Records*]. The Contractor must prepare, and keep up to date, the operation and maintenance manuals in the format and in accordance with other relevant details as stated in the Specification.

The operation and maintenance manuals must be submitted to the Engineer for Review, and the Works are not to be considered completed for the purposes of taking over under Sub-Clause 10.1 [*Taking Over the Works and Sections*] until the Engineer has given (or is deemed to have given) a Notice of No-objection under sub-paragraph (i) of Sub-Clause 4.4.1 [*Preparation and Review*].

Under Sub-Clause 5.6 [*As-Built Records*] of YB 2017 and SB 2017, the basic obligations of the Contractor are similar to those under RB 2017, when the Specification calls for as-built records from the Contractor. However, YB 2017 and SB 2017 go further than the FIDIC 1999 contracts in that the Contractor must submit to the Engineer for Review under Sub-Clause 5.2.2 [*Review by Engineer*]:

(*a*) *the as-built records for the Works or Section (as the case may be) before the commencement of the Tests on Completion; and*
(*b*) *updated as-built records to the extent that any work is executed by the Contractor:*
  (i) *during and/or after the Tests on Completion, before the issue of any Taking-Over Certificate under Sub-Clause 10.1 [Taking Over the Works and Sections]; and*
  (ii) *after taking over under Sub-Clause 10.1 [Taking Over the Works and Sections], before the issue of the Performance Certificate.*

Thus, the updating process must continue until the expiry of the DNP and the rectification of all notified defects.

Sub-Clause 5.7 [*Operation and Maintenance Manuals*] of YB 2017 and SB 2017 follows a similar approach.

The Contractor must prepare, and keep up to date, a complete set of operation and maintenance manuals for the Works (referred to as the 'O&M Manuals') in accordance with the format and other relevant details stated in the Employer's Requirements.

These initial versions of the O&M Manuals must be in sufficient detail for the Employer to:

(*a*) operate, maintain and adjust the Works to ensure that the performance of the Works, Section and/or Plant (as the case may be) continues to comply with the performance criteria specified in the Employer's Requirements and the Schedule of Performance Guarantees; and
(*b*) operate, maintain, dismantle, reassemble, adjust and repair the Plant.

They must also include an inventory of spare parts required for the Employer's future operation and maintenance of the Plant.

Before commencement of the Tests on Completion, the Contractor must submit provisional versions of the O&M Manuals for the Works or Section (as the case may be) to the Engineer under Sub-Clause 5.2.2 [*Review by Engineer*].

However, if any error or defect is found in the provisional O&M Manuals during the Tests on Completion, the Contractor must promptly rectify the error or defect at its own risk and cost.

Thereafter and before the issue of any TOC under Sub-Clause 10.1 [*Taking Over the Works and Sections*], the final O&M Manuals shall be submitted to the Engineer under Sub-Clause 5.2.2 [*Review by Engineer*].

The final version of the O&M Manuals must be the subject of a Notice of No-objection from the Engineer, otherwise it cannot be final. Therefore, any modification of the O&M Manuals subsequent to the Tests on Completion, could entail a Review period of 21 days and a delay in the issue of the TOC.

> **Note:**
>
> The final paragraph of Sub-Clause 5.7 [*Operation and Maintenance Manuals*] states that the final version of the O&M Manuals must be issued before the issue of the TOC. It does not state that the final version must be issued before the date of Taking Over.
>
> Sub-Clause 10.1 [*Taking Over the Works and Sections*] states that prior to taking over, the Engineer must have given (or be deemed to have given) a Notice of No-objection to the **provisional** O&M Manuals. Thus, a requirement for modification of the O&M Manuals after the Tests on Completion could delay the issue of the TOC **but not the date of taking over to be stated in the TOC**.

7.6. Training

There is no obvious link between design and training. However, FIDIC adds provisions related to training to those related to design, which include the requirements for as-built drawings, O&M Manuals and spare parts – all as part of the requirements for the taking over of the Works by the Employer. The relevant sub-clause in RB 2017 is Sub-Clause 4.5 [*Training*]. In YB 2017 and SB 2017, it is Sub-Clause 5.5 [*Training*].

Under RB 2017, if no training by the Contractor of employees of the Employer (and/or other identified personnel) is stated in the Specification, Sub-Clause 4.5 [*Training*] does not apply.

Under YB 2017 and SB 2017, the Contractor's obligation is to train the employees of the Employer (and/or other personnel identified in the Employer's Requirements) to the extent specified.

The timing of the training must be as stated in the Specification/Employer's Requirements (if not stated, as acceptable to the Employer). If the Specification/Employer's Requirements specifiy training which is to be carried out before taking over, the Works are not to be considered completed for the purposes of taking over under Sub-Clause 10.1 [*Taking Over the Works and Sections*] until this training has been completed.

In all cases, it is the responsibility of the Contractor to provide qualified and experienced training staff, training facilities and all training materials as necessary and/or as stated in the Specification/Employer's Requirements.

It is the responsibility of the Employer to provide suitably qualified and experienced trainees at the required time(s).

Hint:

Contractors should be attentive with respect to the details of such training:

- How much training is to be given on Site, how much in the Contractor's home country or at suppliers' premises?
- Who is to pay for travel expenses and accommodation?
- Who is to organise visas?
- Is the training to achieve a guaranteed standard?
- What qualifications or experience are expected of trainees prior to the training?

Smith G
ISBN 978-0-7277-6652-6
https://doi.org/10.1680/fcmh.66526.097

Chapter 8
Project start-up

8.1. Performance, Environmental and Social Performance, and Advance Payment Securities

8.1.1 Introduction

One of the first obligations borne by the Contractor is to provide the Performance Security, which is to ensure the Contractor's proper performance of the Contract. Under RB 2017 and YB 2017, it is to be provided within 28 days of receiving the Letter of Acceptance. Under SB 2017, it is to be provided within 28 days of signature of the Contract Agreement by both Parties. If the Contractor fails to do so, the Employer is permitted to terminate the Contract under Sub-Clause 15.2 [*Termination for Contractor's Default*] and award the Contract to a different bidder. For this purpose, the bid security provided by each bidder is normally to be valid until the successful bidder provides a satisfactory Performance Security.

For projects financed by the World Bank and some other MDBs, in addition to the Performance Security, the Contractor must submit an Environmental and Social (ES) Performance Security, which is designed to ensure that the Contractor fulfils its obligations with respect to ES aspects. It is also to be given within 28 days of receipt of the Letter of Acceptance.

Another security which is to be supplied by the Contractor within the early days of the project is the Advance Payment Guarantee, which is to provide some security to the Employer in return for making an advance payment to the Contractor to facilitate mobilisation and cash-flow.

Unlike the Performance Security and the ES Performance Security, there is no time limit imposed for provision of the Advance Payment Guarantee but unless the Contractor provides the guarantee, it will not be entitled to receive the advance payment.

Moreover, under Sub-Clause 8.1 [*Commencement of Work*] of the Special Provisions imposed by the World Bank and some other MDBs, the provision of the Advance Payment Guarantee is one of the conditions which must be satisfied before the Engineer can fix the Commencement Date. ADB requires the Advance Payment Security to be provided within 28 days after receipt of the Letter of Acceptance, otherwise its provision is not a condition for the fixing of the Commencement Date.

8.1.2 Performance Security

Under Sub-Clause 4.2 [*Performance Security*] the Contractor must provide a Performance Security at its own expense. The Performance Security must be in the amount and currencies stated in the Contract Data. If no amount is stated in the Contract Data, the sub-clause does not apply.

The Contractor must deliver the Performance Security to the Employer and provide a copy to the Engineer. It must be in the form annexed to the Particular Conditions, or in another form agreed by the Employer. The Performance Security must be issued by an entity and from within a country to which the Employer has given consent.

Note:

The form of Performance Security proposed by FIDIC states that it is subject to the Uniform Rules for Demand Guarantees (URDG) 2010 Revision, ICC Publication No. 758. Articles 34 and 35 of these rules define the governing law and the applicable jurisdiction in the event of a dispute related to the guarantee.

Article 34 (a) states that unless otherwise provided in the guarantee, the governing law shall be that of the location of the guarantor's branch or office that issued the guarantee.

Article 35 (a) states that unless otherwise provided in the guarantee, any dispute between the guarantor and the beneficiary relating to the guarantee shall be settled exclusively by the competent court of the country of the location of the guarantor's branch or office that issued the guarantee.

In most cases, the Contractor will provide a guarantee issued by a bank based in the Contractor's home country. If so, Article 34 (a) of URDG 2010 imposes the law of the Contractor's country as the governing law and Article 35 (a) would require the Employer to commence proceedings in a court in the Contractor's country should a dispute arise with respect to the Performance Security. This means that, in the event of a dispute related to the application of the Performance Security, the Employer must commence proceedings in the country of the Contractor (or of the issuing bank if not the country of the Contractor) and be subject to the law of that country.

For this reason, some MDBs insist that the Performance Security must be issued by a bank or other entity in the Employer's country, with the Contractor providing a counter-guarantee when necessary.

The Contractor must ensure that the Performance Security remains valid and enforceable until the issue of the Performance Certificate after the end of the DNP and the rectification of all notified defects for which the contractor is liable, and the Contractor has cleared the Site in compliance with Sub-Clause 11.11 [*Clearance of Site*].

Banks and other issuing organisations rarely issue Performance Securities which have no cut-off date. Thus, the wording of the Performance Security would usually state under FIDIC 1999 contracts that it was valid until a fixed calendar date, for example, 31 December 2023, or the issue of the Performance Certificate whichever occurred earlier. Under FIDIC 2017 contracts, the Performance Security must fix as a latest cut-off date the clearance of the Site in accordance with Sub-Clause 11.11. Whereas the date of issue of the Performance Certificate was relatively easy to establish, the date of completion of the clearance of the Site is likely to be more open to dispute.

If the terms of the Performance Security specify such an expiry date, and the Contractor has not become entitled to receive the Performance Certificate by the date 28 days before the expiry date, the Contractor must extend the validity of the Performance Security until the issue of the Performance Certificate and compliance with Sub-Clause 11.11 [*Clearance of Site*].

It is part of the Engineer's role to monitor the expiry date of the Performance Security, to remind the Contractor of the obligation to extend the validity and to warn the Employer in sufficient time for the Employer to act if the validity is not extended.

Whenever Variations and/or adjustments under Clause 13 [*Variations and Adjustments*] result in an accumulative increase of the Contract Price by more than 20% of the Accepted Contract Amount, the Employer may require the Contractor to increase the amount of the Performance Security by a percentage equal to the accumulated increase. Any Cost incurred by the Contractor as a result of this Employer's request, is to be treated as a Variation under Sub-Clause 13.3.1 [*Variation by Instruction*].

In the case of a decrease in the Contract Price, the Contractor may decrease, subject to the Employer's prior consent, the amount of the Performance Security by a percentage equal to the accumulative decrease. The Contract does not address the situation when the Employer refuses its consent to such a decrease. In the event of the Employer's refusal or failure to give consent, the Contractor will have no choice but to refer the dispute to the DAAB.

The trigger for such increases or decreases in the amount of the Performance Security is more clearly defined than under FIDIC 1999 and the Pink Book 2010, wherein it was said to be **a** Variation (i.e. a single but major Variation) **or** adjustments (i.e. excluding Variations) which exceeded 20% of the Accepted Contract Amount.

8.1.3 Claims under the Performance Security

The Employer must not make a claim under the Performance Security, except for amounts to which the Employer is entitled under the Contract. The circumstances which may give rise to such a claim under RB 2017 are:

(*a*) failure by the Contractor to extend the validity of the Performance Security, in which event the Employer may claim the full remaining amount of the Performance Security;
(*b*) failure by the Contractor to pay the Employer an amount due, as agreed or determined under Sub-Clause 3.7 [*Agreement or Determination*] or agreed or decided under Clause 21 [*Disputes and Arbitration*], within 42 days of the date of the agreement or determination or decision or arbitral award (as the case may be);
(*c*) failure by the Contractor to remedy a default stated in a Notice given under Sub-Clause 15.1 [*Notice to Correct*] within 42 days or other time (if any) stated in the Notice;
(*d*) circumstances which entitle the Employer to terminate the Contract under Sub-Clause 15.2 [*Termination for Contractor's Default*], irrespective of whether a Notice of termination has been given; or
(*e*) if under Sub-Clause 11.5 [*Remedying of Defective Work off Site*] the Contractor removes any defective or damaged Plant from the Site, failure by the Contractor to repair such Plant, return it to the Site, reinstall it and retest it by the date of expiry of the relevant duration stated in the Contractor's Notice (or other date agreed by the Employer).

If the Employer makes a claim for an amount to which it is not entitled, it must indemnify and hold the Contractor harmless against and from all damages, losses and expenses (including legal fees and expenses) to the extent that the Employer was not entitled to make the claim.

If the Employer is paid an amount under the Performance Security, the amount must be taken into account:

(*a*) in the Final Payment Certificate under Sub-Clause 14.13 [*Issue of FPC*]; or
(*b*) if the Contract is terminated, in calculating any payment due to the Contractor under Sub-Clause 15.4 [*Payment after Termination for Contractor's Default*], Sub-Clause 15.7 [*Payment after Termination for Employer's Convenience*], Sub-Clause 16.4 [*Payment after Termination by Contractor*], Sub-Clause 18.5 [*Optional Termination*], or Sub-Clause 18.6 [*Release from Performance under the Law*].

Under Sub-Clause 4.2.2 [*Claims under the Performance Security*] of the Special Provisions imposed by the World Bank and some other MDBs, *'The Employer shall not make a claim under the Performance Security, except for amounts for which the Employer is entitled under the Contract.'*

There is no other mention in the Contract of any entitlement to payment under the Performance Security and almost nothing in the Contract entitles the Employer to amounts from the Contractor without first following the claims procedure set out under Sub-Clause 20.2 [*Claims for Payment and/or EOT*]. Indeed, under Sub-Clause 20.2.7 [*General Requirements*], it is expressly stated that the Employer will only be entitled to claim any payment from the Contractor (or set off against or make any deduction from any amount due to the Contractor) by complying with this Sub-Clause 20.2.

8.1.4 Return of the Performance Security

The Employer must return the Performance Security to the Contractor:

- within 21 days of the issue of the Performance Certificate and the Contractor's compliance with Sub-Clause 11.11 [*Clearance of Site*]; (28 days under the World Bank's Special Provisions); or
- promptly after the date of termination if the Contract is terminated in accordance with Sub-Clause 15.5 [*Termination for Employer's Convenience*], Sub-Clause 16.2 [*Termination by Contractor*], Sub-Clause 18.5 [*Optional Termination*] or Sub-Clause 18.6 [*Release from Performance under the Law*].

Although the date of issue of the Performance Certificate is usually clear, the situation with respect to the deemed issue of the Performance Certificate under Sub-Clause 11.9 [*Performance Certificate*] is less clear, as is the date when the Contractor fulfils its obligations under Sub-Clause 11.11 [*Clearance of Site*].

The relevant part of Sub-Clause 11.9 [*Performance Certificate*] of YB 2017 states:

> *'The Engineer shall issue the Performance Certificate to the Contractor (with a copy to the Employer and to the DAAB) within 28 days after the latest of the expiry dates of the Defects Notification Periods, or as soon thereafter as the Contractor has:*
>
> (*a*) *supplied all the Contractor's Documents, and the Engineer has given (or is deemed to have given) a Notice of No-objection to the as-built records under sub-paragraph (b) of Sub-Clause 5.6 [As-Built Records]; and*
> (*b*) *completed and tested all the Works (including remedying any defects) in accordance with the Contract.*

If the Engineer fails to issue the Performance Certificate within this period of 28 days, the Performance Certificate shall be deemed to have been issued on the date 28 days after the date on which it should have been issued, as required by this Sub-Clause.'

Any Contractor who seeks to recover its Performance Security on the basis that the Performance Certificate is deemed to have been issued and that it has fully complied with its obligations related to site clearance, must be very attentive to the demonstration of the fulfilment of the conditions.

8.1.5 Environmental and Social Performance Security

The Special Provisions imposed by the World Bank and some other MDBs replace the standard wording of the first paragraph of Sub-Clause 4.2 [*Performance Security and ES Performance Security*] with:

'The Contractor shall obtain (at its cost) a Performance Security for proper performance and, if applicable, an Environmental and Social (ES) Performance Security for compliance with the Contractor's ES obligations, in the amounts stated in the Contract Data and denominated in the currency(ies) of the Contract or in a freely convertible currency acceptable to the Employer. If amounts are not stated in the Contract Data, this Sub-Clause shall not apply.'

In the following Sub-Clauses of the General Conditions, the term 'Performance Security' is replaced with 'Performance Security and, if applicable, an Environmental and Social (ES) Performance Security':

2.1 – 'Right of Access to the Site;
14.2 – Advance Payment;
14.6 – Issue of IPC;
14.12 – Discharge;
14.13 – Issue of FPC;
14.14 – Cessation of Employer's Liability;
15.2 – Termination for Contractor's Default;
15.5 – Termination for Employer's Convenience.'

The first paragraph of Sub-Clause 4.2.1. [*Contractor's Obligations*] is replaced with:

'The Contractor shall deliver the Performance Security and, if applicable, an ES Performance Security to the Employer within 28 days after receiving the Letter of Acceptance and shall send a copy to the Engineer. The Performance Security shall be issued by a reputable bank or financial institution selected by the Contractor and shall be in the form annexed to the Particular Conditions, as stipulated by the Employer in the Contract Data, or in another form approved by the Employer. The ES Performance Security shall be issued by a reputable bank selected by the Contractor and shall be in the form annexed to the Particular Conditions, as stipulated by the Employer in the Contract Data, or in another form approved by the Employer.

Thereafter, throughout Sub-Clause 4.2 'Performance Security' is replaced with: 'Performance Security and, if applicable, ES Performance Security.'

In these respects, the Contract Data states:

'The ES Performance Security will be in the form of a 'demand guarantee' in the amount(s) of [insert % figure(s) normally 1% to 3%] of the Accepted Contract Amount and in the same currency(ies) of the Accepted Contract Amount.

[The sum of the total 'demand guarantees' (Performance Security and ES Performance Security) shall normally not exceed 10% of the Accepted Contract Amount.]'

8.1.6 Advance Payment Guarantee

The Contractor must obtain (at the Contractor's own cost) an Advance Payment Guarantee in amounts and currencies equal to the advance payment and must submit it to the Employer (with a copy to the Engineer). This guarantee must be issued by an entity and from within a country to which the Employer has given consent and must be based on the sample form included in the tender documents or on another form agreed by the Employer.

The Contractor must maintain the validity and enforceability of the Advance Payment Guarantee until the advance payment has been fully repaid. However, its amount is usually reduced by the amount repaid by the Contractor as stated in the Interim Payment Certificates.

If the terms of the Advance Payment Guarantee specify its expiry date, and the advance payment has not been repaid by the date 28 days before the expiry date:

- the Contractor must extend the validity of this guarantee until the advance payment has been repaid;
- the Contractor must immediately submit evidence of this extension to the Employer, (with a copy to the Engineer); and
- if the Employer does not receive this evidence 7 days before the expiry date of this guarantee, the Employer will be entitled to claim under the guarantee the amount of advance payment which has not been repaid.

Similar to the Performance Security, it is part of the Engineer's role to monitor the expiry date of the Advance Payment Security, to remind the Contractor of the obligation to extend the validity and to warn the Employer in sufficient time for the Employer to act if the validity is not extended.

8.2. Advance Payment

The amount of the advance payment and the currencies in which it is to be paid must be stated in the Contract Data. If no amount of advance payment is stated in the Contract Data, Sub-Clause 14.2 [*Advance Payment*] does not apply.

When submitting the Advance Payment Guarantee (see above), the Contractor must include an application (in the form of a Statement) for the advance payment.

8.2.1 Advance Payment Certificate

The Engineer must issue an Advance Payment Certificate for the advance payment within 14 days of:

- the Employer receiving both the Performance Security and the Advance Payment Guarantee, in the form stipulated in, and issued by an entity in accordance with, Sub-Clause 4.2.1 [*Contractor's Obligations*] and Sub-Clause 14.2.1 [*Advance Payment Guarantee*] respectively; and
- the Engineer receiving a copy of the Contractor's application for the advance payment under Sub-Clause 14.2.1 [*Advance Payment Guarantee*].

After receiving the Advance Payment Certificate, the Employer must make the advance payment within the period stated in the Contract Data. If no period is stated, payment must be made within 21 days of receipt of the Advance Payment Certificate (Sub-Clause 14.7 [*Payment*]).

8.2.2 Repayment of Advance Payment

The advance payment is described as an interest-free loan for mobilisation (and design, if any). This loan must be repaid by means of deductions in Payment Certificates.

Unless other percentages are stated in the Contract Data:

- deductions commence in the IPC in which the total of all certified interim payments in the same currency as the advance payment (excluding the advance payment and deductions and release of retention money) exceeds 10% of the portion of the Accepted Contract Amount payable in that currency less Provisional Sums; and
- deductions are made at the rate of one quarter (25%) of the amount of each IPC (excluding the advance payment and deductions and release of retention money) in the currencies and proportions of the advance payment, until the advance payment has been fully repaid.

If the advance payment has not been fully repaid before the issue of the Taking-Over Certificate for the Works, or before termination under Clause 15 [*Termination by Employer*], Clause 16 [*Suspension and Termination by Contractor*] or Clause 18 [*Exceptional Events*], the whole of the remaining amount immediately becomes due and payable by the Contractor to the Employer.

8.3. Notice of Commencement

Under Sub-Clause 8.1 [*Commencement of Works*] of RB 2017 and YB 2017, the Engineer must give a Notice to the Contractor stating the Commencement Date. The Notice must be given at least 14 days before the Commencement Date. The Commencement Date must be within 42 days of the Contractor receiving the Letter of Acceptance, unless otherwise stated in the Particular Conditions. Thus the Notice must be given within 28 days of the Contractor receiving the Letter of Acceptance.

Under Sub-Clause 8.1 [*Commencement of Works*] of SB 2017, unless the Particular Conditions state otherwise, the Commencement Date must be within 42 days of the date on which the Contract comes into full force and effect under Sub-Clause 1.6 [*Contract Agreement*] which is the date stated in the Contract Agreement. Unless the Contract Agreement fixes the Commencement Date, the Employer is to give a Notice of the Commencement Date to the Contractor, at least 14 days before the Commencement Date.

The Contractor must commence the execution of the Works on, or as soon as is reasonably practicable after, the Commencement Date and must proceed with due expedition and without delay.

The timeline for activities leading to the Commencement Date under RB 2017 and YB 2017 is shown in Figure 8.1.

It can be seen from the timeline that the Advance Payment is largely unrelated to the Commencement Date and is unlikely to be received by the Contractor before the Commencement Date unless the Contractor provides the Advance Payment Security and its Statement soon after receipt of the Letter of Acceptance.

Theoretically, the Engineer may issue the Notice of Commencement Date before the Contract Agreement has been signed and before the Performance Security has been provided by the Contractor. In practice, this is unlikely to happen and therefore there is a high probability that the Commencement Date will not occur within 42 days of the receipt by the Contractor of the Letter of Acceptance, unless the Engineer is immediately informed by the Employer that an acceptable Performance Security has been provided and the Engineer is equally rapid in issuing the Notice of the Commencement Date.

Notwithstanding a failure to fix a Commencement Date within 42 days of receipt of the Letter of Acceptance, the Contractor may not give Notice of an intention to terminate the Contract under Sub-Clause 16.2 [*Termination by Contractor*] unless no Notice of the Commencement Date has been received within 84 days of receipt of the Letter of Acceptance.

Example of Notice of Commencement Date under RB 2017:

A.N. OTHER & Partners
Consulting Engineers

BETTA CONTRACTORS LTD.

31 January 2018

For the attention of Mr. Alright, Contractor's Representative

Subject: **NOTICE OF COMMENCEMENT DATE**

Dear Sir,

You are hereby notified in accordance with Sub-Clause 8.1 [*Commencement of Works*] of the Contract that the Commencement Date shall be .... [*Insert date not less than 14 days after the date of this Notice*].

The Contractor is to commence the execution of the Works as soon as is reasonably practicable after the Commencement Date, and shall then proceed with the Works with due expedition and without delay.

Yours faithfully,

The Engineer
cc The Employer

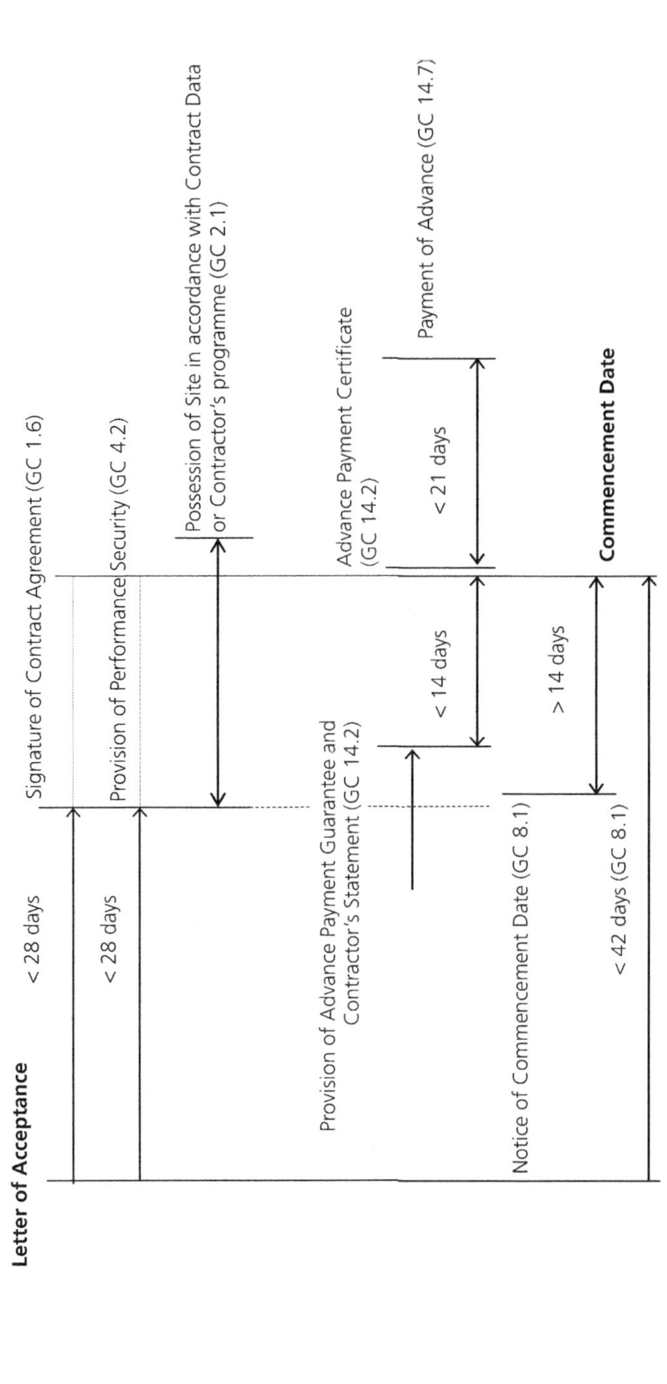

Figure 8.1 Commencement Date under RB 2017 and YB 2017

Under the Special Provisions imposed by the World Bank and some other MDBs, the standard wording of Sub-Clause 8.1 [*Commencement of Works*] is replaced in its entirety by the following:

> *'The Engineer shall give a Notice to the Contractor stating the Commencement Date, not less than 14 days before the Commencement Date. The Notice shall be issued promptly after the Engineer determines the fulfilment of the following conditions:*
>
> (a) *Signature of the Contract Agreement by both Parties and, if required, approval of the Contract by relevant authorities of the Country;*
> (b) *delivery to the Contractor of reasonable evidence of the Employer's financial arrangements (under Sub-Clause 2.4 [Employer's Financial Arrangements]);*
> (c) *except if otherwise specified in the Contract Data, effective access to and possession of the Site given to the Contractor together with such permission(s) under (a) of Sub-Clause 1.13 [Compliance with Laws] as required for the commencement of the Works;*
> (d) *receipt by the Contractor of the Advance Payment under Sub-Clause 14.2 [Advance Payment] provided that the corresponding bank guarantee has been delivered by the Contractor;*
> (e) *constitution of the DAAB in accordance with Sub-Clause 21.1 and Sub-Clause 21.2 as applicable.*
>
> *Subject to Sub-Clause 4.1 on the Management Strategies and Implementation Plans and the C-ESMP and Sub-Clause 4.8 on the health and safety manual, the Contractor shall commence the execution of the Works as soon as is reasonably practicable after the Commencement Date, and shall then proceed with the Works with due expedition and without delay.'*

The impact of this revised wording on the activities leading to the Commencement Date is shown in Figure 8.2.

It can be seen that the process for Commencement is much more complex for projects financed by the MDB. The reason for the complexity is to encourage the Employer to resolve the issues which often beset the start of infrastructure projects in developing countries before the Contractor mobilises its resources. If these issues were not resolved before the Commencement Date, the Contractor would be entitled to claim the costs of resources which had been mobilised but were unable to work.

It should be noted that under Sub-Clause 4.1 of the Special Provisions, even though the application of the Contractor's MSIPs is not a condition to be satisfied before the Commencement Date, the Contractor is not permitted to mobilise resources to Site until the Engineer gives a Notice of No-objection with respect to the Contractor's proposals for managing the ES risks and impacts.

Figure 8.2 Commencement Date under WB Special Provisions

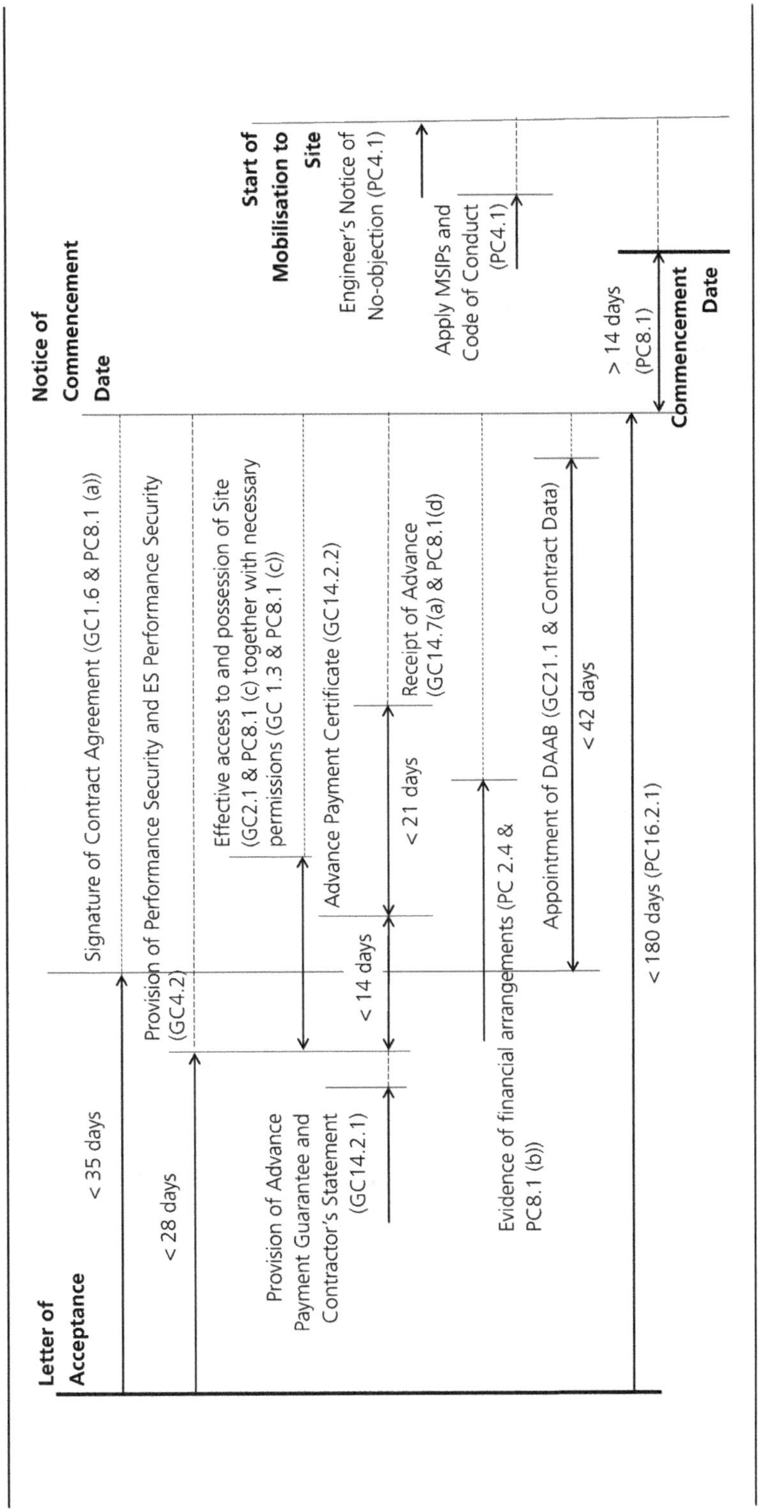

Example of Notice of Commencement Date under RB 2017 for a WB financed project:

A.N. OTHER & Partners
Consulting Engineers

BETTA CONTRACTORS LTD.

31 January 2018

For the attention of Mr. Alright, Contractor's Representative

Subject: **NOTICE OF COMMENCEMENT DATE**

Dear Sir,

You are hereby notified in accordance with Sub-Clause 8.1 [*Commencement of Works*] of the General Conditions of Contract as modified by the Special Provisions, that the Commencement Date shall be .... [*Insert date not less than 14 days after the date of this Notice*].

In this respect, the Engineer confirms that all the below-mentioned conditions have been satisfied:
(i) signature of the Contract Agreement by both Parties, [and approval of the Contract by relevant authorities of the Country];
(ii) delivery to the Contractor of reasonable evidence of the Employer's financial arrangements (under Sub-Clause 2.4 [*Employer's Financial Arrangements*]);
(iii) effective access to and possession of the Site given to the Contractor together with such permission(s) under (a) of Sub-Clause 1.13 [*Compliance with Laws*] as required for the commencement of the Works;
(iv) receipt by the Contractor of the Advance Payment under Sub-Clause 14.2 [*Advance Payment*];
(v) constitution of the DAAB.

Subject to Sub-Clause 4.1 on the Management Strategies and Implementation Plans and the C-ESMP and Sub-Clause 4.8 on the health and safety manual, the Contractor is to commence the execution of the Works as soon as is reasonably practicable after the Commencement Date and shall then proceed with the Works with due expedition and without delay.

Yours faithfully,

The Engineer
cc The Employer

8.4. Means of access and additional working areas

The Employer's obligations with respect to providing the Contractor with the right of access to and possession of the Site are described in Chapter 4.1 above. The Employer's obligations, if any, with respect to the means of access are also addressed.

However, under Sub-Clause 4.15 [*Access Route*] the Contractor is deemed to have satisfied himself, prior to the Base Date, as to the suitability and availability of the access routes to the Site. The Employer does not guarantee the suitability or availability of particular access routes; and all Costs incurred by the Contractor due to their non-suitability or non-availability are generally to be borne by the Contractor.

The exception to this general rule is when the non-suitability or non-availability of an access route arises as a result of changes to that access route by the Employer or a third party after the Base Date. If the Contractor suffers delay and/or incurs Cost due to such changes, the Contractor is entitled to an EOT and/or payment of such Cost subject to compliance with Sub-Clause 20.2 [*Claims for Payment and/or EOT*].

The Contractor must take all necessary measures to prevent any road or bridge from being damaged by the Contractor's vehicles or by the Contractor's Personnel, such as ensuring compliance with legal load and width limits or other restrictions. The Contractor must repair and maintain the access routes and must obtain any permissions or permits which may be required from the relevant authorities for the Contractor's use of the routes and for the installation of signs and directions.

If any third-party claims arise from the Contractor's use or otherwise of any access route, the Employer shall have no responsibility in this respect.

Under Sub-Clause 4.13 [*Rights of Way and Facilities*], the Contractor must bear all costs and charges related to any special and/or temporary rights of way which may be required, including for access to the Site. The Contractor is also responsible for obtaining any additional facilities needed for the Works such as off-Site storage or assembly areas.

8.5. Permits

In general, the Employer is responsible for obtaining the permits, licences and approvals necessary for the Permanent Works whereas the Contractor is responsible for obtaining those required for the execution of the Works (Sub-Clause 1.13 [*Compliance with Laws*]). The Party responsible for obtaining a permit must indemnify the other Party in relation to any failure to do so (unless the other Party contributed to the failure).

If the Contractor suffers delay and/or incurs Cost as a result of the Employer's delay or failure to obtain any permit, the Contractor is entitled, subject to Sub-Clause 20.2 [*Claims for Payment and/or EOT*], to EOT and/or payment of such Cost Plus Profit. If the Employer incurs additional costs as a result of the Contractor's failure, the Employer is entitled, subject to Sub-Clause 20.2 [*Claims for Payment and/or EOT*], to reimbursement of these costs by the Contractor.

With respect to the Contractor's obligation to obtain permits, it should be remembered that the Employer is under an obligation to provide the Contractor with 'reasonable assistance', if the Contractor requests this (Chapter 4.2 above and Sub-Clause 2.2 [*Assistance*]).

8.6. Mobilisation

The Contractor must commence the execution of the Works on, or as soon as is reasonably practicable after the Commencement Date. However, several obligations must be fulfilled in the days which follow the receipt of the Notice of Commencement Date, some of which condition the start of work.

Under Sub-Clause 4.8 [*Health and Safety Obligations*], within 21 days of the Commencement Date and before commencing any construction on the Site, the Contractor must submit the project-specific health and safety manual to the Engineer '*for information*' (Chapter 6.12 above).

Under Sub-Clause 4.9 [*Quality Management and Compliance Verification Systems*] the Contractor must submit for Review by the Engineer a project-specific QM System. The submission must be within

28 days of the Commencement Date. No period for Review is stated. However, if the QM System is defined in the Specification to be a 'Contractor's Document', Sub-Clause 4.4.1 [*Preparation and Review*] will apply and the Review period will be 21 days.

Under Sub-Clause 8.3 [*Programme*], the Contractor must submit to the Engineer its initial programme for the execution of the Works, within 28 days of receiving the Notice of the Commencement Date.

Under Sub-Clause 4.1 [*Contractor's General Obligations*] of the Special Provisions imposed by the World Bank and some other MDBs:

> *'The Contractor shall not carry out mobilization to Site (e.g. limited clearance for haul roads, site accesses and work site establishment, geotechnical investigations or investigations to select ancillary features such as quarries and borrow pits) unless the Engineer gives a Notice of No-objection to the Contractor, a Notice that shall not be unreasonably delayed, to the measures the Contractor proposes to manage the environmental and social risks and impacts, which at a minimum shall include applying the Management Strategies and Implementation Plans (MSIPs) and Code of Conduct for Contractor's Personnel submitted as part of the Bid and agreed as part of the Contract.'*

Under Sub-Clause 5.1 [*General Design Obligations*] of YB 2017, promptly after receiving the Notice of Commencement Date, the Contractor must scrutinise the Employer's Requirements and notify the Engineer of any error or omission (see Chapter 7.2 above).

Under Sub-Clause 14.4 [*Schedule of Payments*], unless the Contract contains a Schedule of Payments specifying the instalments in which the Contract Price will be paid, within 42 days of the Commencement Date the Contractor must submit an initial non-binding estimate of the payments which the Contractor expects to become due during each period of 3 months. Revised estimates must be submitted at intervals of 3 months thereafter. It should be noted that this provision does not oblige the Contractor to provide a 'cash-flow' forecast for the entire Time for Completion but only for the three-month period after preparation of each forecast.

8.7. Checklist for Commencement of Works

Sub-Clause	Action	RB 2017	By	WB Special Provisions	By
GC 1.6 & WB PC 1.6	Sign Contract Agreement	Within 28 days of receipt of Letter of Acceptance	E & C	Within 35 days of receipt of Letter of Acceptance	E & C
GC 2.4 & WB 8.1	Provide evidence of financial arrangements	In bidding documents	E	No deadline	E
GC 4.2	Issue Performance Security	Within 28 days of receipt of Letter of Acceptance	C	Within 28 days of receipt of Letter of Acceptance	C
WB PC 4.2	Issue ES Performance Security			Within 28 days of receipt of Letter of Acceptance	C

Sub-Clause	Action	RB 2017	By	WB Special Provisions	By
GC 4.3	Request consent with respect to Contractor's Representative	Before Commencement Date	C	Before Commencement Date	C
GC 14.2.1 WB PC 8.1	Issue Statement for Advance Payment	No deadline	C	No deadline	C
GC 14.2.2 WB PC 8.1	Issue Advance Payment Guarantee	No deadline	C	No deadline	C
GC 14.2.2 WB PC 8.1	Issue Advance Payment Certificate	Within 14 days of receipt of Advance Payment Guarantee and Statement	Eng	Within 14 days of receipt of Advance Payment Guarantee and Statement	Eng
GC 14.7 (a) WB PC 8.1	Pay Advance Payment	Within 21 days of receipt of Advance Payment Certificate or as indicated in Contract Data	E	Within 21 days of receipt of Advance Payment Certificate or as indicated in Contract Data	E
GC 2.1 WB PC 8.1	Provide right of access to and possession of Site	As indicated in Contract Data or to match Programme and after receipt of Performance Security	E	As indicated in Contract Data or to match Programme and after receipt of Performance Security	E
GC 1.13 (a) WB PC 8.1	Provide permits necessary for commencement of Works which are Employer's responsibility	No deadline	E	No deadline	E
GC 21.1 WB PC 8.1 & 21.1	Appoint DAAB	Within 28 days of receipt of Letter of Acceptance	E & C	Within 42 days of signature of Contract Agreement	E & C
GC 8.1 WB PC 8.1	Issue Notice of Commencement Date	At least 14 days before Commencement Date	Eng	After fulfilment of all conditions listed under PC 8.1 and at least 14 days before Commencement Date	Eng
GC 19.2	Provide Insurances	Before Commencement Date	C	Before Commencement Date	C
GC 8.3	Submit initial programme for execution of Works	Within 28 days of receipt of Notice of Commencement Date	C	Within 28 days of receipt of Notice of Commencement Date	C
GC 4.7	Verify accuracy of items of reference	Within 28 days of Commencement Date	C	Within 28 days of Commencement Date	C

Sub-Clause	Action	RB 2017	By	WB Special Provisions	By
GC 4.8 WB PC 4.8	Submit specific Health & Safety Manual	Within 21 days of Commencement Date and before commencing any construction on Site	C	No deadline	C
WB PC 4.1	Apply MSIPs & Code of Conduct			Before mobilisation	C
GC 4.9	Submit specific QM System	Within 28 days of Commencement Date	C	Within 28 days of Commencement Date	C
WB PC 4.21	Submit security management plan			No deadline	C
GC 5.1	Submit subcontractors' details for consent	More than 42 days before planned commencement of Subcontractor's work	C	More than 42 days before planned commencement of Subcontractor's work	C

E = Employer
Eng = Engineer
C = Contractor

Smith G
ISBN 978-0-7277-6652-6
https://doi.org/10.1680/fcmh.66526.113

Chapter 9
Quality

9.1. Quality Assurance

Under Sub-Clause 4.9 [*Quality Management and Compliance Verification Systems*], the Contractor must establish a quality management system (QM System) and a compliance verification system (Compliance Verification System).

9.1.1 Quality Management System

The QM System must be specifically prepared for the Works and submitted to the Engineer within 28 days of the Commencement Date. Whenever the QM System is updated or revised, a copy must be promptly submitted to the Engineer.

The Engineer may (but is not obliged to) review the QM System and may give a Notice to the Contractor stating the extent, if any, to which it does not comply with the Contract. Within 14 days of receiving such a Notice, the Contractor must rectify the non-compliance. If the Engineer does not give such a Notice within 21 days of the date of submission of the QM System, the Engineer is deemed to have given a Notice of No-objection.

Thereafter, the Engineer may give a Notice to the Contractor, at any time, stating the extent to which the Contractor is failing to correctly implement the QM System. After receiving this Notice, the Contractor must immediately remedy the failure.

The QM System must be in accordance with the requirements of the Specification (if any). It must include three parts:

(i) Details of the Contractor's document management system to ensure that all Notices and other communications under Sub-Clause 1.3 [*Notices and Other Communications*], Contractor's Documents, as-built records (if applicable), operation and maintenance manuals (if applicable), and contemporary records can be traced, with full certainty, to the Works, Goods, work, workmanship or test to which they relate.
(ii) Details of the Contractor's system to ensure proper coordination and management of interfaces between the stages of execution of the Works, and between Subcontractors.
(iii) Details of the Contractor's system for monitoring the submission and Review by the Engineer of the Contractor's Documents.

The Contractor must audit the QM System regularly (and at least once every 6 months) and must submit to the Engineer, within 7 days of completion of the audit, a report which lists the results. When necessary, the report must propose measures to improve and/or rectify the QM System and/or its implementation.

The Contractor (and each JV partner) must similarly give Notice to the Engineer of any failings identified by any external audit imposed as part of the Contractor's quality assurance certification.

9.1.2 Compliance Verification System

The Contractor must implement a Compliance Verification System to demonstrate that the design (if any), Materials, Plant, and workmanship fully comply with the Contract.

The Compliance Verification System must satisfy the requirements of the Specification and must foresee the reporting of the results of all inspections and tests carried out by the Contractor.

The Contractor must prepare and submit to the Engineer a complete set of compliance verification documentation for the Works or Section (as the case may be), fully compiled and collated in the manner described in the Specification or, if not so described, in a manner acceptable to the Engineer. No indication is given of when this documentation must be provided. Nor is there any cross reference to the as-built records which must be provided under Sub-Clause 4.4.2 [*As-Built Records*] as a pre-requisite for taking over under Sub-Clause 10.1 [*Taking Over the Works and Sections*].

9.2. Samples

The Contractor must submit samples of Materials (and relevant information) in three circumstances:

(i) samples specified in the Contract (at the Contractor's cost);
(ii) manufacturer's standard samples of Materials (at the Contractor's cost); and
(iii) additional samples instructed by the Engineer (as a Variation).

The samples (which must be labelled as to origin and their intended use in the Works) are to be submitted for the consent of the Engineer prior to being used in or for the Works. There is no express time limit by which the Engineer must respond. However, the Engineer is bound by Sub-Clause 1.3 [*Notices and Other Communications*] not to unreasonably delay the response. Similarly, although the Contract does not expressly state the grounds for refusing the consent, Sub-Clause 1.3 [*Notices and Other Communications*] imposes the condition that any consent must not be unreasonably withheld, and this puts the Engineer under an implied obligation to justify any objection.

An example of a request for consent for a sample of Material to be used in the Works is as follows:

BETTA CONTRACTORS LTD.

A.N. OTHER & Partners
Consulting Engineers

15 September 2022

For the attention of Mr. Other, the Engineer

Subject: **Request for consent under Sub-Clause 7.2 for the use of water-stop**

Dear Sir,

In accordance with Sub-Clause 7.2 [*Samples*] we hereby request the Engineer's consent to the use of water-stop produced by ACE Construction Materials Ltd. In this respect, you will find attached a standard sample provided by the manufacturer together with the relevant data sheet.

We would appreciate your early response to our request, but should you require any further information, we would be pleased to provide it.

Yours faithfully,

Contractor's Representative

9.3. Inspections

During normal working hours, as stated in the Contract Data, and at all other reasonable times, under Sub-Clause 7.3 [*Inspection*], the Contractor must allow the Employer's Personnel (including the Engineer and assistants):

- full access to all parts of the Site and to all places from which natural Materials are being obtained (i.e. quarries and borrow pits);
- to examine, inspect, measure and test (to the extent stated in the Specification) the Materials, Plant and workmanship, during production and manufacture as well as during construction (at the Site and elsewhere);
- to check the progress of manufacture of Plant and production and manufacture of Materials;
- to make records (including photographs and/or video recordings); and
- to carry out other duties and inspections, as specified in the Conditions of Contract and the Specification.

The Contractor must allow the Employer's Personnel full opportunity to do so, including by providing safe access, facilities, permissions and safety equipment.

The Contractor must give a Notice to the Engineer whenever any Materials, Plant or work is ready for inspection, and before it is to be covered up, put out of sight or packaged for storage or transport.

The Employer's Personnel must then either carry out the examination, inspection, measurement or testing 'without unreasonable delay', or must promptly give a Notice to the Contractor that it will not do so. If the Engineer gives no such Notice and/or the Employer's Personnel do not attend at the time stated in the Contractor's Notice (or at another time agreed with the Contractor), the Contractor may proceed with covering up, putting out of sight or packaging for storage or transport.

The reference to 'unreasonable delay' by the Employer's Personnel is difficult to understand. Firstly, the question must be asked: what delay is reasonable? Secondly, as the Contractor is entitled to proceed in the absence of the Engineer, with or without Notice from the Engineer, the mention of 'unreasonable delay' appears to be irrelevant.

Note that there is no mention of the time by which the Contractor's Notice must be provided. It is implied that the Engineer must be given reasonable warning of when the inspection will be required, so that they can take the appropriate steps.

If the Contractor has failed to give the Notice, the Engineer may instruct the Contractor to uncover the work, Plant or Materials to allow inspection and thereafter reinstate and make good, all at the Contractor's risk and cost.

An example of a Contractor's Notice of readiness for inspection is as follows:

BETTA CONTRACTORS LTD.

A.N. OTHER & Partners
Consulting Engineers

15 September 2022

For the attention of Mr. Other, the Engineer

Subject: **Notice of readiness for inspection under Sub-Clause 7.3**

Dear Sir,

In accordance with Sub-Clause 7.3 [*Inspection*] we hereby notify you that the following is ready for inspection by the Engineer:

Materials, Plant or work to be inspected: ....

Location: ....

Expected date and time of inspection: ....

Should you choose not to carry out the said inspection, we would be grateful to receive your prompt notification.

Yours faithfully,

Contractor's Representative

cc The Employer

9.4. Tests (other than the Tests after Completion)

Under Sub-Clause 7.4 [*Testing by the Contractor*], the Contractor must provide everything necessary for efficient execution of specified tests (apparatus, assistance, documents and other information, temporary supplies of electricity and water, fuel, consumables, labour and suitably qualified, experienced and competent staff).

If requested by the Engineer, the Contractor must submit calibration certificates for test apparatus before carrying out the testing.

The Contractor must give a Notice to the Engineer, stating the time and place for the specified testing of any Plant, Materials and other parts of the Works. This Notice must be given in reasonable time, having regard to the location of the testing, so that the Employer's Personnel may attend.

An example of a Contractor's Notice of readiness for testing is as follows:

BETTA CONTRACTORS LTD.

A.N. OTHER & Partners
Consulting Engineers

15 September 2022

For the attention of Mr. Other, the Engineer

Subject: **Notice of readiness for testing under Sub-Clause 7.4**

Dear Sir,

In accordance with Sub-Clause 7.4 [*Testing by the Contractor*] and Specification Clause [.... *Insert clause no.* ......] we hereby notify you that the following will shortly be ready for testing:

Materials, Plant or work to be tested: ....

Location: ....

Nature of Test: ....

Expected date and time of test: ....

We request that you notify us not less than 72 hours prior to the said date and time of your intention to attend the test.

Yours faithfully,

Contractor's Representative

cc The Employer

The Engineer may, under Clause 13 [*Variations and Adjustments*], vary the location or timing or details of specified tests, or instruct the Contractor to carry out additional tests. If these varied or additional tests show that the tested Plant, Materials or workmanship is not in accordance with the Contract, the Cost and any delay incurred in carrying out this Variation shall be borne by the Contractor. It is implied that if the varied or additional tests show that the Plant, Materials or workmanship are in accordance with the Contract, the Contractor will be entitled to be paid for the work in executing the Variation and to be granted an EOT with respect to any delay incurred.

Following receipt of the Contractor's Notice with respect to the planned test, the Engineer must give a Notice to the Contractor of their intention to attend the tests not less than 72 hours prior to the planned test. If the Engineer does not attend at the time and place stated in the Contractor's Notice, the Contractor may proceed with the tests, unless otherwise instructed by the Engineer. These tests shall then be deemed to have been made in the Engineer's presence. If the Contractor suffers delay and/or incurs Cost from complying with any such instruction or as a result of a delay for which the Employer is responsible, the Contractor will be entitled subject to Sub-Clause 20.2 [*Claims for Payment and/or EOT*] to EOT and/or payment of Cost Plus Profit.

If the Contractor causes any delay to the specified tests (including varied or additional tests) and that delay causes the Employer to incur costs, the Employer will be entitled subject to Sub-Clause 20.2 [*Claims for Payment and/or EOT*] to payment of these costs by the Contractor.

The Contractor must promptly submit to the Engineer duly certified reports of the tests. When the specified tests have been passed, the Engineer must either endorse the Contractor's test certificate, or issue a test certificate to the Contractor, to that effect. If the Engineer has not attended the tests, they will be deemed to have accepted the readings as accurate.

9.5. Defects and Remedial Work

If any inspection or test shows a non-compliance with the Contract, Sub-Clause 7.5 [*Defects and Rejection*] applies.

In the event of a non-compliance, the Engineer must give a Notice to the Contractor describing the item of Plant, Materials, design or workmanship that has been found to be defective. The Contractor must then promptly prepare and submit a proposal for the necessary remedial work.

An example of an Engineer's Notice of a defect is as follows:

A.N. OTHER & Partners
Consulting Engineers

BETTA CONTRACTORS LTD.

15 October 2022

For the attention of Mr. Alright, Contractor's Representative

Subject: **Notice of a Defect under Sub-Clause 7.5**

Dear Sir,

In accordance with Sub-Clause 7.5 [*Defects and Rejection*] you are hereby notified that the following work is defective and not in accordance with the Specification:

Description of work found to be defective: ....
....

Location: ....

Nature of defect: ....

You are requested to promptly submit for my Review a proposal for the necessary remedial work.

Yours faithfully,

The Engineer

cc The Employer

The Engineer may (but is not obliged to) Review this proposal and give a Notice to the Contractor stating the extent to which the proposed work, if carried out, would not remedy the non-compliance. After receiving such a Notice the Contractor must promptly submit a revised proposal to the Engineer. If the Engineer gives no such Notice within 14 days of receiving the Contractor's proposal (or revised proposal), the Engineer is deemed to have given a Notice of No-objection.

If the Contractor fails to promptly submit a proposal (or revised proposal) for remedial work, or fails to carry out the proposed remedial work to which the Engineer has given (or is deemed to have given) a Notice of No-objection, the Engineer may:

- instruct the Contractor under sub-paragraph (a) and/or (b) of Sub-Clause 7.6 [*Remedial Work*] as explained below; or
- reject the Plant, Materials, workmanship or Contractor's design (if any) by Notice to the Contractor, with reasons, in which case, sub-paragraph (a) of Sub-Clause 11.4 [*Failure to Remedy Defects*] will apply.

Under sub-paragraph (a) of Sub-Clause 11.4 [*Failure to Remedy Defects*] the Employer may carry out the remedial work or have it carried out by others (including any retesting), at the Contractor's cost. The Employer will be entitled subject to Sub-Clause 20.2 [*Claims for Payment and/or EOT*] to payment by the Contractor of the costs reasonably incurred by the Employer in remedying the defect or damage; but the Contractor will have no responsibility for this work.

After remedying defects in any Plant, Materials, design or workmanship, if the Engineer requires any such items to be retested, the tests must be repeated in accordance with Sub-Clause 7.4 [*Testing by the Contractor*] at the Contractor's risk and cost. If the rejection and retesting cause the Employer to incur additional costs, the Employer will be entitled, subject to Sub-Clause 20.2 [*Claims for Payment and/or EOT*], to reimbursement of these costs by the Contractor.

At any time before the issue of the Taking-Over Certificate for the Works, under Sub-Clause 7.6 [*Remedial Work*] the Engineer may instruct the Contractor to:

- repair or remedy or remove and replace any Plant or Materials which are not in accordance with the Contract;
- repair or remedy, or remove and re-execute, any other work which is not in accordance with the Contract; and
- carry out any remedial work which is urgently required for the safety of the Works, whether because of an accident, unforeseeable event or otherwise.

The Contractor must comply with the instruction as soon as practicable and not later than any time which is specified in the instruction, or immediately, if urgency is specified under sub-paragraph (c) above.

The Contractor must bear the cost of all remedial work required under Sub-Clause 7.6 [*Remedial Work*], except to the extent that any work under sub-paragraph (c) above is attributable to:

- any act by the Employer or the Employer's Personnel; or
- an Exceptional Event.

If the work is necessary due to an act by the Employer or the Employer's Personnel and the Contractor suffers delay and/or incurs Cost in carrying out such work, the Contractor will be entitled, subject to Sub-Clause 20.2 [*Claims for Payment and/or EOT*] to EOT and/or payment of the Cost Plus Profit.

If the work is necessary due to an Exceptional Event, Sub-Clause 18.4 [*Consequences of an Exceptional Event*] will apply (see Chapter 15).

If the Contractor fails to comply with the Engineer's instruction given under Sub-Clause 7.6 [*Remedial Work*], the Employer may employ and pay other persons to carry out the work. Except to the extent that the Contractor would have been entitled to payment for the instructed remedial work, the Employer will be entitled, subject to Sub-Clause 20.2 [*Claims for Payment and/or EOT*], to payment by the Contractor of all costs arising from this failure.

Note:

The Employer's right is to be reimbursed '*all costs*' arising from the failure and not only the costs incurred by employing others to carry out the remedial work. Moreover, this entitlement is without prejudice to any other rights the Employer may have, under the Contract or otherwise.

However, the Employer's right to reimbursement is offset against the amount to which the Contractor would have been entitled for defect-free work. The Employer is only entitled to the additional cost incurred due to the Contractor's failure to remedy the defect. The Employer cannot refuse to pay the Contractor because of the defect and deduct the costs of remedial work carried out by others. This would equate to a double punishment for the Contractor.

Smith G
ISBN 978-0-7277-6652-6
https://doi.org/10.1680/fcmh.66526.121

Chapter 10
Timing

10.1. Contractor's Programme

FIDIC Contracts do not normally include a programme as a contract document. This can be seen from Sub-Clause 1.5 [*Priority of Documents*] which lists the documents which form the contract. The list does not mention a programme. There is reference to Schedules but this means 'lists' such as the Dayworks Schedule which sets out hourly rates for labour and Equipment or the Schedule of cost indexation mentioned under Sub-Clause 13.7 [*Adjustments for Changes in Cost*].

The reason for not including a programme as a contract document is that contract documents cannot be changed without an amendment to the contract being agreed by the Parties. Such a requirement would make the programme unusable as a tool for managing the project. It must be possible to modify a programme on a regular basis to take account of actual progress, changes of logic and so forth. Indeed, Sub-Clause 8.3 [*Programme*] and Sub-Clause 4.20 [*Progress Reports*] impose specific requirements for the Contractor's programme to be updated each month, or whenever any programme ceases to reflect actual progress or is otherwise inconsistent with the Contractor's obligations (see below).

Thus, instead of including the Contractor's programme as a contract document, FIDIC contracts require the Contractor to submit a detailed programme soon after award of the contract. Under Sub-Clause 8.3 [*Programme*], the Contractor must submit an initial programme for the execution of the Works to the Engineer within 28 days after receiving the Notice of the Commencement Date under Sub-Clause 8.1 [*Commencement of Works*]. As the Notice of the Commencement Date must be issued at least 14 days prior to the Commencement Date, the Contractor will have no more than 14 days after the Commencement Date in which to provide the detailed programme.

Unlike earlier FIDIC contracts, FIDIC 2017 requires the programme to be prepared using the programming software stated in the Specification (if the programming software is not stated, it must be acceptable to the Engineer). It is important, during the preparation of the bidding documents, to specify the software so that bidders can take the requirements into account. If this is not done, much time may be lost during the early days of the project, arguing about the choice of the software.

FIDIC 2017 is also more specific than earlier FIDIC contracts about the information to be provided in the initial programme and each revised programme. The following must now be included:

(*a*) the Commencement Date and the Time for Completion, of the Works and of each Section (if any);
(*b*) the date on which the right of access to and possession of (each part of) the Site is to be given to the Contractor in accordance with the time (or times) stated in the Contract Data, or if not so stated, the dates at which the Contractor requires the Employer to give such right of access to and possession of (each part of) the Site;

(*c*) the order in which the Contractor intends to carry out the Works, including the anticipated timing of each stage of design (if any), preparation and submission of Contractor's Documents, procurement, manufacture, inspection, delivery to Site, construction, erection, installation, work to be undertaken by any nominated Subcontractor (as defined in Sub-Clause 5.2 [*Nominated Subcontractors*]) and testing;
(*d*) the Review periods for any submissions stated in the Specification or required under the Conditions of Contract;
(*e*) the sequence and timing of inspections and tests specified in, or required by, the Contract;
(*f*) for a revised programme: the sequence and timing of the remedial work (if any) to which the Engineer has given a Notice of No-objection under Sub-Clause 7.5 [*Defects and Rejection*] and/or the remedial work (if any) instructed under Sub-Clause 7.6 [*Remedial Work*];
(*g*) all activities (to the level of detail stated in the Specification), logically linked and showing the earliest and latest start and finish dates for each activity, the float (if any) and the critical path(s);
(*h*) the dates of all locally recognised days of rest and holiday periods (if any);
(*i*) all key delivery dates of Plant and Materials;
(*j*) for a revised programme and for each activity: the actual progress to date, any delay to such progress and the effects of such delay on other activities (if any); and
(*k*) a supporting report which includes:

(i) a description of all the major stages of the execution of the Works;
(ii) a general description of the methods which the Contractor intends to adopt in the execution of the Works;
(iii) details showing the Contractor's reasonable estimate of the number of each class of Contractor's Personnel, and of each type of Contractor's Equipment, required on the Site, for each major stage of the execution of the Works;
(iv) if a revised programme: identification of any significant change(s) to the previous programme submitted by the Contractor; and
(v) the Contractor's proposals to overcome the effects of any delay(s) on progress of the Works.

The initial programme and any updates must be submitted to the Engineer in one paper copy, one electronic copy and additional paper copies (if any) as stated in the Contract Data. To be of use to the Engineer, the electronic copy should be in native format rather than a PDF file.

Within 21 days of receiving the initial programme; or within 14 days of receiving a revised programme, the Engineer must Review the submitted programme and may give a Notice to the Contractor stating the extent to which it does not comply with the Contract or ceases to reflect actual progress or is otherwise inconsistent with the Contractor's obligations. The Engineer is not permitted to comment on the logic or the outputs upon which the programme is based unless the Contract imposes constraints which have not been taken into account by the Contractor. The Engineer does not approve the programme or give consent to it.

If the Engineer gives no such Notice of a failure to comply with the Contract within the said period, the Engineer is deemed to have given a Notice of No-objection and the initial programme or revised programme (as the case may be) becomes the Programme.

The Contractor must proceed in accordance with the Programme, subject to the Contractor's other obligations under the Contract. The Employer's Personnel are entitled to rely on the Programme when planning their activities.

If, at any time, the Engineer gives a Notice to the Contractor that the Programme fails (to the extent stated) to comply with the Contract or ceases to reflect actual progress or is otherwise inconsistent with the Contractor's obligations, the Contractor must submit a revised programme to the Engineer within 14 days of receipt of the Notice.

Nothing in any programme, the Programme or any supporting report is to be taken as a Notice under the Contract. Nor does anything in any programme, the Programme or any supporting report relieve the Contractor of any obligation to give Notice.

10.2. Time for Completion

Sub-Clause 1.1.84 states that the 'Time for Completion' is the time stated in the Contract Data for completing the Works or a Section (as the case may be) under Sub-Clause 8.2 [*Time for Completion*]. This Time for Completion is calculated from the Commencement Date and may be extended under Sub-Clause 8.5 [*Extension of Time for Completion*].

Under Sub-Clause 8.2 [*Time for Completion*], the Contractor must complete the whole of the Works, and each Section (if any), within the Time for Completion for the Works or Section (as the case may be).

This obligation to complete within the Time for Completion includes completion of all work which is stated in the Contract as being required for the Works or Section to be considered to be completed for the purposes of taking over under Sub-Clause 10.1 [*Taking Over the Works and Sections*], such as successful Tests on Completion, submission of As-Built Records, operation and maintenance manuals and so forth.

10.3. Advance Warnings

Under Sub-Clause 8.4 [*Advance Warning*], each Party is to advise the other Party and the Engineer, and the Engineer is to advise the Parties, in advance of any known or probable future events or circumstances which may:

- adversely affect the work of the Contractor's Personnel;
- adversely affect the performance of the Works when completed;
- increase the Contract Price; and/or
- delay the execution of the Works or a Section (if any).

It is to be noted that there is no requirement for a Notice and there is no time limit for providing such advice. No consequences of a failure to advise are specified. However, once such an event or circumstance has been identified, the Engineer may request that the Contractor submit a proposal under Sub-Clause 13.3.2 [*Variation by Request for Proposal*] to avoid or minimise the effects of the event or circumstance.

It is important to differentiate between a Notice of Claim under Sub-Clause 20.2.1 [*Notice of Claim*] and advice under Sub-Clause 8.4 [*Advance Warning*]. Whereas there is no time limit for the advice and no sanction for failing to advise under Sub-Clause 8.4 [*Advance Warning*], a Notice of Claim under

Sub-Clause 20.2.1 [*Notice of Claim*] must be submitted within 28 days of the date on which the claiming Party became aware, or should have become aware, of the event or circumstance giving rise to the cost, loss, delay or other circumstance for which it is intended to claim. If the Notice of Claim is not submitted within this period of 28 days, the claiming Party loses its rights to claim (except in exceptional circumstances). The advice relates to future events or circumstances whether known or probable. The Notice of Claim relates to events or circumstances that have already happened.

10.4. Monthly reports

Closely linked to the requirement for submission of revised programmes is the requirement for the Contractor to submit monthly progress reports to the Engineer under Sub-Clause 4.20 [*Progress Reports*].

One paper original, one electronic copy and additional paper copies (if any) as stated in the Contract Data are to be submitted in the format stated in the Specification (if not stated, in a format acceptable to the Engineer).

Reports must be submitted within 7 days of the last day of the month to which each report relates. The first report must cover the first month following the Commencement Date. Reporting must continue until the Date of Completion of the Works or the date of completion of any outstanding work which was listed in the Taking-Over Certificate.

Under sub-paragraph (c) of Sub-Clause 14.3 [*Application for Interim Payment*], the progress report is one of the supporting documents that must be provided with the Contractor's Statement applying for interim payment. The period of 28 days during which the Engineer must issue the corresponding IPC under Sub-Clause 14.6 [*Issue of IPC*] begins when the Engineer has received the Statement and supporting documents, including the progress report. Any delay in submitting the progress report will thus delay the issue of the IPC. The author has also seen a case where the Engineer refused to issue an IPC because the progress report did not contain the progress details mentioned under Sub-Clause 4.20 [*Progress Reports*].

Unless otherwise stated in the Specification, each progress report must include:

- charts, diagrams and detailed descriptions of progress, including each stage of (design by the Contractor, if any) Contractor's Documents, procurement, manufacture, delivery to Site, construction, erection and testing;
- photographs and/or video recordings showing the status of manufacture and of progress;
- details of the manufacture of each main item of Plant and Materials;
- the details described in Sub-Clause 6.10 [*Contractor's Records*];
- copies of quality management documents, inspection reports, test results, and compliance verification documentation (including certificates of Materials);
- a list of Variations, and any Notices given (by either Party) under Sub-Clause 20.2.1 [*Notice of Claim*];
- health and safety statistics, including details of any hazardous incidents and activities relating to environmental aspects and public relations; and

- comparisons of actual and planned progress, with details of any events or circumstances which may adversely affect the completion of the Works in accordance with the Programme and the Time for Completion, and the measures being (or to be) adopted to overcome delays.

The last item in the above list is particularly significant with respect to the programming of the Works. In effect, at least once per month, the Contractor is required to identify the critical path (and near-critical paths) and potential delays to completion and to take measures to minimise such delays.

It should be noted that nothing stated in any progress report constitutes a Notice.

It should also be noted that the requirements of Sub-Clause 6.10 [*Contractor's Records*] are more onerous than similar requirements stated in earlier FIDIC contracts. For every activity shown in the Programme, at each work location and for each day of work, the Contractor must include in each progress report, records of:

- the occupations and actual working hours of each class of Contractor's Personnel;
- the type and actual working hours of each item of the Contractor's Equipment;
- the types of Temporary Works used;
- the types of Plant installed in the Permanent Works; and
- the quantities and types of Materials used.

10.5. Delay Damages

Under Sub-Clause 8.8 [*Delay Damages*], if the Contractor fails to complete the Works or Section within the relevant Time for Completion, the Employer is entitled to payment of Delay Damages by the Contractor, subject to Sub-Clause 20.2 [*Claims for Payment and/or EOT*]. The Delay Damages are calculated at the rate stated in the Contract Data, for every day which elapses between the expiry of the relevant Time for Completion and the actual Date of Completion of the Works or Section. However, the Contract Data may fix the maximum amount of Delay Damages.

These Delay Damages are the only damages due from the Contractor for the Contractor's failure to comply with Sub-Clause 8.2 [*Time for Completion*], other than in the event of termination under Sub-Clause 15.2 [*Termination for Contractor's Default*] before completion of the Works, or in the event of fraud, gross negligence, deliberate default or reckless misconduct by the Contractor. An example of such deliberate default could be where the Contractor deliberately delays completion of the Works to put economic or political pressure on the Employer to settle its Claims.

The payment of Delay Damages does not release the Contractor from the obligation to complete the Works, or from any other obligations.

The legal position with respect to Delay Damages can vary from one country or jurisdiction to another. In some countries, the amount must correspond to a reasonable estimate of the loss which is anticipated to be suffered by the Employer due to late completion. In some countries, some evidence of actual loss is required to justify the application of the Delay Damages. In some countries, the law imposes additional formalities with respect to a claim for Delay Damages. It is therefore essential for both Parties to seek legal advice with respect to the Delay Damages provisions before entering into the Contract.

10.6. Intermediate Milestones

It is perhaps because of the variety of legal viewpoints that standard FIDIC Contracts do not foresee intermediate milestones which allow the Employer to claim Delay Damages if the Contractor is late in achieving the milestone. In some countries, it would be necessary for the Employer to be able to demonstrate that the amount of Delay Damages corresponded to a reasonable estimate of the loss which was anticipated to be suffered by the Employer, due to the Contractor's failure to meet the milestone. In many cases, a failure to meet a milestone date does not directly cause the Employer to suffer loss.

The author has seen contracts which have attempted to use Sections as a way of overcoming this obstacle to the use of intermediate milestones. Thus, for a design and build project, completion of design was defined as a Section, with its own Time for Completion and its own Delay Damages. The main difficulty with this approach is that when the design is completed, the Engineer must issue a TOC for the design and the DNP for the design begins. This is probably not what the Employer was seeking. From the Contractor's point of view, the drawback of this approach was that it was difficult to determine when the Section was complete, given that design might be taken to include preparation and submission of as-built documents, which inevitably occurs towards completion of the Works.

To avoid problems such as these, the Notes on the Preparation of Special Provisions offer guidance should the Employer seek to have certain parts of the Works completed within specific periods without being obliged to take over those parts after completion. FIDIC states that such parts should be clearly described in the Specification/Employer's Requirements as 'Milestones'.

The guidance includes two new definitions: 'Milestone' and 'Milestone Certificate'.

Sub-Clause 1.1.?	***'Milestone'*** *means a part of the Plant and/or a part of the Works stated in the Contract Data (if any), and described in detail in the Specification as a Milestone, which is to be completed by the time for completion stated in Sub-Clause 4.24 [Milestone Works] but is not to be taken over by the Employer after completion.*
Sub-Clause 1.1.?	***'Milestone Certificate'*** *means the certificate issued by the Engineer under Sub-Clause 4.24 [Milestones].*

The Contract Data is to include a definition of each Milestone, its Time for Completion expressed as a number of days starting from the Commencement Date and the amount of Delay Damages (if any) expressed as a percentage of the final Contract Price per day of delay together with the maximum amount of Delay Damages for Milestones.

Wording is suggested for the new Sub-Clause 4.24 [*Milestones*]:

> *'If no Milestones are specified in the Contract Data, this Sub-Clause shall not apply.*
>
> *The Contractor shall complete the works of each Milestone (including all work which is stated in the Specification as being required for the Milestone to be considered complete) within the time for completion of the Milestone, as stated in the Contract Data, calculated from the Commencement Date.*

The Contractor shall include, in the initial programme and each revised programme, under sub-paragraph (a) of Sub-Clause 8.3 [Programme], the time for completion for each Milestone.

Sub-paragraph (d) of Sub-Clause 8.4 [Advance Warning] and Sub-Clause 8.5 [Extension of Time for Completion] shall apply to each Milestone, such that 'Time for Completion' under Sub-Clause 8.5 shall be read as the time for completion of a Milestone under this Sub-Clause.

The Contractor may apply, by Notice to the Engineer, for a Milestone Certificate not earlier than 14 days before the works of a Milestone which will, in the Contractor's opinion, be complete. The Engineer shall, within 28 days after receiving the Contractor's Notice:

(*a*) *issue the Milestone Certificate to the Contractor, stating the date on which the works of the Milestone were completed in accordance with the Contract, except for any minor outstanding work and defects (as shall be listed in the Milestone Certificate); or*
(*b*) *reject the application, giving reasons and specifying the work required to be done and defects required to be remedied by the Contractor to enable the Milestone Certificate to be issued.*

The Contractor shall then complete the work referred to in sub-paragraph (b) of this Sub-Clause before issuing a further Notice of application under this Sub-Clause.

If the Engineer fails either to issue the Milestone Certificate or to reject the Contractor's application within the above period of 28 days, and if the works of a Milestone are complete in accordance with the Contract, the Milestone Certificate shall be deemed to have been issued on the date which is 14 days after the date stated in the Contractor's Notice of application.

If Delay Damages for a Milestone are stated in the Contract Data, and if the Contractor fails to complete the works of the Milestone within the time for completion of the Milestone (with any extension under this Sub-Clause):

(i) *the Contractor shall, subject to Sub-Clause 20.1 [Claims], pay Delay Damages to the Employer for this default;*
(ii) *such Delay Damages shall be the amount stated in the Contract Data, for every day which shall elapse between the time for completion of the Milestone (with any extension under this Sub-Clause) and the date stated in the Milestone Certificate;*
(iii) *these Delay Damages shall be the only damages due from the Contractor for such default; and*
(iv) *the total amount of Delay Damages for all Milestones shall not exceed the maximum amount stated in the Contract Data (this shall not limit the Contractor's liability for Delay Damages in any case of fraud, gross negligence, deliberate default or reckless misconduct by the Contractor).'*

Finally, FIDIC recommends that if payment to the Contractor is to be tied to completion of each Milestone, such payments should be specified in a Schedule of Payments forming part of the Contract and consideration should be given to amending Sub-Clause 14.4 [*Schedule of Payments*] to expressly refer to the Milestone payments.

The suggested wording of Sub-Clause 4.24 [*Milestones*] is very similar to the wording of Sub-Clause 10.1 [*Taking Over the Works and Sections*] and Sub-Clause 8.8 [*Delay Damages*].

The most significant difference between the wording of Sub-Clause 10.1 [*Taking Over the Works and Sections*] and the proposed wording for Sub-Clause 4.24 [*Milestones*] is that the former mentions minor outstanding work and defects '*which will not substantially affect the safe use of the Works or Section for their intended purpose*', whereas the latter does not. The decision with respect to whether minor outstanding work should block the issue of the Milestone Certificate is therefore more subjective than the decision with respect to outstanding work at the time of taking over.

In reality, the most difficult aspects of using Milestones, is to define precisely each Milestone and what needs to be done for the Milestone to be considered complete for the purposes of the Milestone Certificate and for payment.

10.7. Extensions of Time

The most common defence against a claim for Delay Damages is that the Contractor is entitled to an Extension of Time (EOT) because the delay to completion was caused by an event or circumstance for which the Contractor is not liable under the Contract, but by an Employer's delay or a neutral event.

Under Sub-Clause 8.5 [*Extension of Time for Completion*] of RB 2017 and YB 2017, the Contractor is entitled, subject to Sub-Clause 20.2 [*Claims for Payment and/or EOT*], to an EOT if and to the extent that completion for the purposes of Sub-Clause 10.1 [*Taking Over the Works and Sections*] is or will be delayed by any of the following causes:

(*a*) a Variation (except that there shall be no requirement to comply with Sub-Clause 20.2 [*Claims for Payment and/or EOT*]);
(*b*) a cause of delay giving an entitlement to EOT under another sub-clause of the Conditions of Contract;
(*c*) exceptionally adverse climatic conditions;
(*d*) unforeseeable shortages in the availability of Personnel or Goods (or Employer-Supplied Materials, if any) caused by epidemic or governmental actions; or
(*e*) any delay, impediment or prevention caused by or attributable to the Employer, the Employer's Personnel, or the Employer's other contractors on the Site.

In this context, '*exceptionally adverse climatic conditions*' means adverse climatic conditions at the Site which are unforeseeable having regard to climatic data made available by the Employer under Sub-Clause 2.5 [*Site Data and Items of Reference*] and/or climatic data published in the Country for the geographical location of the Site.

Items (c) and (d) in the above list are not found in the corresponding list in SB 2017.

Under RB 2017, the Contractor is also entitled to EOT, subject to Sub-Clause 20.2 [*Claims for Payment and/or EOT*], if the quantity of any item of work measured for payment in accordance with Clause 12 [*Measurement and Valuation*] is greater than the quantity of the item in the BOQ or other Schedule by more than 10% and such an increase in quantity causes a delay to completion. The reference to Sub-Clause 20.2 [*Claims for Payment and/or EOT*] means that the Contractor must carefully monitor the quantities executed, in order to submit a Notice of Claim as soon as the measured quantity exceeds the quantity in the BOQ by 10%.

When addressing such a claim for the impact of the increased quantity, the Engineer may also review the impact on the critical path of the measured quantities of other items of work which are less than the corresponding quantities in the Bill of Quantities or other Schedule by more than 10%. However, the net effect of such a review cannot result in a net reduction in the Time for Completion.

When determining each EOT under Sub-Clause 20.2 [*Claims for Payment and/or EOT*], the Engineer must review previous determinations and may increase, but not decrease, the total EOT. In other words, once an EOT has been given, it cannot be taken away.

The Contractor has no entitlement to an EOT except for the reasons listed under Sub-Clause 8.5 [*Extension of Time for Completion*]. It is possible, however, for the Parties to negotiate an EOT as part of an amicable settlement. Whatever the basis for an EOT, the effect of the EOT is to release the Contractor from any liability for Delay Damages during the period of the EOT. There is a mistaken understanding in some countries that after the original Time for Completion, unless there is an EOT, the Contractor is no longer bound to complete the Works, nor has the Employer the possibility to pay the Contractor. Therefore, to '*keep the Contract alive*', the Contractor is given an EOT, but the Employer still seeks to claim Delay Damages for this period. Such an arrangement is unnecessary. The Contractor does not contract to work for a fixed duration (i.e. the Time for Completion) but contracts to complete the Works and remedy any defects, almost without regard to the time needed to achieve this.

Finally, it should be noted that Sub-Clause 8.5 [*Extension of Time for Completion*] does not mention payment. An award of an EOT does not automatically entitle the Contractor to claim its additional time-related Costs (sometimes called 'prolongation Costs') or any other Costs. If the Contractor seeks additional payment based on any of the circumstances listed under this sub-clause, it must identify another sub-clause (or legal grounds) which gives an entitlement to additional payment.

For example, in relation to an EOT due to Unforeseeable shortages in the availability of personnel or Goods caused by governmental actions, it might be possible to claim the additional Costs under Sub-Clause 13.6 [*Adjustments for Changes in Laws*]. However, if the shortage was caused by an epidemic, there is no sub-clause which permits the Contractor to claim additional payment. The additional time-related Costs arising from an EOT under Sub-Clause 8.5 (a) due to a Variation, must be included in the evaluation of the Variation. For an EOT arising under Sub-Clause 8.5 (b) from an entitlement under another sub-clause of the Conditions of Contract, the said sub-clause may give an entitlement to reimbursement of additional Costs, including those which are time-related.

10.8. Concurrent delay

FIDC 2017 Contracts contain a new feature under Sub-Clause 8.5 [*Extension of Time for Completion*] not found in earlier contracts. It is the treatment of concurrent delay.

It states:

> *'If a delay caused by a matter which is the Employer's responsibility is concurrent with a delay caused by a matter which is the Contractor's responsibility, the Contractor's entitlement to EOT shall be assessed in accordance with the rules and procedures stated in the Special Provisions (if not stated, as appropriate taking due regard of all relevant circumstances).'*

The Notes on the Preparation of Special Provisions state that FIDIC drafted this provision in this manner because there was no standard set of rules and/or procedures for dealing with such circumstances. FIDIC recognises that different rules or procedures may apply in different legal jurisdictions. It also acknowledges, however, that the approach given in the 'Delay and Disruption Protocol' published by the Society of Construction Law (UK)[5] is increasingly being adopted internationally. Therefore, both in the Notes on the Preparation of Special Provisions and in the FIDIC 2017 Contracts Guide [2022], FIDIC strongly recommends that when preparing the Special Provisions, the Employer seek advice from a professional with extensive experience in construction programming, analysis of delays and assessment of EOT in the context of the governing Laws.

Given that Sub-Clause 8.5 [*Extension of Time for Completion*] does not impose a method of treating concurrent delay but suggests that in the absence of specific mention in the Special Provisions, the 'appropriate' method must be used, it might have been better not to include the topic in the Conditions. It is likely that many Employers will see the provision as an invitation to impose Special Provisions which restrict the Contractor's entitlement to EOT in the presence of concurrent delay. If bidders see such restrictions, they will increase their bid prices to cover the risk. In any event, the invitation to impose specific rules and procedures with respect to concurrent delay is likely to stimulate disputes.

10.9. Delays Caused by Authorities

One sub-clause of the Conditions which may entitle the Contractor to claim an EOT but not additional payment, is Sub-Clause 8.6 [*Delays Caused by Authorities*].

In this context, 'authorities' means relevant legally constituted public authorities or private utility entities in the Country.

If the Contractor has diligently followed the procedures laid down by a public authority or private utility entity but that authority or entity delays or disrupts the Contractor's work to an Unforeseeable extent, the delay or disruption will be considered as a cause of delay under sub-paragraph (b) of Sub-Clause 8.5 [*Extension of Time for Completion*]. As such, the Contractor's right to claim will be subject to Sub-Clause 20.2 [*Claims for Payment and/or EOT*].

Thus to make a claim under this sub-clause, the Contractor must demonstrate that it diligently followed the relevant procedures, that the authority or private entity nevertheless caused delay or disruption and that such delay or disruption was Unforeseeable, that is, the work of the Contractor was delayed or disrupted to an extent not reasonably foreseeable by an experienced contractor prior to the Base Date. In other words, if the electricity authority has a reputation for being slow to divert existing cables, the Contractor would be unable to claim for the delay encountered.

---

[5] https://www.scl.org.uk/sites/default/files/SCL_Delay_Protocol_2nd_Edition_Final.pdf

With respect to additional payment, the FIDIC Contracts Guide [2000] suggested, in relation to a similar sub-clause in the FIDIC 1999 Contracts, that under some circumstances, the Contractor might be reimbursed its additional Costs: '*This Sub-Clause ... makes no mention of the financial consequences, because they would depend upon the particular circumstances.*' However, there was no contractual basis for such reimbursement. Neither the FIDIC 2017 Contracts Guide [2022] nor the Notes on the Preparation of Special Provisions for the FIDIC 2017 Contracts make such a suggestion.

10.10. Suspensions instructed by the Engineer/Employer

At any time, the Engineer may, under Sub-Clause 8.9 [*Employer's Suspension*] of RB 2017 or YB 2017 (the Employer under SB 2017), instruct the Contractor to suspend activity on part or all of the Works. The instruction must state the date and the reason for the suspension.

Although it is not mentioned, it would be preferable for the instruction to also provide information concerning the likely duration of the suspension because, during the suspension, the Contractor must protect and secure the relevant part of or all the Works against any deterioration, loss or damage. The choice of appropriate protection measures will depend on the likely duration.

If the cause of the suspension is the responsibility of the Contractor, Sub-Clauses 8.10 [*Consequences of Employer's Suspension*], 8.11 [*Payment for Plant and Materials after Employer's Suspension*] and 8.12 [*Prolonged Suspension*] will not apply. This means that the Contractor will not be entitled to an EOT or additional payment due to the suspension.

Moreover, the Contractor is not entitled to an EOT or to payment of the Cost incurred in making good any deterioration, loss or damage caused by the Contractor's failure to protect the suspended part of or all the Works.

However, if the cause of the suspension is not the responsibility of the Contractor, and the Contractor suffers delay and/or incurs Cost from complying with an Engineer's instruction to suspend work under Sub-Clause 8.9 [*Employer's Suspension*] and/or from resuming work under Sub-Clause 8.13 [*Resumption of Work*], the Contractor is entitled under Sub-Clause 8.10 [*Consequences of Employer's Suspension*] to EOT and/or payment of such Cost Plus Profit, subject to compliance with Sub-Clause 20.2 [*Claims for Payment and/or EOT*]. The cost of protection measures during the suspension would form part of the Costs for which the Contractor is entitled to be paid.

The Contractor will also be entitled to payment for Plant and/or Materials which have not been delivered to Site, if:

(*a*) the work on the Plant, or delivery of Plant and/or Materials, has been suspended for more than 28 days and:

- (i) the Plant and/or Materials were scheduled, in accordance with the Programme, to have been completed and ready for delivery to the Site during the suspension period; and
- (ii) the Contractor provides the Engineer with reasonable evidence that the Plant and/or Materials comply with the Contract; and

(*b*) the Contractor has marked the Plant and/or Materials as the Employer's property in accordance with the Engineer's instructions.

There is no requirement for the Contractor to provide security such as a bank guarantee to cover the payment, notwithstanding that the Plant and/or Materials may not yet be in the Country.

The payment is to cover the 'value' of the Plant and/or Materials as at the date of suspension stated in the suspension instruction. FIDIC gives no indication either in the Notes on the Preparation of Special Provisions or the FIDIC 2017 Contracts Guide [2022] of how the 'value' is to be assessed. However, once payment has been made, ownership of the Plant and/or Materials is transferred to the Employer under sub-paragraph (b) of Sub-Clause 7.7 [*Ownership of Plant and Materials*]. Therefore, 'value' should logically include the cost paid by the Contractor to the manufacturer or supplier, the costs of transport, insurance and storage, plus a mark-up to cover the Contractor's overheads and profit.

Despite the transfer of ownership, the Contractor remains responsible for the protection and care of the Plant and/or Materials under Sub-Clause 8.9 [*Employer's Suspension*] and Sub-Clause 17.1 [*Responsibility for Care of the Works*]. The Contractor must also maintain insurance cover for the Plant and/or Materials under Sub-Clause 19.2 [*Insurance to be provided by the Contractor*].

If the suspension under Sub-Clause 8.9 [*Employer's Suspension*] has continued for more than 84 days, the Contractor may submit a Notice to the Engineer under Sub-Clause 8.12 [*Prolonged Suspension*] (the Employer under SB 2017) requesting permission to proceed. If the Engineer fails to respond by a Notice under Sub-Clause 8.13 [*Resumption of Work*] within 28 days of receiving the Contractor's Notice, the Contractor may agree to a further suspension.

In this case, the Parties may agree the EOT and/or Cost Plus Profit (if the Contractor incurs Cost) and/or payment for suspended Plant and/or Materials arising from the total period of suspension. As the duration of the total period of suspension will be uncertain at this time, the agreement must include a method for finally ascertaining the EOT, the Cost Plus Profit and the payment for suspended Plant and/or Materials.

It is noticeable that the FIDIC 2017 Contracts Guide [2022] links the Contractor's acceptance of a continued suspension to an agreement on the EOT and financial aspects, which from the Contractor's viewpoint would be desirable. However, the wording of sub-paragraph (a) of Sub-Clause 8.12 [*Prolonged Suspension*] places the Contractor's agreement to the prolonged suspension before the possible discussion of the EOT and financial consequences.

As an alternative (and if the Parties fail to reach such agreement on the consequences of the total period of suspension), the Contractor may, after giving a (second) Notice to the Engineer, treat the affected part of the Works as an omission from the Contract (as if instructed under Sub-Clause 13.3.1 [*Variation by Instruction*]). The Notice will have immediate effect and will release the Contractor from any further obligation to protect, store and secure the Works under Sub-Clause 8.9 [*Employer's Suspension*]. If the suspension affects the whole of the Works, the Contractor may give a Notice of termination to the Employer under sub-paragraph 16.2.1 (h) of Sub-Clause 16.2 [*Termination by Contractor*].

Under Sub-Clause 8.13 [*Resumption of Work*], the Contractor must resume work as soon as practicable after receiving a Notice from the Engineer to proceed with the suspended work, unless the Contractor has already served Notice of termination under Sub-Clause 8.12 [*Prolonged Suspension*] and Sub-Clause 16.2 [*Termination by Contractor*].

The meaning of '*as soon as practicable*' will depend on the circumstances. A full remobilisation after a long suspension may take months if, for example, a manufacturer of Plant has begun work for another client and cannot resume work on the suspended order until the replacement work has been completed.

At the time stated in the Notice to resume work (or immediately after the Contractor receives this Notice, if no time is stated), the Contractor and the Engineer must carry out a joint inspection of the Works and the Plant and/or Materials affected by the suspension. The Engineer is to record any deterioration, loss, damage or defect in the Works and/or Plant and/or Materials which has occurred during the suspension and provide this record to the Contractor. The Contractor must promptly make good all such deterioration, loss, damage or defect so that the Works, when completed, comply with the Contract.

10.11. Rate of Progress

If it becomes apparent that the Contractor has fallen or will fall behind the Programme (or the initial programme), or that actual progress is too slow to complete the Works or a Section (if any) within the relevant Time for Completion, other than as a result of a cause which entitles the Contractor to an EOT, the Engineer may instruct the Contractor under Sub-Clause 8.7 [*Rate of Progress*] to submit a revised programme describing the steps which the Contractor proposes to take in order to catch up and to complete the Works or a Section (if any) within the relevant Time for Completion. Such steps may include working longer hours, increasing the numbers of Contractor's Personnel and/or Equipment, using air freight instead of shipping by sea and so forth and are to be at the Contractor's risk and cost.

Unless the Engineer gives a Notice to the Contractor stating otherwise, the Contractor must adopt the proposed measures. If, as a result, the Employer incurs additional costs, such as for an increase in the number of Engineer's assistants, the Employer will be entitled, subject to Sub-Clause 20.2 [*Claims for Payment and/or EOT*], to reimbursement of these costs by the Contractor. This reimbursement will be in addition to any Delay Damages, to which the Employer might be entitled, if the Time for Completion is not met.

It should be noted that if the cause of the delay was a matter which entitled the Contractor to an EOT, and the instruction from the Engineer to take steps to catch up was issued without dealing with the EOT, it will be in effect an instruction to accelerate (i.e. to go faster than the Contract requires). The Contractor is compelled by Sub-Clause 3.5 [*Engineer's Instructions*] to comply with the instruction but, before doing so, should immediately give Notice to the Engineer, with reasons, that the instruction represents a Variation. If the Engineer does not respond within 7 days of receiving this Notice, by giving a Notice confirming, reversing or varying the instruction, the Engineer is deemed to have revoked the instruction. If the Engineer confirms the instruction by Notice, the Contractor must comply.

10.12. Acceleration

Under Sub-Clause 8.7 [*Rate of Progress*], the Engineer has the power to instruct acceleration measures to reduce delays resulting from causes listed under Sub-Clause 8.5 [*Extension of Time for Completion*], that is, to reduce delays for which the Contractor is entitled to an EOT.

Such instruction to take acceleration measures is to be treated as a Variation under Sub-Clause 13.3.1 [*Variation by Instruction*] and entitles the Contractor to an adjustment of the Contract Price.

Somewhat confusingly, the FIDIC 2017 Contracts Guide [2022] states that the Engineer/Employer has no entitlement or authority under the Contract to instruct the Contractor to complete the Works or Section before the expiry of the Time for Completion. It goes on to state that if the Employer wishes the Contractor to be so instructed, it must be agreed with the Contractor as a supplemental agreement (i.e. an amendment to the Contract).

Yet, the provision clearly states that the instruction is to take measures to reduce delays arising from causes which entitle the Contractor to an EOT. The instruction cannot reduce or eliminate the Contractor's entitlement to an EOT due to those causes of the delay. Thus, the Engineer is required to agree or determine the EOT, notwithstanding the instruction to take acceleration measures to reduce the delay, and the Contractor will not be liable for Delay Damages if the acceleration measures do not have the desired effect, provided that the Works are completed before the expiry of the extended Time for Completion.

If the Employer wishes the Contractor to complete the Works or Section earlier than the expiry date of the new Time for Completion and be liable for Delay Damages if it fails to do so, this must be negotiated with the Contractor and confirmed by an agreement.

Smith G
ISBN 978-0-7277-6652-6
https://doi.org/10.1680/fcmh.66526.135

Chapter 11
Completion of the works

11.1. Tests on Completion

'*Tests on Completion*' are defined under Sub-Clause 1.1.83 of RB 2017 (Sub-Clause 1.1.85 of YB 2017 and Sub-Clause 1.1.75 of SB 2017):

Sub-Clause 1.1.83	***'Tests on Completion'*** *means the tests which are specified in the Contract or agreed by both Parties or instructed as a Variation, and which are carried out under Clause 9 [Tests on Completion] before the Works or a Section (as the case may be) are taken over under Clause 10 [Employer's Taking Over].*

Not all projects require Tests on Completion. However, they are very common on projects which involve mechanical and/or electrical plant such as those executed under YB 2017 and SB 2017. For this reason, the provisions with respect to Tests on Completion are more extensive in YB 2017 and SB 2017, than in RB 2017. In all three contracts, Sub-Clause 9.1 [*Contractor's Obligations*] sets out the administrative procedure.

The first step is the submission of a detailed test programme by the Contractor, not less than 42 days before the intended date for starting the Tests on Completion. This programme must include details of the timing and required resources. Following receipt of the proposed test programme, the Engineer (Employer under SB 2017) may (but is not obliged to) Review it and give Notice to the Contractor if the test programme does not comply with the Contract. The Notice must state the extent of the non-compliance. If the Engineer (Employer under SB 2017) gives no such Notice within 14 days after receiving the test programme (or revised test programme), the Engineer (Employer under SB 2017) is deemed to have given a Notice of No-objection. If the Engineer/Employer issues a Notice of non-compliance within the said period of 14 days, the Contractor has a further period of 14 days in which to rectify the non-compliance and to submit the revised programme for Review. The Contractor must not commence the Tests on Completion until a Notice of No-objection has been given (or is deemed to have been given) by the Engineer/Employer. Therefore, it is important for the Contractor to avoid multiple resubmissions.

The second step by the Contractor is the submission of a Notice to the Engineer (Employer under SB 2017) not less than 21 days before the date after which the Contractor will be ready to carry out each of the Tests on Completion. Prior to the said date, the Contractor must also submit any as-built documents required under Sub-Clause 4.4.2 [*As-Built Records*] and operation and maintenance manuals (O&M Manuals), required under Sub-Clause 4.4.3 [*Operation and Maintenance Manuals*] (Sub-Clause 5.6 and 5.7 in YB 2017 and SB 2017). Under YB 2017 and SB 2017, only provisional copies of the O&M Manuals are required prior to the start of the Tests on Completion and the final versions may be given after the Tests on Completion but before the issue of the TOC.

It should be noted that the Contractor must submit a Notice with respect to the date of readiness for each of the Tests on Completion. This may be simple for a road project for which the only Test on Completion is a load test on a bridge. However, giving Notice more than 21 days before the date of each test on more complex electrical and mechanical Works such as a power station can be much more difficult to manage. In such cases, it might be better for the parties to agree that Notice will only be required from the Contractor for the start of the first test shown on the test programme or the start of each phase of testing.

In this respect, YB 2017 and SB 2017 define three main phases:

(1) pre-commissioning tests (on or off the Site, as appropriate), including appropriate inspections and ('dry' or 'cold') functional tests to demonstrate that each item of the Works or Section can safely undertake the next phase of testing;
(2) commissioning tests, including the operational tests specified in the Employer's Requirements to demonstrate that the Works or Section can be operated safely and as specified in the Employer's Requirements, under all available operating conditions; and
(3) trial operation (to the extent possible under available operating conditions) to demonstrate that the Works or Section perform reliably and in accordance with the Contract.

During trial operation, when the Works or Section (as the case may be) have reached a stable condition, the Contractor must give a Notice that the Works are ready for any other Tests on Completion, including performance tests.

Performance tests are to demonstrate that the Works (or Section) comply with the performance criteria specified in the Employer's Requirements and with the Schedule of Performance Guarantees. Such trial operation, including performance testing, does not constitute taking over under Clause 10 [*Employer's Taking Over*]. Any product produced by the trial operation (and any revenue or other benefit resulting from it) are to be the property of the Employer.

The Contractor must commence the Tests on Completion within 14 days of the date stated in the Contractor's Notice, or on such day or days as the Engineer (Employer under SB 2017) instructs, and must proceed in accordance with the Contractor's test programme for which a Notice of No-objection has been given (or is deemed to have been given).

Nevertheless, the testing is to be carried out in accordance with Sub-Clause 7.4 [*Testing by the Contractor*]. Thus, the Contractor is required to submit a further Notice to the Engineer stating the time and place for the specified testing. This Notice must be given in reasonable time, having regard to the location of the testing, for the Employer's Personnel to attend. The Engineer must then give a Notice to the Contractor of not less than 72 hours of their intention to attend the tests. If the Engineer does not attend at the time and place stated in the Contractor's Notice under Sub-Clause 7.4, the Contractor may proceed with the tests, unless instructed otherwise.

Under YB 2017 and SB 2017, each of the three test phases mentioned above is to commence only after the Works or Section have passed the previous phase. As soon as the Works or Section have, in the Contractor's opinion, passed each phase, the Contractor must submit a certified report of the results of

Figure 11.1 Procedure for Tests on Completion

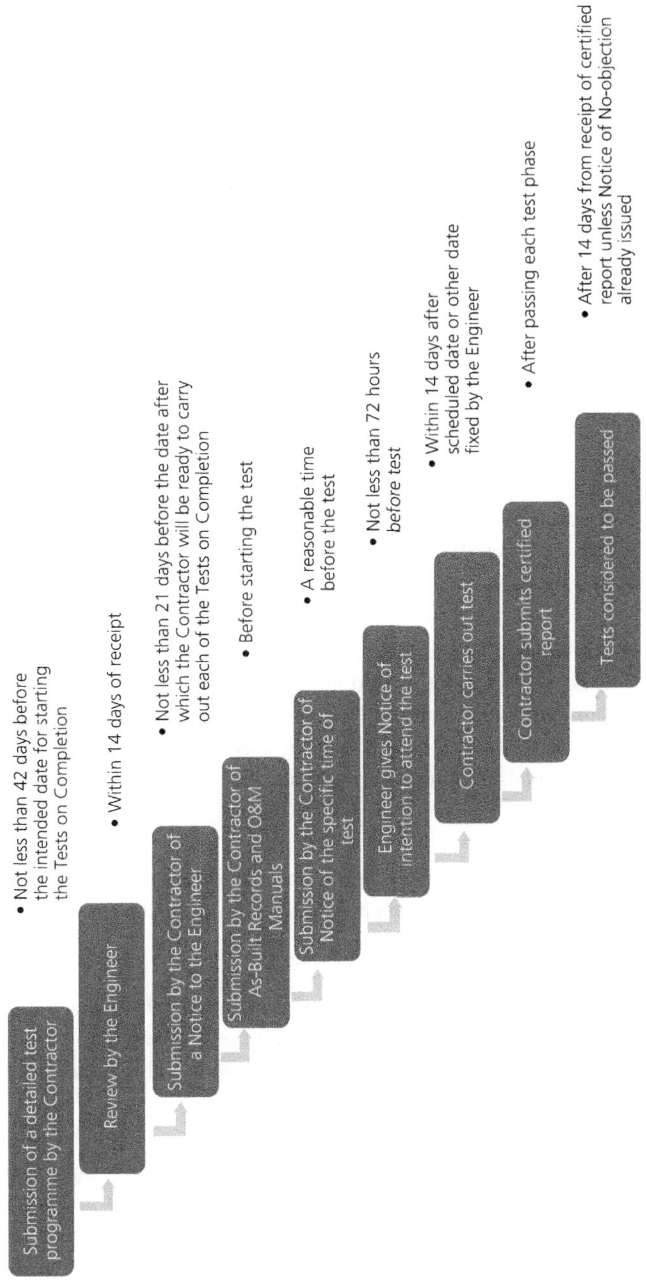

these tests to the Engineer (Employer under SB 2017). The Engineer/Employer must Review each such report and may give a Notice to the Contractor stating the extent to which the results of the tests do not comply with the Contract. If no such Notice is issued within 14 days of receiving the results of the tests, the Engineer (Employer under SB 2017) is deemed to have given a Notice of No-objection.

According to the FIDIC 2017 Contracts Guide [2022], it is this Notice of No-objection which confirms the '*passing of the Tests on Completion*'. If this is so, at least one period of up to 14 days is added to the critical path between the end of the Tests on Completion and the TOC. If the same interpretation applies to the passing of each phase of testing, there will be a 14-day 'gap' between completion of each phase and the start of the following phase, which will add 42 days to the critical path during which the Contractor's resources will largely be idle.

Under Sub-Clause 9.3 [*Retesting*], if the Works or a Section fail to pass the Tests on Completion, Sub-Clause 7.5 [*Defects and Rejection*] applies.

The Engineer or the Contractor may require these failed tests, and the Tests on Completion on any related work, to be repeated under the same terms and conditions. Such repeated tests are treated as Tests on Completion.

Under Sub-Clause 7.5 [*Defects and Rejection*] the Engineer may give a Notice to the Contractor describing the item of Plant, Materials, design or workmanship that has been found to be defective. The Contractor must then promptly prepare and submit a proposal for the necessary remedial work.

The Engineer may Review this proposal and may give a Notice to the Contractor stating the extent to which the proposed work, if carried out, would not result in the Plant, Materials, Contractor's design (if any) or workmanship complying with the Contract. After receiving such a Notice, the Contractor must promptly submit a revised proposal. If the Engineer gives no such Notice within 14 days of receiving the Contractor's proposal (or revised proposal), the Engineer is deemed to have given a Notice of No-objection.

If the Contractor fails to promptly submit a proposal (or revised proposal) for remedial work, or fails to carry out the proposed remedial work to which the Engineer has given (or is deemed to have given) a Notice of No-objection, the Engineer may:

(*a*) instruct the Contractor under sub-paragraph (a) and/or (b) of Sub-Clause 7.6 [*Remedial Work*]; or
(*b*) reject the Plant, Materials, Contractor's design (if any) or workmanship by giving a Notice to the Contractor, with reasons, in which case sub-paragraph (a) of Sub-Clause 11.4 [*Failure to Remedy Defects*] shall apply.

After remedying defects in any Plant, Materials, design (if any) or workmanship, if the Engineer requires the items to be retested, the tests shall be repeated at the Contractor's risk and cost. If the rejection and retesting cause the Employer to incur additional costs, the Employer shall be entitled, subject to Sub-Clause 20.2 [*Claims for Payment and/or EOT*], to payment of these costs by the Contractor.

Figure 11.2 Three phases of Tests on Completion

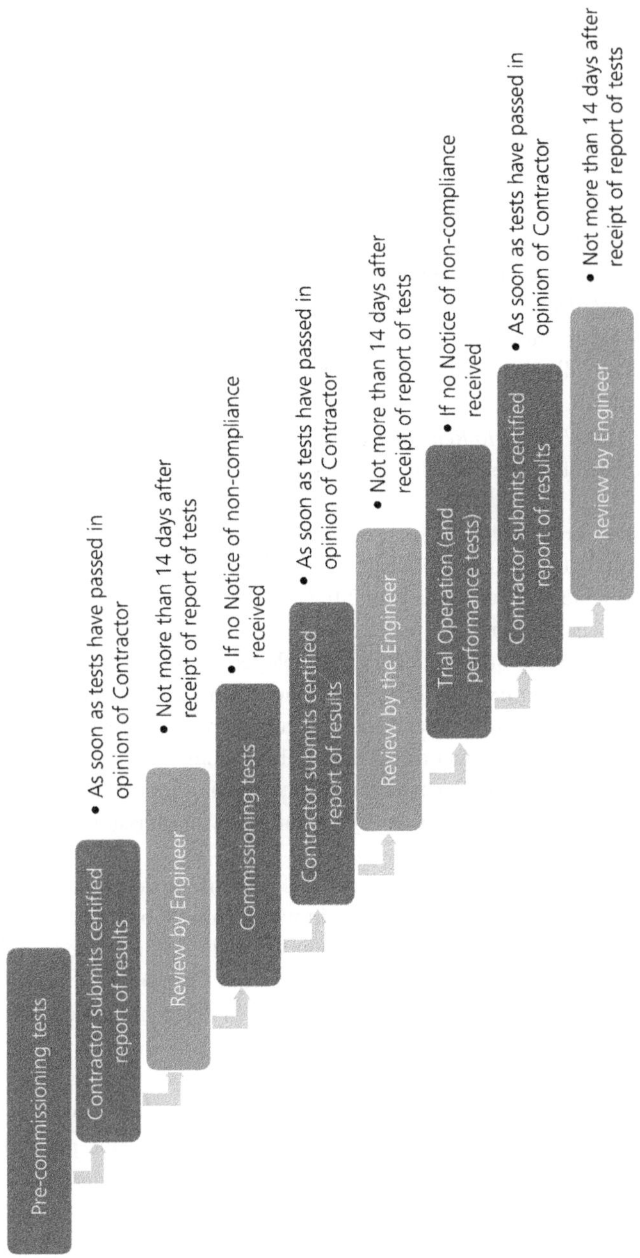

If the Works (or a Section) fail to pass the Tests on Completion repeated under Sub-Clause 9.3 [*Retesting*], the Engineer is entitled under Sub-Clause 9.4 [*Failure to Pass Tests on Completion*] to:

(*a*) order further repetition of the Tests on Completion under Sub-Clause 9.3 [*Retesting*];
(*b*) reject the Works if the failure is sufficient to deprive the Employer of substantially the whole benefit of the Works; in which event the Employer will be entitled to the remedies under sub-paragraph (d) of Sub-Clause 11.4 [*Failure to Remedy Defects*];
(*c*) reject the Section if the effect of the failure is that the Section cannot be used for its intended purpose(s); in which event the Employer will be entitled to the remedy under sub-paragraph (c) of Sub-Clause 11.4 [*Failure to Remedy Defects*]; or
(*d*) issue a Taking-Over Certificate, if the Employer so requests; in which event the Employer will be entitled subject to Sub-Clause 20.2 [*Claims for Payment and/or EOT*] to payment by the Contractor or a reduction in the Contract Price under sub-paragraph (b) of Sub-Clause 11.4 [*Failure to Remedy Defects*]. This entitlement shall be without prejudice to any other rights the Employer may have, under the Contract or otherwise. The Contractor must nevertheless proceed in accordance with all other obligations under the Contract.

Under Sub-Clause 11.4 [*Failure to Remedy Defects*], if the Contractor fails to remedy a defect which is the Contractor's responsibility the Employer may (at the Employer's sole discretion):

(*a*) carry out the work or have the work carried out by others (including any retesting), at the Contractor's cost, but the Contractor shall have no responsibility for this work. The Employer will be entitled, subject to Sub-Clause 20.2 [*Claims for Payment and/or EOT*], to payment by the Contractor of the costs reasonably incurred by the Employer in remedying the defect;
(*b*) accept the defective work, in which case the Employer will be entitled, subject to Sub-Clause 20.2 [*Claims for Payment and/or EOT*], to a reduction in the Contract Price which will be in full satisfaction of this failure only and shall be in the amount as shall be appropriate to cover the reduced value to the Employer as a result of this failure;
(*c*) require the Engineer to treat any part of the Works which cannot be used for its intended purpose(s) under the Contract by reason of this failure as an omission, as if instructed under Sub-Clause 13.3.1 [*Variation by Instruction*]; or
(*d*) terminate the Contract as a whole with immediate effect if the defect deprives the Employer of substantially the whole benefit of the Works, in which case, Sub-Clause 15.2 [*Termination for Contractor's Default*] will not apply. The Employer shall then be entitled, subject to Sub-Clause 20.2 [*Claims for Payment and/or EOT*], to recover from the Contractor all sums paid for the Works, plus financing charges and any costs incurred in dismantling the same, clearing the Site and returning Plant and Materials to the Contractor.

It should be remembered that the definition of Tests on Completion is not limited to the tests specified in the Contract but includes those instructed as a Variation. Moreover, the possibility of the Engineer instructing additional or varied tests under Clause 13 [*Variations and Adjustments*] is expressly mentioned under Sub-Clause 7.4 [*Testing by the Contractor*] to which reference is made under Sub-Clause 9.1 [*Contractor's Obligations*]. Whether or not the Contractor is to be paid for these varied or additional tests will depend on the results of the tests. If they show that the tested Plant, Materials or workmanship is not in accordance with the Contract, the Cost and any delay incurred in carrying out the Variation must be borne by the Contractor.

Figure 11.3 Repeat testing

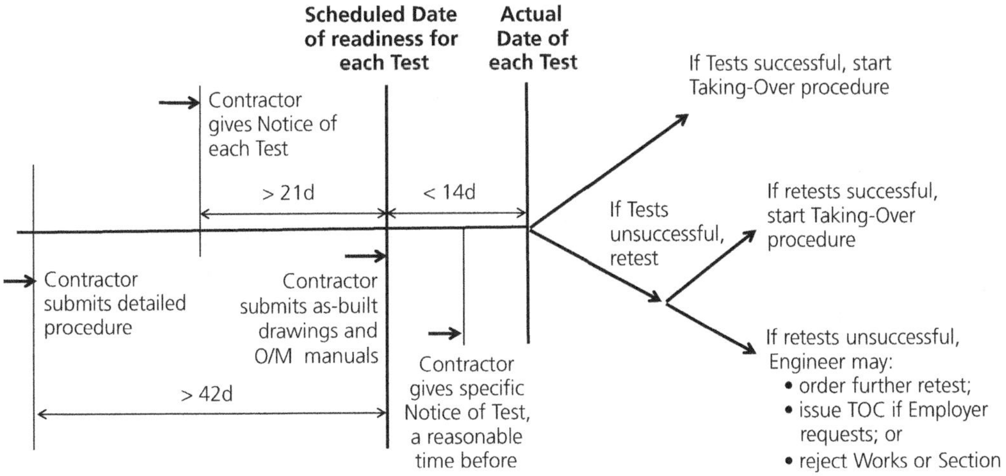

If the Contractor has given a Notice under Sub-Clause 9.1 [*Contractor's Obligations*] that the Works (or Section) are ready for Tests on Completion, and these tests are unduly delayed by the Employer's Personnel or by a cause for which the Employer is responsible, Sub-Clause 10.3 [*Interference with Tests on Completion*] applies.

Under Sub-Clause 10.3 [*Interference with Tests on Completion*] of RB 2017, if the Employer's Personnel or a cause for which the Employer is responsible, prevents the Contractor from carrying out the Tests on Completion for more than 14 days (either a continuous period, or multiple periods which total more than 14 days):

(*a*) the Contractor must give a Notice to the Engineer describing such prevention;
(*b*) the Employer will be deemed to have taken over the Works (or Section) on the date when the Tests on Completion would otherwise have been completed; and
(*c*) the Engineer must immediately issue a Taking-Over Certificate for the Works (or Section).

After receipt of this Taking-Over Certificate, the Contractor must carry out the Tests on Completion as soon as practicable and, in any case, before the expiry date of the DNP. The Engineer must give a Notice to the Contractor not less than 14 days before the date after which the Contractor may carry out each of the Tests on Completion.

If the Contractor suffers delay and/or incurs Cost as a result of being prevented from carrying out the Tests on Completion, the Contractor is entitled, subject to Sub-Clause 20.2 [*Claims for Payment and/or EOT*], to EOT and/or payment of such Cost Plus Profit.

Under YB 2017, Sub-Clause 10.3 [*Interference with Tests on Completion*] also applies to delay to any performance test during trial operation due to the necessary operating conditions not being available.

Figure 11.4 Delayed testing

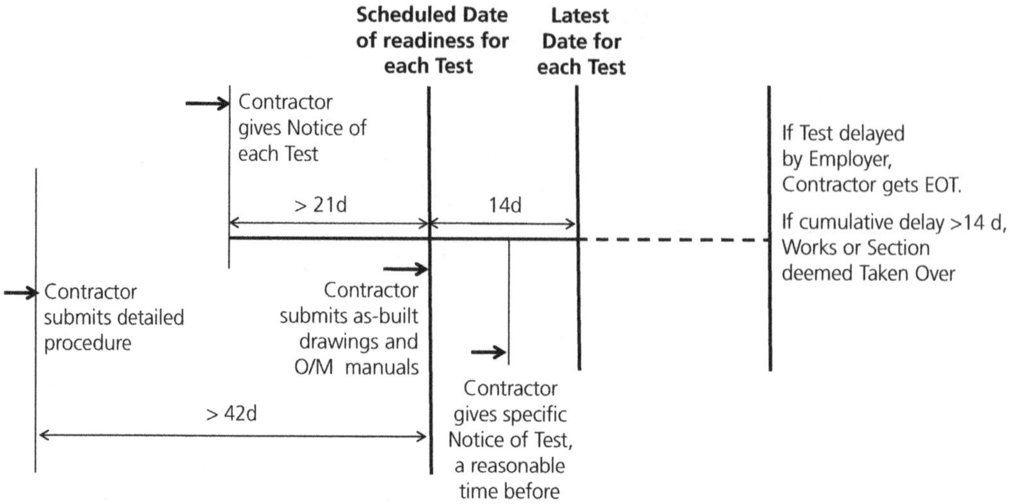

Under SB 2017, Sub-Clause 10.3 [*Interference with Tests on Completion*] also contains the reference to the performance test but the remainder of the text is slightly different from RB 2017 and YB 2017:

> *'If the Contractor is prevented, for more than 14 days (either a continuous period, or multiple periods which total more than 14 days), from carrying out the Tests on Completion by the Employer's Personnel or by a cause for which the Employer is responsible (including any performance test that is not possible due to available operating conditions during trial operation):*
>
> (*a*) *the Contractor shall carry out the Tests on Completion as soon as practicable and, in any case, before the expiry date of the relevant DNP; and*
> (*b*) *if the Contractor suffers delay and/or incurs Cost as a result of being so prevented, the Contractor shall be entitled subject to Sub-Clause 20.2 [Claims for Payment and/or EOT] to EOT and/or payment of such Cost Plus Profit.'*

If the delay in conducting the Tests on Completion is due to the Contractor, the Engineer may give a Notice to the Contractor requiring the Contractor to carry out the tests within 21 days. In this case, the Contractor may choose the day or days for carrying out the tests within this period of 21 days subject to Notice to the Engineer, not less than 7 days prior to the chosen date or dates.

If the Contractor fails to carry out the Tests on Completion within this period of 21 days:

(*a*) the Employer's Personnel may proceed with the tests after a second Notice is given to the Contractor;
(*b*) the Contractor may attend and witness these tests;
(*c*) within 28 days of these tests being completed, the Engineer must send a copy of the test results to the Contractor; and

(*d*) if the Employer incurs additional costs as a result of such testing, the Employer is entitled, subject to Sub-Clause 20.2 [*Claims for Payment and/or EOT*], to payment by the Contractor of the costs reasonably incurred.

These Tests on Completion are deemed to have been carried out in the presence of the Contractor, and the Contractor cannot contest the results.

11.2. Taking Over

The Contractor may apply for a Taking-Over Certificate by giving a Notice to the Engineer (Employer under SB 2017) not more than 14 days before the Works will be complete, in the Contractor's opinion, and ready for taking over. The Contractor may also apply in the same way for a Taking-Over Certificate for any Section.

Within 28 days of receiving the Contractor's Notice, the Engineer (Employer under SB 2017) must either:

(*a*) issue the Taking-Over Certificate to the Contractor, stating the date on which the Works or Section were completed in accordance with the Contract, except for any minor outstanding work and defects (as listed in the Taking-Over Certificate) which will not substantially affect the safe use of the Works or Section for their intended purpose (either until or whilst this work is completed and these defects are remedied); or
(*b*) reject the application by giving a Notice to the Contractor, with a list of work to be done, defects to be rectified, reasons or documents to be provided, prior to reapplying for the Taking-Over Certificate.

If the Engineer (Employer under SB 2017) does not issue the Taking-Over Certificate or rejects the Contractor's application within this period of 28 days, and if the conditions described in sub-paragraphs (a) to (d) below have been fulfilled, the Works or Section will be deemed to have been completed in accordance with the Contract on the fourteenth day after the Engineer (Employer under SB 2017) received the Contractor's Notice applying for the Taking-Over Certificate. The Taking-Over Certificate will be deemed to have been issued at such a date.

Under Sub-Clause 10.1 [*Taking Over the Works and Sections*] of RB 2017, the Works (or Section) must be taken over by the Employer when:

(*a*) the Works (or Section) have been completed in accordance with the Contract, including the passing of the Tests on Completion, except for any minor outstanding work and defects which will not substantially affect the safe use of the Works (or Section);
(*b*) the Engineer has given (or is deemed to have given) a Notice of No-objection to the as-built records if applicable;
(*c*) the Engineer has given (or is deemed to have given) a Notice of No-objection to the O&M Manuals if applicable;
(*d*) the Contractor has carried out the training under Sub-Clause 5.5 [*Training*] if applicable; and
(*e*) a Taking-Over Certificate for the Works has been issued or is deemed to have been issued.

YB 2017 and SB 2017 are similar, except that the O&M Manuals may be provisional versions; and under SB 2017, it is the Employer who issues the Notices of No-objection.

Example of a Taking-Over Certificate:

> **A.N. OTHER & Partners**
>
> Consulting Engineers
> BETTA CONTRACTORS LTD.
>
> 31 December 2022
>
> **For the attention of Mr. Alright, Contractor's Representative**
>
> Subject: **Taking-Over Certificate issued under Sub-Clause 10.1 [*Taking Over the Works and Sections*]**
>
> Dear Sir,
>
> In accordance with Sub-Clause 10.1 [*Taking Over the Works and Sections*] you are hereby notified that the Works [*Section* ...] were [was] completed in accordance with the Contract except for the minor outstanding work and defects as indicated in the attached list and were [was] taken over by the Employer on [insert date].
>
> [You are hereby instructed to complete the outstanding work and remedying of the defects by the dates indicated.]
>
> Yours faithfully,
>
> The Engineer
>
> cc The Employer

Under Sub-Clause 9.4 [*Failure to Pass Tests on Completion*], Sub-Clause 10.2 [*Taking Over of Parts of the Works*] and Sub-Clause 10.3 [*Interference with Tests on Completion*], the provisions with respect to Taking Over are slightly different. The relevant provisions under Sub-Clause 9.4 [*Failure to Pass Tests on Completion*] and Sub-Clause 10.3 [*Interference with Tests on Completion*] have been addressed under Chapter 11.1 above.

Under Sub-Clause 10.2 [*Taking Over of Parts of the Works*] of RB 2017 and YB 2017, the Employer may require the Engineer to issue a Taking-Over Certificate for any part of the Permanent Works but is not generally obliged to do so.

However, the Employer must not use any part of the Works unless and until the Engineer has issued a Taking-Over Certificate for that part, except if such use is temporary and was either foreseen in the Specification or is with the prior agreement of the Contractor.

If the Employer does use any part of the Works before the Taking-Over Certificate is issued, the Contractor must give a Notice to the Engineer identifying the part and describing such use. That part then becomes a '*Part*' as defined in Sub-Clause 1.1.58 and:

(*a*) that Part is deemed to have been taken over by the Employer as from the date on which it was first used;
(*b*) from that date, the Contractor will no longer be liable for the care of such Part and responsibility for repairing damage, for security, insurance and so forth is passed to the Employer; and
(*c*) the Engineer must immediately issue a Taking-Over Certificate for that Part, and any outstanding work to be completed (including Tests on Completion) and/or defects to be remedied must be listed in this certificate.

The definition of '*Part*' as stated in RB 2017 is as follows:

Sub-Clause 1.1.58	***'Part'*** *means a part of the Works or part of a Section (as the case may be) which is used by the Employer and deemed to have been taken over under Sub-Clause 10.2 [Taking Over Parts].*

This definition does not reflect the possibility of the Employer choosing to have a Taking-Over Certificate issued for a part. For this reason, the 2022 Reprint modifies this definition (see Chapter 19).

After the Engineer has issued a Taking-Over Certificate for a Part, the Contractor must be given the earliest opportunity to carry out the outstanding work (including Tests on Completion) and/or remedial work for any defects listed in the certificate. The Contractor must carry out this work as soon as practicable and, in any case, before the expiry date of the relevant DNP.

Example of a Taking-Over Certificate following deemed taking over:

A.N. OTHER & Partners
Consulting Engineers

BETTA CONTRACTORS LTD.

31 December 2022

For the attention of Mr. Alright, Contractor's Representative

Subject: **Taking-Over Certificate issued under Sub-Clause 10.2 [*Taking Over Parts*]**

Dear Sir,

You are hereby notified that the following Part of the Works was put to use by the Employer on [insert date] (other than as a temporary measure, which is either stated in the Specification or with the prior agreement of the Contractor).

Description of the Part:

....

In accordance with Sub-Clause 10.2 [*Taking Over Parts*], the said Part is deemed to have been taken over by the Employer at the said date.

[Notwithstanding the said Taking-Over, you are required to complete the minor outstanding work and defects as indicated in the attached list as well as to execute and complete the Tests on Completion in accordance with Sub-Clause 9.1 [*Tests on Completion*].]

Yours faithfully,

The Engineer

cc The Employer

If the Contractor incurs Cost as a result of the Employer taking over and/or using a Part, the Contractor will be entitled, subject to Sub-Clause 20.2 [*Claims for Payment and/or EOT*], to payment of such Cost Plus Profit. Sub-Clause 10.2 [*Taking Over of Parts of the Works*] says nothing about the possible delay to completion of the remaining Works caused by the Employer taking over and/or using a Part. However, the Contractor would be entitled to an EOT under sub-paragraph (e) of Sub-Clause 8.5 [*Extension of Time for Completion*], subject always to Sub-Clause 20.2 [*Claims for Payment and/or EOT*], if the taking over of the Part causes a delay to completion of the remainder of the Works.

If any Part of the Works is taken over under Sub-Clause 10.2 [*Taking Over of Parts of the Works*], the remaining Works or Section are not to be taken over until the conditions described in sub-paragraphs (a) to (e) of Sub-Clause 10.1 [*Taking Over the Works and Sections*] have been fulfilled. This may be workable if the Part represents a minor portion of the whole Works. However, it would be possible for the Employer to begin using the whole of the Works before the TOC is issued, in which circumstances, it would be wrong to deny deemed taking over on the basis that all the conditions listed under Sub-Clause 10.1 have not been fulfilled.

If the Engineer issues a Taking-Over Certificate for any part of the Works, or if the Employer is deemed to have taken over a Part under sub-paragraph (a) of Sub-Clause 10.2 [*Taking Over Parts*], for any period of delay after the date at which the part or Part was taken over, the Delay Damages for completion of the remainder of the Works must be reduced. It is also stated in the last paragraph of Sub-Clause 10.2 [*Taking Over of Parts of the Works*] that the Delay Damages for the remainder of the Section (if any) in which this Part is included must also be reduced.

The remainder of the paragraph then deals with a Part (i.e. a part of the Works which is deemed to be taken over because of prior use by the Employer) without mention of a part for which the Engineer has issued a Taking-Over Certificate solely at the Employer's discretion. This appears to be an error in drafting. Logically, the same reduction in Delay Damages should apply with respect to any part of the Works for which the Engineer issues a Taking-Over Certificate, without there being any requirement for prior use of the part by the Employer. (Indeed, the 2022 Reprint has addressed this problem – see Chapter 19.)

The reduction is to be calculated as the proportion which the value of the Part (except the value of any outstanding works and/or defects to be remedied) bears to the value of the Works (or Section) as a whole.

The Engineer must proceed under Sub-Clause 3.7 [*Agreement or Determination*] to agree or determine this reduction (and for the purpose of Sub-Clause 3.7.3 [*Time limits*], the date the Engineer receives the Contractor's Notice under Sub-Clause 10.2 [*Taking Over Parts*] is the date of commencement of the time limit for agreement under Sub-Clause 3.7.3 [*Time limits*]). The reduction in the Delay Damages only applies to the daily rate of Delay Damages and does not affect the maximum amount of these damages.

The Contractor's Notice under Sub-Clause 10.2 [*Taking Over Parts*] only applies to a Part that is used by the Employer prior to taking over. There is no requirement for a Contractor's Notice with respect to a part of the Works for which the Engineer issues a Taking-Over Certificate at the discretion of the Employer. This means that there is no trigger for the commencement of the time limit for agreement under Sub-Clause 3.7.3 [*Time limits*] of a reduction of the Delay Damages related to such a part.

An Employer who wishes to claim Delay Damages without taking account of a part of the Works for which a Taking-Over Certificate has been issued should seek legal advice before doing so.

It should be noted that Sub-Clause 10.2 [*Taking Over of Parts of the Works*] in SB 2017, differs from the corresponding provisions in RB 2017 and YB 2017. Under SB 2017, unless the Employer's Requirements foresee the possibility of taking over a part of the Works or such taking over is agreed between the Parties, such taking over or use of a part of the Works by the Employer is not permitted.

11.3. Tests after Completion

For some kinds of Works, it is necessary to conduct Tests after Completion, particularly to ensure proper performance. Such is the case, for example, for sewage treatment plants which can only be properly tested when the plant is connected to the sewerage system and sewage is available.

Sub-Clause 1.1.84 **YB 2017** **Sub-Clause 1.1.74** **SB 2017**	***'Tests after Completion'*** *means the tests (if any) which are stated in the Specification and which are carried out in accordance with the Special Provisions after the Works or a Section (as the case may be) are taken over under Clause 10 [Employer's Taking Over].*

Sub-Clause 12 of YB 2017 and SB 2017 applies if such Tests after Completion are specified in the Employer's Requirements.

In such a case, Sub-Clause 12.1 [*Procedure for Tests after Completion*] requires the Employer to:

(*a*) provide all necessary electricity, water, sewage (if applicable), equipment, fuel, consumables, instruments, labour, materials and suitably qualified, experienced and competent staff for the Tests after Completion to be executed efficiently and properly; and

(*b*) carry out the Tests after Completion in accordance with:

(i) the Employer's Requirements,

(ii) the O&M Manuals to which the Engineer has given (or is deemed to have given) a Notice of No-objection, under Sub-Clause 5.7 [*Operation and Maintenance Manuals*], and

(iii) such guidance as the Contractor may be required to give during the course of these tests; and
(iv) in the presence of such Contractor's Personnel as either Party may reasonably request.

The Tests after Completion must be conducted at the times stated in the Employer's Requirements or, if not stated, as soon as is reasonably practicable after the Works (or Section) have been taken over.

The Engineer must give a Notice to the Contractor, not less than 21 days before the date on which the Tests after Completion will be carried out, together with details of their location. This Notice must include a test programme showing the estimated timing for each of the tests. Unless otherwise agreed with the Contractor, these tests must be carried out on the stated date.

The results of the Tests after Completion are to be compiled and evaluated by both Parties. In so doing, account is to be taken of the effect of the Employer's prior use of the Works.

If the Contractor does not attend at the time and place stated in the Engineer's Notice (or otherwise agreed with the Contractor), the Employer may proceed with the Tests after Completion. These will be deemed to have been conducted in the Contractor's presence, and the Contractor will be unable to contest the results.

If the Contractor incurs Cost as a result of any unreasonable delay by the Employer in carrying out the Tests after Completion, the Contractor shall be entitled under Sub-Clause 12.2 [*Delayed Tests*] and subject to Sub-Clause 20.2 [*Claims for Payment and/or EOT*] to payment of such Cost Plus Profit.

If, for reasons not attributable to the Contractor, a Test after Completion cannot be completed during the DNP (or any other period agreed by both Parties), then the Works (or Section) shall be deemed to have passed this Test after Completion.

Subject to Sub-Clause 12.4 [*Failure to Pass Tests after Completion*], if the Works (or a Section) fail to pass the Tests after Completion:

(*a*) the Contractor must execute all work required to remedy the defects, following receipt of a Notice given by (or on behalf of) the Employer under sub-paragraph (b) of Sub-Clause 11.1 [*Completion of Outstanding Work and Remedying Defects*] and
(*b*) after such remedying of any defect or damage, Sub-Clause 11.6 [*Further Tests after Remedying Defects*] will apply.

Sub-Clause 11.6 [*Further Tests after Remedying Defects*] requires the Contractor to give a Notice to the Engineer, within 7 days of completion of the remedial work, describing the remedied Works, Section, Part and/or Plant and the proposed repeated tests under Clause 12 [*Tests after Completion*].

Within 7 days of receipt of the Contractor's Notice, the Engineer must give a Notice to the Contractor, in which the Engineer:

- agrees to the proposed testing or
- gives instructions with respect to the repeated tests, as necessary to demonstrate that the remedied Works, Section, Part and/or Plant comply with the Contract.

If the Contractor fails to give such a Notice within the 7 days, the Engineer may give a Notice to the Contractor within 14 days of completion of the remedial work, issuing instructions with respect to the repeated tests as necessary to demonstrate that the remedied Works, Section, Part and/or Plant comply with the Contract

The repeated tests must be carried out in the same manner as that applicable to the previous tests, except that the repeated tests shall be at the risk and cost of the Party liable for the cost of the remedial work, under Sub-Clause 11.2 [*Cost of Remedying Defects*].

If and to the extent that this failure and retesting are due to any of the matters listed as the Contractor's responsibility under sub-paragraphs (a) to (d) of Sub-Clause 11.2 [*Cost of Remedying Defects*] and cause the Employer to incur additional costs, the Employer will be entitled subject to Sub-Clause 20.2 [*Claims for Payment and/or EOT*] to payment of these costs by the Contractor.

If the Works (or Section) fail to pass any or all of the Tests after Completion and applicable Performance Damages are set out in the Schedule of Performance Guarantees, the Employer will be entitled, subject to Sub-Clause 20.2 [*Claims for Payment and/or EOT*], to payment of these Performance Damages. Payment of these Performance Damages will be in full satisfaction of this failure and the Contractor will be entitled to full payment for the Works (or Section), less the amount of the Performance Damages. If the Contractor pays these Performance Damages to the Employer during the DNP, then the Works (or Section) will be deemed to have passed these Tests after Completion.

Figure 11.5 Tests after Completion

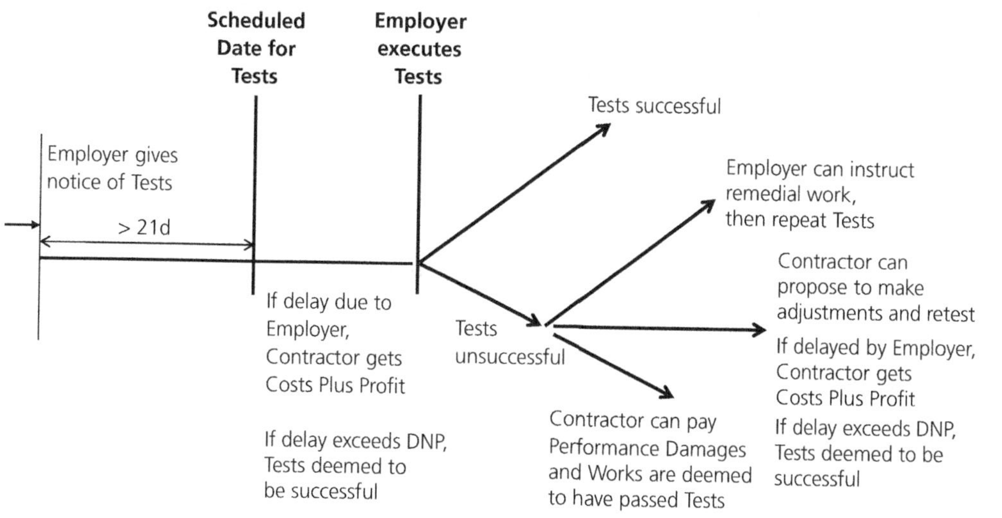

If the Works, (or Section), fail to pass a Test after Completion and, by giving a Notice to the Employer, the Contractor proposes to make adjustments or modifications to the Works (or Section):

(*a*) the Contractor may be instructed by a Notice from the Employer that right of access to the Works (or Section) cannot be given until a time that is convenient to the Employer, which time shall be reasonable
(*b*) the Contractor remains liable to carry out the adjustments or modifications and to satisfy this Test, within a reasonable period of receiving such Notice from the Employer and
(*c*) if the Contractor does not receive such a Notice from the Employer during the relevant DNP, the Contractor will be relieved of the obligation to make the adjustments or modifications and the Works (or Section) will be deemed to have passed this Test after Completion.

If the Contractor incurs additional Cost as a result of any unreasonable delay by the Employer in permitting access to the Works (or Section) either to investigate the causes of a failure to pass a Test after Completion or to carry out any adjustments or modifications, the Contractor will be entitled, subject to Sub-Clause 20.2 [*Claims for Payment and/or EOT*], to payment of any such Cost Plus Profit.

11.4. Defects Notification Period

The '*Defects Notification Period*' or '*DNP*' is defined as follows:

Sub-Clause 1.1.27 RB & YB 2017 Sub-Clause 1.1.24 SB 2017	***'Defects Notification Period'*** *or* ***'DNP'*** *means the period for notifying defects and/or damage in the Works or a Section or a Part (as the case may be) under Sub-Clause 11.1 [Completion of Outstanding Work and Remedying Defects], as stated in the Contract Data (if not stated, one year), and as may be extended under Sub-Clause 11.3 [Extension of the Defects Notification Period]. The period is calculated from the Date of Completion of the Works or Section or Part.*
Sub-Clause 1.1.24 RB	***'Date of Completion'*** *means the date stated in the Taking-Over Certificate issued by the Engineer; or, if the last paragraph of Sub-Clause 10.1 [Taking Over the Works and Sections] applies, the date on which the Works or Section are deemed to have been completed in accordance with the Contract; or Sub-Clause 10.2 [Taking Over Parts] or Sub-Clause 10.3. [Interference with Tests on Completion] applies, the date on which the Works or Section or Part are deemed to have been taken over by the Employer.*

The purpose of the DNP is to ensure that any work which was incomplete at the time of Taking Over is properly completed and that any defects or damage discovered following the Taking Over are rectified by the Contractor so that the Works and Contractor's Documents (and any Section and/or Part) are in the condition required by the Contract (fair wear and tear excepted) at the end of the relevant Defects Notification Period or as soon as practicable thereafter. In this context '*fair wear and tear*' refers to the normal deterioration that occurs during the use of the Works which is not caused by any design error by the Contractor or by poor workmanship, substandard Materials or Plant.

11.5. Completion of Outstanding Work and Remedying Defects

To achieve this purpose, Sub-Clause 11.1 [*Completion of Outstanding Work and Remedying Defects*] requires the Contractor to complete any work which was outstanding on the relevant Date of Completion within the time(s) stated in the Taking-Over Certificate or such other reasonable time as instructed by the Engineer (Employer under SB 2017), and to remedy defects or damage, of which a Notice is given to the Contractor by (or on behalf of) the Employer on or before the expiry date of the relevant DNP. Provided the Contractor receives notification of the defect or damage during the DNP, the Contractor must execute the necessary remedial work, even after expiry of the DNP.

If a defect appears (or if the Works fail to pass any Tests after Completion) or damage occurs during the relevant DNP, a Notice must be given to the Contractor by or on behalf of the Employer before the expiry of the DNP.

Promptly after the Contractor's receipt of the Notice:

(*a*) the Contractor and the Employer's Personnel must jointly inspect the defect or damage;
(*b*) the Contractor must prepare and submit a proposal for the necessary remedial work; and
(*c*) the second, third and fourth paragraphs of Sub-Clause 7.5 [*Defects and Rejection*] apply, that is:

  (i) the Engineer (Employer under SB 2017) may Review the proposal and give Notice to the Contractor stating the extent to which the proposed work, if carried out, would not achieve the desired result;
  (ii) the Engineer (Employer under SB 2017) may instruct the Contractor to repair, remedy or replace any Plant, Materials or workmanship that does not comply with the Contract, failing which he may (at the Employer's discretion) employ others to carry out the work at the Contractor's expense (unless the Contractor would have been entitled to be paid for the repair work).

After completion of the remedial work, the Engineer (Employer under SB 2017) may require retesting at the Contractor's risk and cost.

11.6. Cost of Remedying Defects

Under Sub-Clause 11.2 [*Cost of Remedying Defects*], all work by the Contractor with respect to notified defects or damage is to be executed at the Contractor's risk and cost, but only if and to the extent that the work is attributable to a failure by the Contractor to comply with an obligation under the Contract. It is for this reason, that '*fair wear and tear*' is excluded from the DNP provisions. Moreover, the Contractor is not obliged to repair at its own cost any damage caused by the Employer or by third parties, unless such damage results from failure by the Contractor to fully comply with its obligations such as to provide correct As-Built Records and/or O&M Manuals.

If the Contractor considers that the work is attributable to a cause for which the Contractor is not responsible, the Contractor must promptly give a Notice to the Engineer (Employer under SB 2017) and the Engineer (Employer's Representative under SB 2017) must proceed under Sub-Clause 3.7 [*Agreement or Determination*] to agree or determine the cause (and the start of the time limit for agreement under Sub-Clause 3.7.3 [*Time limits*] is the date of the Contractor's Notice). If it is agreed or determined that the work is attributable to a cause which is not the Contractor's responsibility, Sub-Clause 13.3.1 [*Variation by Instruction*] applies as if the remedial work had been instructed by the Engineer (Employer under SB 2017).

11.7. Extension of Defects Notification Period

Under Sub-Clause 11.3 [*Extension of Defects Notification Period*] and subject to Sub-Clause 20.2 [*Claims for Payment and/or EOT*], the Employer is entitled to an extension of the DNP for the Works, or a Section or a Part, if and to the extent that the Works, Section, Part or a major item of Plant cannot be used for the intended purpose(s) after taking over, because of a defect or damage which is the Contractor's responsibility under Sub-Clause 11.2 [*Cost of Remedying Defects*]. Thus, if a Section cannot be used for its intended purpose for three months, the Employer is entitled to an extension of the DNP of three months, subject to the Employer's compliance with Sub-Clause 20.2 [*Claims for Payment and/or EOT*]. In particular, the Employer must submit a Notice of Claim within 28 days of the date when it became aware or should have become aware of the defect which prevented the use of the Works, Section, Part or major item of Plant for the intended purpose, followed by an interim claim within 84 days of such date.

It should be noted that the Employer is not entitled to renewal of the DNP for an item found to be defective during the DNP and for which the Contractor must execute remedial work.

Multiple extensions can be claimed with respect to a recurring defect or a series of defects, which prevent the use of the Works, Section, Part or major item of Plant up to a cumulative period of 2 years after the expiry of the DNP stated in the Contract Data.

If delivery and/or erection of Plant and/or Materials was suspended under Sub-Clause 8.9 [*Employer's Suspension*] due to a cause which was not the Contractor's responsibility, or under Sub-Clause 16.1 [*Suspension by Contractor*], the Contractor's obligations under Clause 11 shall not apply to any defects or damage occurring more than two years after the expiry of the DNP for the Works (of which the Plant and/or Materials form part) would otherwise have expired.

The FIDIC 2017 Contracts Guide [2022] suggests that a few days before the expiry of the DNP, the Engineer/Employer and the Contractor should arrange to carry out a joint inspection of the entire Works. This is good advice in most cases but if the project has been subject to suspensions, care is needed to ensure that such inspection is not carried out after the Contractor's obligations under Clause 11 are no longer applicable.

Example:

The original Time for Completion for construction of a power station was 17 months. The DNP was to be 12 months, that is, from month 18 to month 29 inclusive. Due to multiple, major delays in payment, the Contractor suspended work under Sub-Clause 16.1 [*Suspension by Contractor*]. The suspensions and subsequent remobilisation totalled 27 months. The Contractor claimed and was awarded extensions of time which also totalled 27 months. The Contractor completed the Works within the revised Time for Completion, that is, by the end of month 44. The DNP was therefore to run from month 45 to month 56, inclusive.

Shortly before the end of the DNP, in month 55, the Employer discovered a defect in an item of Plant and instructed the Contractor to rectify the defect and retest.

Due to the limit fixed under Sub-Clause 11.3 [*Extension of Defects Notification Period*], the Contractor's obligations under Sub-Clause 11 with respect to remedial work ended 24 months after the DNP would have expired but for the suspension, that is, at the end of month 53. The Contractor was therefore entitled under the Contract to refuse to rectify the defect.

11.8. Failure to remedy defects

If the Contractor delays the remedying of any defect or damage under Sub-Clause 11.1 [*Completion of Outstanding Works and Remedying Defects*], the Employer or other person on the Employer's behalf, may issue a Notice to the Contractor fixing a date on or by which the defect or damage is to be remedied. The Notice must allow the Contractor a reasonable time to remedy the defect or damage.

If the Contractor fails to comply with the said Notice and the remedial work was to be executed at the cost of the Contractor under Sub-Clause 11.2 [*Cost of Remedying Defects*], the Employer may choose between:

(*a*) carrying out the work or having the work carried out by others (including any retesting) in the manner required under the Contract and at the Contractor's cost. In this case, the Employer shall be entitled subject to Sub-Clause 20.2 [*Claims for Payment and/or EOT*] to payment by the Contractor of the costs reasonably incurred by the Employer in remedying the defect or damage, but the Contractor shall have no responsibility for this work;
(*b*) accepting the damaged or defective work, in which case the Employer will be entitled, subject to Sub-Clause 20.2 [*Claims for Payment and/or EOT*], to a reduction in the Contract Price. The reduction shall be in full satisfaction of this failure only and shall be an amount which is appropriate to cover the reduced value to the Employer, as a result of the damage or defect;
(*c*) requiring the Engineer to treat any part of the Works which cannot be used for its intended purpose(s) under the Contract by reason of this failure, as an omission instructed under Sub-Clause 13.3.1 [*Variation by Instruction*]; or
(*d*) terminating the Contract as a whole, with immediate effect if the defect or damage deprives the Employer of substantially the whole benefit of the Works. In this case, Sub-Clause 15.2 [*Termination for Contractor's Default*] will not apply and the Employer will then be entitled subject to Sub-Clause 20.2 [*Claims for Payment and/or EOT*] to recover from the Contractor the full amount paid to the Contractor, plus financing charges and any costs incurred in dismantling the Works, clearing the Site and returning Plant and Materials to the Contractor.

The exercise of discretion by the Employer with respect to options (c) and (d) above is without prejudice to any other rights the Employer may have, under the Contract or otherwise.

11.9. Remedying of Defective Work off Site

If, during the DNP, the Contractor considers that any defect or damage in any Plant cannot be remedied expeditiously on the Site, they are permitted by Sub-Clause 11.5 [*Remedying of Defective Work off Site*] to give to the Employer a Notice with reasons, requesting consent to remove from Site the defective or damaged Plant for the purposes of repair. The said sub-clause lists the details which must be included in the Notice. But the Contractor must also provide any further details that the Employer may reasonably require.

After receipt of the Employer's consent, the Contractor may remove from the Site the defective or damaged items of Plant. The Employer's consent may be conditional on the Contractor increasing the amount of the Performance Security by the full replacement cost of the defective or damaged Plant.

11.10. Further Tests after Remedying Defects

Within 7 days of remedying any defect or damage, the Contractor must give a Notice to the Engineer (Employer under SB 2017) under Sub-Clause 11.6 [*Further Tests after Remedying Defects*] describing the remedied Works, Section, Part and/or Plant and the proposed repetition of any Tests on Completion.

Within 7 days (14 days under SB 2017) of receiving this Notice, the Engineer (Employer under SB 2017) must give a Notice to the Contractor either agreeing to the proposed testing or issuing instructions with respect to the repeated tests.

If the Contractor fails to give such a Notice within the 7 days, the Engineer (Employer under SB 2017) may give a Notice to the Contractor within 14 days after the defect or damage is remedied, instructing the Contractor with respect to the required tests.

All such repeated tests are to be carried out at the risk and cost of the Party liable for the cost of the remedial work.

11.11. Right of Access after Taking Over

The Contractor must have the right of access to the Works as is reasonably required to comply with the Contractor's obligations related to the DNP until the date 28 days after the issue of the Performance Certificate. However, this right of access may be limited by the Employer's reasonable security restrictions.

Under YB 2017 and SB 2017, the Contractor must also be allowed access to the operation, maintenance, and performance records of the Works.

Whenever such access is required, the Contractor must submit a Notice to the Employer, describing the parts of the Works (and/or records under YB 2017 and SB 2017) to be accessed, the reasons for such access, and the Contractor's preferred date for access.

This Notice must be given reasonably in advance of the preferred date, taking due regard of all relevant circumstances including the Employer's security restrictions. Within 7 days of receiving the Contractor's Notice, the Employer must reply to the Contractor by Notice, consenting to the Contractor's request, or proposing reasonable alternative date(s) with reasons. If the Employer fails to give this Notice within the 7 days, the Employer shall be deemed to have given consent to the Contractor's access on the preferred date stated in the Contractor's Notice.

If the Contractor incurs additional Cost as a result of any unreasonable delay by the Employer in permitting access to the Works, the Contractor will be entitled, subject to Sub-Clause 20.2 [*Claims for Payment and/or EOT*], to payment of such Cost Plus Profit.

11.12. Searching for the cause of a defect (root cause investigation)

If instructed by the Engineer (Employer under SB 2017), the Contractor must search for the cause of any defect, under the direction of the Engineer (Employer under SB 2017). The Contractor must carry out the search on the date(s) stated in the instruction or other date(s) agreed with the Engineer (Employer under SB 2017). Such a search is sometimes referred to as a 'root cause investigation'.

Unless the defect is the Contractor's responsibility and to be remedied at the Contractor's cost, the Contractor will be entitled, subject to Sub-Clause 20.2 [*Claims for Payment and/or EOT*], to payment of the Cost Plus Profit of the search.

If the Contractor fails to carry out the search instructed by the Engineer (Employer under SB 2017), the search may be carried out by the Employer's Personnel. The Contractor must be given a Notice of the date when the search will be carried out and the Contractor may attend at its own cost.

If the defect is to be remedied at the cost of the Contractor, the Employer shall be entitled, subject to Sub-Clause 20.2 [*Claims for Payment and/or EOT*], to payment by the Contractor of the costs of the search reasonably incurred by the Employer.

11.13. Performance Certificate

Performance of the Contractor's obligations under the Contract is not to be considered to have been completed until the Engineer (Employer under SB 2017) has issued the Performance Certificate to the Contractor under Sub-Clause 11.9 [*Performance Certificate*], stating the date on which the Contractor fulfilled the Contractor's obligations under the Contract. The Performance Certificate is for the entire Works and there are no separate certificates for Sections or Parts, even if the Sections or Parts were taken over (or deemed to be taken over) at different dates and had different Defects Notification Periods. Only the Performance Certificate is deemed to constitute acceptance of the Works.

The Engineer (Employer under SB 2017) must issue the Performance Certificate to the Contractor (with a copy to the Employer and to the DAAB) within 28 days of the latest of the expiry dates of the Defects Notification Periods, or as soon thereafter as the Contractor has supplied all the Contractor's Documents, remedied any notified defects and tested all the Works in accordance with the Contract.

If the Engineer (Employer under SB 2017) fails to do so within this period of 28 days, the Performance Certificate is deemed to have been issued on the date 28 days after the date on which it should have been issued under Sub-Clause 11.9 [*Performance Certificate*].

After the issue of the Performance Certificate, each Party remains liable under Sub-Clause 11.10 [*Unfulfilled Obligations*] for the fulfilment of any obligation which remained unperformed at that time and the Contract is deemed to remain in force for this purpose. Examples of such unperformed obligations are the release of the Performance Security and the final payment by the Employer and clearance of the Site by the Contractor.

However, in relation to Plant, the Contractor shall not be liable for any defects or damage occurring more than two years after expiry of the DNP for the Plant except if such limitation is prohibited by law, or in any case of fraud, gross negligence, deliberate default or reckless misconduct.

With respect to clearance of the Site, Sub-Clause 11.11 [*Clearance of Site*] requires the Contractor to remove from the Site any remaining Contractor's Equipment, surplus material, wreckage, rubbish, Temporary Works and so forth promptly after the issue of the Performance Certificate. The Contractor is required to reinstate all parts of the Site which were affected by the Contractor's activities during the execution of the Works but which are not occupied by the Permanent Works and to leave the Site and the Works in the condition stated in the Specification (or, if not stated, in a clean and safe condition).

If the Contractor fails to comply with these obligations within 28 days of the issue of the Performance Certificate, the Employer may sell (to the extent permitted by applicable Laws) or otherwise dispose of any remaining items and/or may reinstate and clean the Site (as may be necessary) at the Contractor's cost. The amount obtained from the sale may be offset by the Employer against the costs reasonably incurred in relation to the sale or disposal and reinstating and/or cleaning the Site. The Employer will be entitled, subject to Sub-Clause 20.2 [*Claims for Payment and/or EOT*], to payment by the Contractor of the remainder of the costs after the deduction of the proceeds of the sale.

It should be remembered that under Sub-Clause 4.2.1 [*Contractor's Obligations*] the Performance Security must remain valid and enforceable until the issue of the Performance Certificate and the fulfilment of the Contractor's obligations under Sub-Clause 11.11 [*Clearance of Site*]. Under Sub-Clause 4.2.3 [*Return of the Performance Security*] the Employer is entitled to keep the Performance Security until 21 days after the issue of the Performance Certificate or the clearance of the Site, whichever is the earlier.

Example of a Performance Certificate:

A.N. OTHER & Partners
Consulting Engineers

BETTA CONTRACTORS LTD.

15 April 2023

For the attention of Mr. Alright, Contractor's Representative

Subject: **Performance Certificate**

Dear Sir,

In accordance with Sub-Clause 11.9 [*Performance Certificate*] I hereby certify that the Contractor fulfilled its obligations to execute and complete the Works and remedy any defects therein on [insert date].

Nevertheless, the Contractor is reminded of its continuing obligations under Sub-Clause 11.10 [*Unfulfilled Obligations*] and, in particular, under Sub-Clause 11.11 [*Clearance of Site*].

Yours faithfully,

The Engineer

cc The Employer

11.14. Legal liability for defects

The default position under FIDIC Contracts is that the DNP should be 12 months from the date of Taking Over unless a different duration is stated in the Contract Data. The Contractor must remedy all defects and damage which are notified to him during this period. There is no contractual obligation to remedy defects and damage which are only notified to the Contractor after expiry of the DNP. However, this does not mean that the Contractor is not legally liable for defects discovered after the DNP.

This is recognised to some extent under Sub-Clause 11.10 [*Unfulfilled Obligations*] wherein it is stated that, in relation to Plant, the Contractor shall not be liable for any defects or damage occurring more than 2 years after expiry of the DNP for the Plant except if prohibited by law or in any case of fraud, gross negligence, deliberate default or reckless misconduct. Thus, FIDIC has attempted to fix a time limit for the Contractor's liability with respect to defective Plant (which is longer than the contractual liability related to the DNP). But, in doing so, FIDIC recognises that such a limit may not be valid under the applicable law. Indeed, the laws in some countries fix a much longer period of liability than the DNP. In many countries, it is commonplace for the Contractor to be legally liable for major defects for 10 years after Taking Over (decennial liability). In some cases, the law requires the Contractor to purchase insurance to cover this risk. In some other countries, a Contractor may be held liable for '*latent defects*' (those not reasonably discoverable during construction or during the DNP) for 6 or 12 years.

It is therefore important for Contractors to seek legal advice with respect to liability for defects under the applicable law, preferably during the bidding phase.

Smith G
ISBN 978-0-7277-6652-6
https://doi.org/10.1680/fcmh.66526.159

Chapter 12
Payments

12.1. Contract Price

The standard Contract Agreement contains the following:

> *'The Employer hereby covenants to pay the Contractor, in consideration of the execution and completion of the Works and the remedying of defects therein, the Contract Price at the times and in the manner prescribed by the Contract.'*

Under Sub-Clause 14.1 [*The Contract Price*] of RB 2017, the Contract Price is defined as the value of the Works in accordance with Sub-Clause 12.3 [*Valuation of the Works*] and subject to adjustments, additions (including Cost or Cost Plus Profit to which the Contractor is entitled under the Conditions) and/or deductions in accordance with the Contract. Sub-Clause 12.3 [*Valuation of the Works*] begins by stating that the Engineer shall value each item of work by applying the agreed or determined measurement and the appropriate rate or price for the item specified in the Bill of Quantities.

Thus, one of the fundamental characteristics of an RB contract is clearly described. It is that the Contractor is to be paid based on measured quantities of work executed in conjunction with the rates and prices detailed in the Bill of Quantities.

Indeed, sub-paragraph (c) of Sub-Clause 14.1 [*The Contract Price*] emphasises that any quantities which are set out in the Bill of Quantities or other Schedule(s) are estimated quantities and are not to be taken as the actual and correct quantities of the Works for the purposes of Clause 12 [*Measurement and Valuation*] or which the Contractor is required to execute. This means that the Contractor will not be paid on the basis of the indicated quantities and that there is no guarantee that the Contractor will be required to execute all the indicated quantities.

The corresponding definition of the Contract Price in YB 2017 states that the Contract Price shall be the lump sum Accepted Contract Amount and be subject to adjustments, additions (including Cost or Cost Plus Profit to which the Contractor is entitled under these Conditions) and/or deductions in accordance with the Contract.

Sub-Clause 1.1.1 RB & YB 2017	***'Accepted Contract Amount'*** *means the amount accepted in the Letter of Acceptance for the execution of the Works in accordance with the Contract.*

Together, these two provisions describe one of the fundamental differences between RB 2017 and YB 2017. Under the latter, the Contractor is to be paid the lump sum stated in the Letter of Acceptance, subject only to certain adjustments, additions and/or deductions permitted by the Contract.

Sub-paragraphs (c) and (d) of Sub-Clause 14.1 [*The Contract Price*] of YB 2017 emphasise that any quantities which may be set out in a Schedule are estimated quantities and are not to be taken as the actual and correct quantities of the Works which the Contractor is required to execute; any such quantities or price data which may be set out in a Schedule shall be used for the purposes stated in the Schedule and may be inapplicable for other purposes. No mention is made of measurement or of payment based on the price data that may be contained within such a Schedule.

Under Sub-Clause 14.1 [*The Contract Price*] of SB 2017, payment for the Works is to be made on the basis of the lump sum Contract Price stated in the Contract Agreement, subject to adjustments, additions (including Cost or Cost Plus Profit to which the Contractor is entitled under the Conditions) and/or deductions in accordance with the Contract. This is very similar to Sub-Clause 14.1 [*The Contract Price*] of YB 2017.

Under all three contracts, the Contractor must pay all taxes, duties and fees required to be paid by the Contractor under the Contract, and the Contract Price is not to be adjusted for any of these costs except as stated in Sub-Clause 13.6 [*Adjustments for Changes in Laws*].

In the Notes on the Preparation of Special Provisions, FIDIC provides an example of a sub-clause for exemption from duties, should the Contractor not be required to pay import duties on Goods imported into the Country, by the Contractor. The suggested sub-clause is in addition to the necessary modification of Sub-Clause 14.1 [*The Contract Price*] to exclude such duties.

The proposed sub-clause is as follows:

> *'All Goods imported by the Contractor into the Country shall be exempt from customs and other import duties, if the Employer's prior written approval is obtained for import. The Employer shall endorse the necessary exemption documents prepared by the Contractor for presentation in order to clear the Goods through Customs, and shall also provide the following exemption documents:*
>
> > *(describe the necessary documents, which the Contractor will be unable to prepare).*
>
> *If exemption is not then granted, the customs duties payable and paid shall be reimbursed by the Employer.*
>
> *All imported Goods, which are not incorporated in or expended in connection with the Works, shall be exported on completion of the Contract.*
>
> *If not exported, the Goods will be assessed for duties as applicable to the Goods involved in accordance with the Laws of the Country.*
>
> *However, exemption may not be available for:*
>
> > *(a) Goods which are similar to those locally produced, unless they are not available in sufficient quantities or are of a different standard to that which is necessary for the Works; and*

(*b*) *any element of duty or tax inherent in the price of goods or services procured in the Country, which shall be deemed to be included in the Accepted Contract Amount.*

Port dues, quay dues and, except as set out above, any element of tax or duty inherent in the price of goods or services shall be deemed to be included in the Accepted Contract Amount.'

The Special Provisions imposed by World Bank and some other MDBs require the Employer to choose between two alternative additions to Sub-Clause 14.1 [*The Contract Price*] to address the topic of import duties and taxes on temporarily imported Equipment:

[Note to the Employer: include one of the following two alternative texts as applicable]
The following is added at the end of the sub-clause:

[Alternative 1]
'Notwithstanding the provisions of subparagraph (b), Contractor's Equipment, including essential spare parts therefor, imported by the Contractor for the sole purpose of executing the Contract shall be exempt from the payment of import duties and taxes upon importation.'

[Alternative 2]
'Notwithstanding the provisions of subparagraph (b), Contractor's Equipment, including essential spare parts therefore, imported by the Contractor for the sole purpose of executing the Contract shall be temporarily exempt from the payment of import duties and taxes upon initial importation, provided the Contractor shall post with the customs authorities at the port of entry an approved export bond or bank guarantee, valid until the Time for Completion plus six months, in an amount equal to the full import duties and taxes which would be payable on the assessed imported value of such Contractor's Equipment and spare parts, and callable in the event the Contractor's Equipment is not exported from the Country on completion of the Contract. A copy of the bond or bank guarantee endorsed by the customs authorities shall be provided by the Contractor to the Employer upon the importation of individual items of Contractor's Equipment and spare parts. Upon export of individual items of Contractor's Equipment or spare parts, or upon the completion of the Contract, the Contractor shall prepare, for approval by the customs authorities, an assessment of the residual value of the Contractor's Equipment and spare parts to be exported, based on the depreciation scale and other criteria used by the customs authorities for such purposes under the provisions of the applicable Laws. Import duties and taxes shall be due and payable to the customs authorities by the Contractor on (a) the difference between the initial imported value and the residual value of the Contractor's Equipment and spare parts to exported and (b) on the initial imported value of the Contractor's Equipment and spare parts remaining in the Country after completion of the Contract. Upon payment of such dues within 28 days of being invoiced, the bond or bank guarantee shall be reduced or released accordingly; otherwise, the security shall be called in the full amount remaining.'

Under the final sub-paragraph of Sub-Clause 14.1 [*The Contract Price*] of RB 2017, within 28 days of the Commencement Date, the Contractor must provide the Engineer with a proposed breakdown of each lump sum price (if any) in the Schedules (such as the Bill of Quantities). The Engineer may take account of the breakdown when preparing Payment Certificates but is not bound by it.

12.2. Preparation of Statements

The payment process begins with an application for payment from the Contractor, referred to by FIDIC as a Statement. Under RB 2017 and YB 2017, the Engineer checks the Statement and issues a Payment Certificate showing the amount which the Engineer fairly considers to be due. The first such Payment Certificate is the '*Advance Payment Certificate*'; the last Payment Certificate is the '*Final Payment Certificate*'. All Payment Certificates issued between the first and last Payment Certificates are known as '*Interim Payment Certificates*'. Following receipt of a Payment Certificate, the Employer must pay the amount certified within the period fixed by the Contract.

Under SB 2017, the process is slightly different in that there is no Engineer to issue a Payment Certificate. Instead, the Contractor's Statement is received and reviewed by the Employer, who must then issue a Notice of payment and pay accordingly. The exception to this process is the advance payment, for which there is no requirement for a Notice from the Employer.

As stated above, under all the FIDIC 2017 contracts, the Contractor must apply for the advance payment and the application must be in the form of a Statement (Sub-Clause 14.2.2 [*Advance Payment*]. The format of Statements is set out under Sub-Clause 14.3 [*Application for Interim Payment*].

Under Sub-Clause 14.3 [*Application for Interim Payment*], the Contractor must submit a Statement to the Engineer (Employer under SB 2017) after the end of the period of payment stated in the Contract Data (if not stated, after the end of each month) and each Statement must:

(*a*) be in a form acceptable to the Engineer (Employer under SB 2017);
(*b*) be submitted in one paper original, one electronic copy and additional paper copies (if any) as stated in the Contract Data; and
(*c*) show in detail the amounts which the Contractor considers are due, with supporting documents which must include sufficient detail for the Engineer (Employer under SB 2017) to investigate these amounts, together with the monthly progress report under Sub-Clause 4.20 [*Progress Reports*].

Although the Engineer/Employer has the power to influence the form of the Statement, this power is very limited as Sub-Clause 14.3 lists the contents of the Statement and the order of the items to be included.

Sub-Clause 14.3 [*Application for Interim Payment*] of RB 2017 demands that the Statement include the following items, valued cumulatively and expressed in the different currencies in which the Contract Price is payable, and in the following sequence:

(*a*) the estimated contract value of the Works executed (and of the Contractor's Documents produced) up to the end of the payment period (including Variations but excluding items described in sub-paragraphs (*b*) to (*j*) below);

(*b*) any amounts to be added and/or deducted for changes in Laws under Sub-Clause 13.6 [*Adjustments for Changes in Laws*] and for changes in Cost under Sub-Clause 13.7 [*Adjustments for Changes in Cost*];
(*c*) any amount to be deducted for retention, calculated by applying the percentage of retention stated in the Contract Data to the total of the amounts under sub-paragraphs (*a*), (b) and (f) of this Sub-Clause, until the amount so retained by the Employer reaches the limit of Retention Money (if any) stated in the Contract Data;
(*d*) any amounts to be added and/or deducted for the advance payment and repayments under Sub-Clause 14.2 [*Advance Payment*];
(*e*) any amounts to be added and/or deducted for Plant and Materials under Sub-Clause 14.5 [*Plant and Materials intended for the Works*];
(*f*) any other additions and/or deductions which have become due under the Contract or otherwise, including those under Sub-Clause 3.7 [*Agreement or Determination*];
(*g*) any amounts to be added for Provisional Sums under Sub-Clause 13.4 [*Provisional Sums*];
(*h*) any amount to be added for release of Retention Money under Sub-Clause 14.9 [*Release of Retention Money*];
(*i*) any amount to be deducted for the Contractor's use of utilities provided by the Employer under Sub-Clause 4.19 [*Temporary Utilities*]; and
(*j*) the deduction of amounts certified in all previous Payment Certificates.

The result of the above will correspond to the amount due for payment by the Employer for the period, as assessed by the Contractor.

As can be seen from the above, items (a) and (c) contain cross references to other items in the list. As a result, if the stipulated sequence is modified, the mathematical result will be incorrect, unless care is taken to renumber the references to the correct items.

Item (*f*) corresponds to amounts claimed by the Contractor, including those for which the Engineer (Employer's Representative under SB 2017) has issued a determination in favour of the Contractor under Sub-Clause 3.7 [*Agreement or Determination*] (Sub-Clause 3.5 under SB 2017).

> **Note:**
>
> Item (f) is not limited to amounts for which the Engineer (Employer's Representative under SB 2017) has issued a determination (or recorded an agreement). It may also include amounts claimed but rejected by the Engineer (Employer's Representative under SB 2017).

Example of a Contractor's Statement:

CONTRACTOR'S STATEMENT No.							
PERIOD ENDING							
PROJECT:							
EMPLOYER:							
ENGINEER:							
CONTRACTOR:							
			CUMUL PREVIOUS PERIOD		CUMUL THIS PERIOD		THIS PERIOD
1	ESTIMATED VALUE OF WORKS EXECUTED INCL. VARIATIONS						
2	ADD/DEDUCT:						
		ADJUSTMENTS FOR CHANGES IN LAWS					
		ADJUSTMENTS FOR CHANGES IN COST					
3	DEDUCT:						
		RETENTION					
4	ADD:						
		ADVANCE PAYMENT					
	DEDUCT:						
		REIMBURSEMENT OF ADVANCE PAYMENT					
5	ADD:						
		PAYMENT FOR PLANT AND/OR MATERIALS SHIPPED OR DELIVERED					
	DEDUCT:						
		PAYMENT FOR PLANT AND/OR MATERIALS INCORPORATED IN WORKS					

6	ADD/DEDUCT:						
		OTHER AMOUNTS DUE INCLUDING UNDER SUB-CLAUSE 3.7					
7	ADD/DEDUCT:						
		AMOUNTS DUE UNDER PROVISIONAL SUMS					
8	ADD/DEDUCT:						
		RELEASE OF RETENTION					
9	DEDUCT:						
		AMOUNTS FOR USE OF UTLILTIES					
10	DEDUCT:						
		AMOUNT PREVIOUSLY CERTIFIED					
	AMOUNT TO BE CERTIFIED FOR PAYMENT THIS CERTIFICATE						
	Date:						
Signature: Contractor's Representative:							

If the Contract includes a Schedule of Payments specifying the instalments in which the Contract Price will be paid, then Sub-Clause 14.4 [*Schedule of Payments*] requires that unless stated otherwise:

(*a*) the instalments quoted in the Schedule of Payments are to be treated as the estimated contract values for the purposes of sub-paragraph (i) of Sub-Clause 14.3 [*Application for Interim Payment*];
(*b*) Sub-Clause 14.5 [*Plant and Materials intended for the Works*] shall not apply; and
(*c*) if:

  (i) these instalments are not defined by reference to the actual progress achieved in the execution of the Works and
  (ii) actual progress is found by the Engineer to differ from that on which the Schedule of Payments was based

then the Engineer (Employer's Representative under SB 2017) may agree or determine revised instalments under Sub-Clause 3.7 [*Agreement or Determination*] (and for this purpose the date when the difference under sub-paragraph (ii) above was found by the Engineer (Employer under SB 2017) is to be the date of commencement of the time limit for agreement under Sub-Clause 3.7.3 [*Time limits*]).

Such revised instalments must take account of the extent to which progress differs from that on which the Schedule of Payments was based.

If the Contract does not include a Schedule of Payments, the estimated contract value of the Works executed is to be assessed on a different basis (see below). Nevertheless, the Contractor must submit non-binding estimates of the payments which the Contractor expects to become due during the following three months. The first estimate must be submitted within 42 days of the Commencement Date and revised estimates must be submitted at intervals of 3 months until the issue of the Taking-Over Certificate for the Works.

Such Schedules of Payment are often used under YB 2017 and SB 2017 but are not used frequently under RB 2017. They can be of three types.

The first type states the fixed monetary amount to be paid to the Contractor for each month of the Time for Completion (or other period). However, this approach assumes that the Contractor will finish on time and if actual progress is less than planned (or more than planned), as stated above, Sub-Clause 14.1 [*Schedule of Payments*] of YB 2017 and SB 2017 permit the Engineer (Employer's Representative under SB 2017) to adjust the Schedule of Payments to the extent to which progress differs from that on which the Schedule of Payments was based.

The second approach is to tie the instalments shown in the Schedule of Payments to the forecast progress, such that the cumulative amount payable to the Contractor as a percentage of the estimated final Contract Price matches the cumulative progress achieved. The key to the successful use of this approach is early agreement on a simple method of measuring the progress achieved.

The third approach is to tie payments to achievement of progress milestones. This requires careful definition of the payment milestones and prior agreement on the supporting documents required to demonstrate achievement of the milestone. Moreover, the milestones should not be so few and the time between them so large that the Contractor's cash flow will suffer while waiting for the next payment.

If the Contract does not contain a Schedule of Payments specifying the instalments to be used for the purposes of preparing the Contractor's Statements (which is the usual case under RB 2017), the Works must be measured, and valued for payment, in accordance with Clause 12 [*Measurement and Valuation*].

Sub-Clause 12.1 [*Works to be Measured*] of RB 2017 sets out the process for recording and agreeing measurements. The Engineer must give at least 7 days' Notice to the Contractor of when and where measurements will be conducted. Unless otherwise agreed with the Contractor, the measurement on Site shall be conducted on this date and the Contractor's Representative must attend (or send another qualified representative) to assist the Engineer and to endeavour to reach agreement. The Contractor's Representative must also provide any particulars requested by the Engineer.

If the Contractor does not attend at the time and place stated in the Engineer's Notice or otherwise agreed, the measurement made by the Engineer is deemed to have been made in the Contractor's presence and the Contractor cannot contest the accuracy of the measurement.

Any part of the Permanent Works that is to be measured from records must be identified in the Specification and usually such records are prepared by the Engineer. The Notes on the Preparation of Special Provisions recommend that if the Contractor is to be responsible for any such records, this should be specified in the Contract (either by the Employer in the Specification or by the Contractor in the Tender). The FIDIC 2017 Contracts Guide [2022] goes further than this and recommends that in such cases, the standard wording of the first paragraph of the sub-clause should be amended to reflect that the Contractor will prepare the records and that the Contractor's valuation will apply subject to the correction by the Engineer of any errors that they identify. Such correction is to be made under Sub-Clause 3.7 [*Agreement or Determination*].

If the Engineer is responsible for preparing the records for any part of the Works, they must give at least 7 days' Notice to the Contractor of when and where the Contractor's Representative must attend to examine and agree the records with the Engineer. If the Contractor does not attend at the time and place stated in the Engineer's Notice or otherwise agreed, the Contractor is deemed to have accepted the records as accurate.

If the Engineer and the Contractor are unable to agree the measurement (either on Site or from records), the Contractor must give a Notice to the Engineer setting out the reasons why the Contractor considers the measurement on Site or records are inaccurate. The Notice must be given within 14 days of attending the measurement on Site or examining the measurement records. If the Engineer does not receive such a Notice within the said period, the Contractor is deemed to have accepted the measurement as accurate.

After receiving such a Contractor's Notice, unless at that time the disputed measurement is already subject to the last paragraph of Sub-Clause 13.3.1 [*Variation by Instruction*], the Engineer must proceed under Sub-Clause 3.7 [*Agreement or Determination*] to agree or determine the measurement. Moreover, for the purpose of Sub-Clause 3.7.3 [*Time limits*], the date on which the Engineer receives the Contractor's Notice will be the date of commencement of the time limit for agreement under Sub-Clause 3.7.3.

Given that the process of agreeing or determining the measurement under Sub-Clause 3.7 [*Agreement or Determination*] may take 84 days or longer, the Engineer must assess a provisional measurement for the purposes of Interim Payment Certificates, until such time as the measurement is agreed or determined.

It is important for the Contract to define the method of measurement which will be applied, either by naming a standard publication in the Contract Data or by including the relevant information in the Bill of Quantities or other applicable Schedules. It is not uncommon for such information to be provided within the Specification.

Sub-Clause 12.2 [*Method of Measurement*] states that, except as otherwise mentioned in the Contract, measurement will be based on the '*net actual quantity of each item of the Permanent Works and no allowance shall be made for bulking, shrinkage or waste*'. This is the default position and only applies if no method of measurement is defined in the Contract Data or elsewhere in the Contract. It is nevertheless an important principle which is often misunderstood. If this principle is applied, for example, to excavation of a trench for installation of a 150 mm dia. pipe, it means that the Contractor is paid for excavating a trench 150 mm wide, rather than the actual volume of the trench.

Under Sub-Clause 12.3 [*Valuation of the Works*], the Engineer must value each item of work by applying the agreed or determined measurement and the appropriate rate or price for the item, unless stated otherwise in the Contract. In most cases, the appropriate rate or price for the item will be the rate or price specified for that item in the Bill of Quantities (or other Schedule). If there is no such item in the Bill of Quantities, the Engineer may use a rate or price specified for similar work.

Any item of work which is identified in the Bill of Quantities (or other Schedule), but for which no rate or price has been inserted, is deemed to be included in other rates and prices in the Bill of Quantities (or other Schedule), and the Contractor is not entitled to have a rate or price inserted after award of the Contract.

The fourth paragraph of Sub-Clause 12.3 [*Valuation of the Works*] states that a new rate or price will be appropriate for an item of work if:

(*a*) *'the item is not identified in, and no rate or price for this item is specified in, the Bill of Quantities or other Schedule and no specified rate or price is appropriate because the item of work is not of similar character, or is not executed under similar conditions, as any item in the Contract;*

(*b*)

- (i) *the measured quantity of the item is changed by more than 10% from the quantity of this item in the Bill of Quantities or other Schedule,*
- (ii) *this change in quantity multiplied by the rate or price specified in the Bill of Quantities or other Schedule for this item exceeds 0.01% of the Accepted Contract Amount,*
- (iii) *this change in quantity directly changes the Cost per unit quantity of this item by more than 1%, and*
- (iv) *this item is not specified in the Bill of Quantities or other Schedule as a 'fixed rate item', 'fixed charge' or similar term referring to a rate or price which is not subject to adjustment for any change in quantity; and/or*

(*c*) *the work is instructed under Clause 13 [Variations and Adjustments] and sub-paragraph (a) or (b) above applies.'*

It is important to note the '*and/or*' at the end of sub-paragraph (*b*) (iv). It means that a new rate or price will be appropriate in three distinct circumstances.

The first is where the item of work is not included in the Bill of Quantities or other Schedule and no rate or price is applicable because the item of work differs from all the items specified in the Bill of Quantities (or other Schedule) either because it is of a different character or it is executed under different circumstances.

The second circumstance is where there is a significant difference between the quantity of an item measured for payment and the quantity mentioned for that item in the Bill of Quantities (or other Schedule).

In this case, a new rate may be fixed if four conditions are satisfied. This circumstance and the operation of this sub-paragraph is examined in more detail in Chapter 13.9 below.

The third circumstance is where the work results from a Variation instructed or approved under Clause 13 [*Variations and Adjustments*] and either of the first two circumstances arises.

The first circumstance does not exist in RB 1999. The FIDIC Contracts Guide [2000] stated that if an item of work was described in the Contract (that is, shown on the Drawings or mentioned in the Specification) but not mentioned in the Bill of Quantities, the cost of the work was deemed to be included in other items, unless the specified method of measurement clearly required that such work be measured. If the method of measurement did so require, a new item was to be added to the Bill of Quantities. No indication was given of a contractual basis for such an addition.

With respect to the wording under RB 2017, the FIDIC 2017 Contracts Guide [2022] states that no new item needs to be added to the Bill of Quantities or other Schedule if the method of measurement does not clearly require that the work be measured or does not include principles of measurement for the work and the work is as described in the Contract, that is, not arising from a Variation. This is contrary to the wording of sub-paragraph (*a*) which does not give any exceptions and imposes a new rate for any work which is not covered by an item in the Bill of Quantities and no specified rate or price is applicable because of the different character or circumstances.

Thus, Employers must be more careful than in the past to ensure that the Bill of Quantities includes items which cover a wide range of characteristics and circumstances. If they do not, Contractors will seek to identify work which is not expressly mentioned in the Bill of Quantities and ask for new items to be inserted.

If the Engineer and the Contractor are unable to agree the appropriate rate or price, the Contractor must give a Notice to the Engineer setting out the reasons why the Contractor disagrees with the Engineer's assessment. After receiving such a Contractor's Notice, unless at that time the Engineer has already started the process of agreement or determination under Sub-Clause 3.7 [*Agreement or Determination*] and the last paragraph of Sub-Clause 13.3.1 [*Variation by Instruction*], the Engineer must:

- proceed under Sub-Clause 3.7 [*Agreement or Determination*] to agree or determine the appropriate rate or price; and
- for the purpose of Sub-Clause 3.7.3 [*Time limits*], the date on which the Engineer receives the Contractor's Notice is the date of commencement of the time limit for agreement.

Pending the agreement or determination of an appropriate rate or price, the Engineer must assess a provisional rate or price for the purposes of Interim Payment Certificates. Experience shows that in some countries, the local system makes it difficult for the Engineer to do this. However, unless the applicable law forbids such use of provisional rates or prices, the Engineer is bound to follow the Contract; if they do not, the Employer will be in breach of the Contract.

Under the Special Provisions imposed by the World Bank and some other MDBs for use with FIDIC 2017, Sub-Clause 3.2 [*Engineer's Duties and Authority*] requires the Engineer to obtain the consent in writing of the Employer before taking any action under Sub-Clause 13.1 [*Right to Vary*] and Sub-Clause 13.2 [*Value Engineering*].

Therefore, it might be considered that the Engineer is unable to assess a provisional rate or price without the Employer's consent.

However, the Special Provision does not replace the General Condition but adds to it. This can be seen from the third paragraph of Sub-Clause 3.2 [*Engineer's Duties and Authority*] of the General Conditions which states: '*If the Engineer is required to obtain the consent of the Employer before exercising a specified authority, the requirements shall be as stated in the Particular Conditions.*' The Special Provision fulfils this requirement and there is no indication that it is intended to replace the original sub-clause.

Therefore, all other requirements of Sub-Clause 3.2 of the General Conditions remain valid, including the fourth paragraph which states that '*whenever the Engineer exercises a specified authority for which the Employer's consent is required, then (for the purposes of the Contract) such consent shall be deemed to have been given*'.

Thus, regardless of whether the Employer consented to an instruction or approval by the Engineer under Sub-Clause 13.1 [*Right to Vary*] or Sub-Clause 13.2 [*Value Engineering*], once such instruction or approval has been issued by the Engineer, the Employer's consent is deemed to have been given and the Engineer is bound under Sub-Clause 12.3 [*Valuation of the Works*] to fix provisional rates or prices for the purposes of Interim Payment Certificates, pending the agreement or determination of appropriate rates and prices.

12.3. Retention

Sub-Clause 1.1.69 **RB 2017** **Sub-Clause 1.1.70** **YB 2017** **Sub-Clause 1.1.60** **SB 2017**	***'Retention Money'*** *means the accumulated retention moneys which the Employer retains under Sub-Clause 14.3 [Application for Interim Payment] and pays under Sub-Clause 14.9 [Release of Retention Money].*

The Retention Money is deducted from Interim Payments at the percentage stated in the Contract Data, until the total amount of the Retention Money reaches the maximum amount which is also stated in the Contract Data. The percentage is applied to the sum of the value of Works executed (including Contractor's Documents) plus amounts added or deducted for changes in laws and changes in cost under Sub-Clauses 13.6 [*Adjustments for Changes in Laws*] and Sub-Clause 13.7 [*Adjustments for Changes in Cost*], plus any other additions and/or deductions which have become due under the Contract or otherwise, including those under Sub-Clause 3.7 [*Agreement or Determination*].

The purpose of the Retention Money is to permit the Employer to constitute a fund to pay for remedial work by others, if the Contractor refuses to remedy defects both during construction (before Taking Over) and subsequently, during the DNP (after Taking Over).

After the issue of the Taking-Over Certificate for the Works, the Contractor must include in a Statement the release of the first half of the Retention Money. If a Taking-Over Certificate is issued for a Section, the Contractor must include in a Statement the release of the relevant percentage of the first half of the Retention Money. The relevant percentage for such a Section is stated in the Contract Data. However, no such early release of the Retention Money for the Section is permitted if no percentage is stated in the Contract Data. No such early release of Retention Money is foreseen for a Part of the Works.

After the latest of the expiry dates of the Defects Notification Periods, the Contractor must include the second half of the Retention Money in a Statement promptly after such latest date. If a Taking-Over Certificate was (or was deemed to have been) issued for a Section, the Contractor must include the relevant percentage of the second half of the Retention Money in a Statement promptly after the expiry date of the DNP for the Section.

In the next IPC after the Engineer receives any such Statement, the Engineer must certify the release of the corresponding amount of Retention Money. However, when certifying any release of Retention Money under Sub-Clause 14.6 [*Issue of IPC*], if any work remains to be executed under Clause 11 [*Defects after Taking Over*], the Engineer may withhold certification of the estimated cost of this work until it has been executed. SB 2017 is similar except that the Employer must release the Retention (less the estimated cost of the work remaining to be executed) without any certification of the amount.

Figure 12.1 Release of Retention Money under RB 2017 & YB 2017

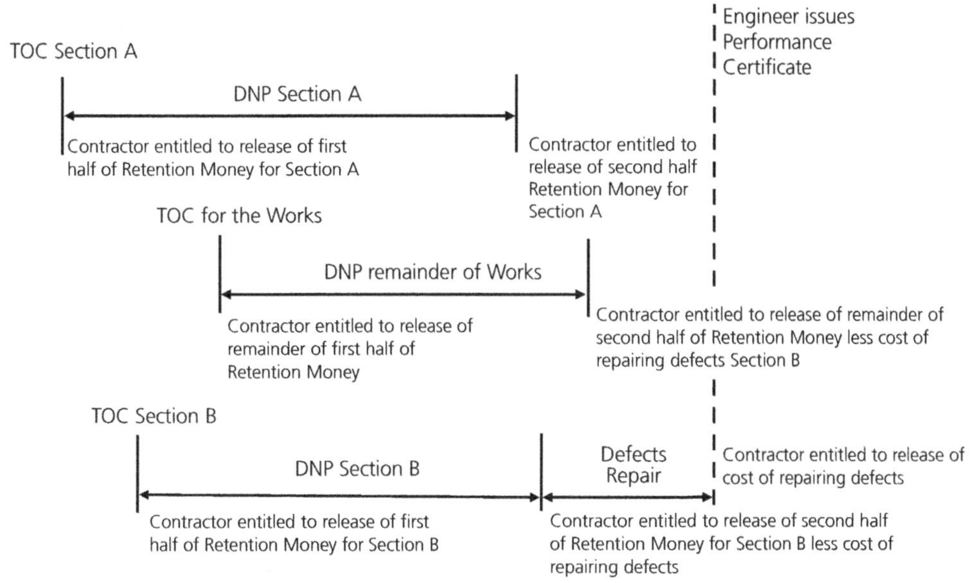

Under the Special Provisions imposed by the World Bank and some other MDBs for use with FIDIC 2017, Sub-Clause 14.9 [*Release of Retention Money*] permits the Contractor to substitute a guarantee in the place of the second half of the Retention Money. On receipt by the Employer of a guarantee in the required form and for the required amount and currencies, the Engineer must certify payment of the second half of the Retention Money.

However, if the Performance Security (and, if applicable, an ES Performance Security required under Sub-Clause 4.2) is in the form of a demand guarantee, and the amount guaranteed when the TOC is issued, is more than half of the Retention Money, then the Retention Money guarantee is not required, and the Contractor is entitled to have all the Retention Money released.

If the amount guaranteed under the Performance Security and, if applicable, an ES Performance Security, when the TOC is issued, is less than half of the Retention Money, the Retention Money guarantee will only be required for the difference between half of the Retention Money and the amount guaranteed under the Performance Security (and, if applicable, an ES Performance Security). In other words, the Retention Money Guarantee is only required to top up the amount of the Performance Security (and, if applicable, the ES Performance Security) so that the total amount guaranteed is equivalent to half of the Retention Money.

As the amount of the Performance Security is, in most cases, more than half the Retention Money, on most projects financed by the World Bank and some other MDBs, the Contractor will be entitled to the release of all the Retention Money soon after the issue of the TOC.

12.4. Advances for Materials on Site

If the Contract Data includes lists of Plant and/or Materials for which advance payment is to be made when shipped and/or delivered, Sub-Clause 14.5 [*Plant and Materials intended for the Works*] is applicable. If no such lists are included, the sub-clause does not apply.

If Sub-Clause 14.5 [*Plant and Materials intended for the Works*] is applicable, the Contractor must include, under sub-paragraph (v) of Sub-Clause 14.3 [*Application for Interim Payment*]

(*a*) an amount to be added for Plant and Materials which have been shipped or delivered (as the case may be) to the Site for incorporation in the Permanent Works; and
(*b*) an amount to be deducted when the contract value of such Plant and Materials is included as part of the Permanent Works under sub-paragraph (i) of Sub-Clause 14.3 [*Application for Interim Payment*].

Thus, the Contractor is entitled to an advance payment when the listed Plant and/or Materials are either shipped to Site or delivered to Site and the advance payment is reimbursed when the Contractor is paid for incorporating the Plant and/or Materials in the Works.

The amount of the advance to be paid is the equivalent of 80% of the value of the Plant and Materials as agreed or determined by the Engineer (Employer's Representative under SB 2017) under Sub-Clause 3.7 [*Agreement or Determination*], taking account of the contract value and the documents submitted by the Contractor. However, several conditions must be satisfied before the Contractor is entitled to such a payment (see below). For the purpose of Sub-Clause 3.7.3 [*Time limits*] the date these conditions are fulfilled shall be the date of commencement of the time limit for agreement under Sub-Clause 3.7.3 [*Time limits*]. The currencies for this certified sum are to be the same as those in which payment will become due when the contract value is included under sub-paragraph (i) of Sub-Clause 14.3 [*Application for Interim Payment*].

In order to be entitled to such an advance payment, the Contractor must:

(*a*) keep satisfactory records (including the orders, receipts, details of Costs and of use of the Plant and Materials) which are to be available for inspection by the Engineer;
(*b*) submit evidence demonstrating that the Plant and Materials comply with the Contract (such as test certificates under Sub-Clause 7.4 [*Testing by the Contractor*] and/or compliance verification documentation under Sub-Clause 4.9.2 [*Compliance Verification System*]); and
(*c*) submit a statement of the Cost of acquiring and shipping or delivering (as the case may be) the Plant and Materials to the Site, supported by satisfactory evidence.

Moreover, the relevant Plant and Materials must:

(*a*) correspond to those listed in the Contract Data for payment when shipped;
(*b*) have been shipped to the Country and be '*en route*' to the Site, in accordance with the Contract; and
(*c*) be described in a clean shipped bill of lading or other evidence of shipment, which has been submitted to the Engineer (Employer under SB 2017) together with:

- evidence of payment of freight and insurance
- any other documents reasonably required by the Engineer and
- a written undertaking by the Contractor that the Contractor will deliver to the Employer (prior to submitting the next Statement) a bank guarantee in a form and issued by an entity to which the Employer gives consent, in amounts and currencies equal to the amount due under the sub-clause. This guarantee must be in a similar form to that of the Advance Payment Guarantee issued under Sub-Clause 14.2.1 [*Advance Payment Guarantee*] and must be valid until the Plant and Materials are properly stored on Site and protected against loss, damage or deterioration.

Alternatively, the relevant Plant and Materials must:

(*a*) correspond to those listed in the Contract Data for payment when delivered to the Site; and
(*b*) have been delivered to and be properly stored on the Site, and protected against loss, damage or deterioration, and appear to be in accordance with the Contract.

If the requested payment relates to Plant and/or Materials shipped to Site, the Engineer will have no obligation to certify any payment until the Employer has received the relevant bank guarantee.

At the time when payment of the contract value of the Plant and/or Materials is included in the IPC, the IPC must also include the applicable amount to be deducted which shall be equivalent to, and in the same currencies and proportions as, the advance payment for the relevant Plant and Materials.

The difficulty with this sub-clause is that the Engineer (Employer's Representative under SB 2017) must complete the procedure under Sub-Clause 3.7 [*Agreement or Determination*] before certifying payment, which could take 84 days. The corresponding sub-clause in FIDIC 1999 contracts did not require an agreement or determination, and the Engineer was allowed to proceed directly to certification if the relevant documents had been submitted. The new requirement for an agreement or determination brings no apparent benefit to the process. On the contrary, by extending the processing period, it significantly reduces the benefit of the advance payment, to the Contractor.

Sub-Clause 14.5 [*Plant and Materials intended for the Works*] does not mention Sub-Clause 7.7 [*Ownership of Plant and Materials*] but the two sub-clauses are closely linked. Under Sub-Clause 7.7 [*Ownership of Plant and Materials*], Plant and Materials become the property of the Employer at whichever is the earlier of the following times (subject to the mandatory requirements of the Laws of the Country)

(*a*) when the item of Plant or Materials is delivered to the Site
(*b*) when the Contractor is paid the value of the Plant and Materials under Sub-Clause 8.11 [*Payment for Plant and Materials after Employer's Suspension*] or
(*c*) when the Contractor is paid the amount determined for the Plant and Materials under Sub-Clause 14.5 [*Plant and Materials intended for the Works*].

Thus, when the Contractor is paid the advance under Sub-Clause 14.5 [*Plant and Materials intended for the Works*] for Materials which have been delivered to Site, the Employer is not at risk because, under Sub-Clause 7.7 [*Ownership of Plant and Materials*], the Materials belong to the Employer and an unpaid supplier would not be entitled to recover them (unless the applicable law states otherwise).

Similarly, when the Contractor is paid the advance under Sub-Clause 14.5 [*Plant and Materials intended for the Works*] for Plant which has been shipped to Site, the Employer is not at risk because, under Sub-Clause 7.7 [*Ownership of Plant and Materials*], the Materials belong to the Employer, even though the Contractor has been paid only 80% of the value of the Plant and also because the Employer holds a bank guarantee.

12.5. Interim Payment Certificate

Under Sub-Clause 14.6 [*Issue of IPC*], of RB 2017 and YB 2017, within 28 days of receiving a Statement and supporting documents, the Engineer must issue an IPC to the Employer, with a copy to the Contractor. The IPC must state the amount which the Engineer fairly considers to be due, including any additions and/or deductions which have become due under Sub-Clause 3.7 [*Agreement or Determination*] or under the Contract or otherwise. It must be accompanied by detailed supporting particulars which explain any difference between a certified amount and the corresponding amount in the Statement.

However, no amount is to be certified or paid to the Contractor until the Employer has received the Performance Security in the form, and issued by an entity, in accordance with Sub-Clause 4.2.1 [*Contractor's obligations*], and until the Contractor has appointed the Contractor's Representative in accordance with Sub-Clause 4.3 [*Contractor's Representative*].

Moreover, prior to the issue of the Taking-Over Certificate for the Works, the Engineer may withhold an IPC if the amount which would be certified (after retention and other deductions) would be less than the minimum amount stated in the Contract Data. In such a case, the Engineer must promptly give a Notice to the Contractor informing him.

An IPC must not be withheld for any other reason, although:

(*a*) If anything supplied or any work done by the Contractor is not in accordance with the Contract, the estimated cost of rectification or replacement may be withheld until rectification or replacement has been completed.
(*b*) If the Contractor was or is failing to perform any work, service or obligation in accordance with the Contract, the value of this work or obligation may be withheld until the work or obligation has been performed. In this event, the Engineer must promptly give a Notice to the Contractor describing the failure and providing detailed supporting particulars of the value withheld.
(*c*) If the Engineer finds any significant error or discrepancy in the Statement or supporting documents, the amount of the IPC may take account of the extent to which this error or discrepancy has prevented or prejudiced proper investigation of the amounts in the Statement until such error or discrepancy is corrected in a subsequent Statement.

For each amount so withheld, the supporting particulars accompanying the IPC must detail the Engineer's calculation of the amount withheld and state the reasons for the withholding.

Although the 28-day period for issuing the IPC only begins after the Engineer has received the Statement and the supporting particulars, the Engineer must not delay the issue of the IPC while awaiting a missing document. In this respect, the Engineer is obliged by Sub-Clause 1.3 [*Notices and other Communications*] not to unreasonably withhold or delay the issue of the IPC. The Contractor should be asked to provide the missing document and if it is not provided within a reasonable time, given the 28-day period for issuing the IPC, the Engineer would be justified in not certifying for payment the work for which the document was relevant.

The author has experienced the situation where the Engineer wrote to inform the Contractor that as the Contractor had repeatedly complained about late payment, the Engineer would seek more supporting particulars in relation to future payments, to allow the Employer more time to pay. Such behaviour is not only unprofessional but fraudulent. As far as possible, the Engineer should set out from the start of the project the type of supporting particulars that will be required, so that the Contractor can take appropriate action to minimise the certification and payment periods.

Sub-Clause 14.6 [*Interim Payments*] of SB 2017 sets out a similar process to that under RB 2017 and YB 2017 except the IPC from the Engineer is replaced by a Notice of interim payment from the Employer.

It is an important feature of FIDIC contracts that no payment, except the Final Payment, is definitive and no Payment Certificate is deemed to indicate the Engineer's acceptance, approval, consent or Notice of No-objection to any Contractor's Document or to (any part of) the Works. In consequence, in any Payment Certificate, the Engineer may make a correction or modification to any previous Payment Certificate. In some countries this principle is not well understood as payments are not allowed unless the Employer or certifier have approved the work. In consequence, payments are wrongly withheld or delayed.

If the Contractor considers that an IPC (interim payment under SB 2017) does not include any amount to which the Contractor is entitled, such amounts must be identified in the next Statement (*'identified amounts'*). The Engineer (Employer under SB 2017) must then make the appropriate correction or modification (if any) in the next IPC (interim payment under SB 2017). Thereafter, if the Contractor is not satisfied that this next IPC (interim payment under SB 2017) includes the identified amounts (and the identified amounts do not concern a matter for which the Engineer (Employer's Representative under SB 2017) is already carrying out their duties under Sub-Clause 3.7 [*Agreement or Determination*]), the Contractor may give a Notice referring this matter to the Engineer (Employer's Representative under SB 2017) for agreement or determination under Sub-Clause 3.7 [*Agreement or Determination*]. In this case, the date at which the Engineer (Employer's Representative under SB 2017) receives this Notice will be the date of commencement of the time limit for agreement under Sub-Clause 3.7.3 [*Time limits*].

It can be seen from the above timeline that it could take almost five months from when the Contractor first includes a contested amount in a Statement to the issue of the determination on the matter. If, after receiving the determination, the Contractor still believes that payment is due, they must refer the dispute to the DAAB.

Figure 12.2 Typical sequence of payment events (RB 2017 & YB 2017)

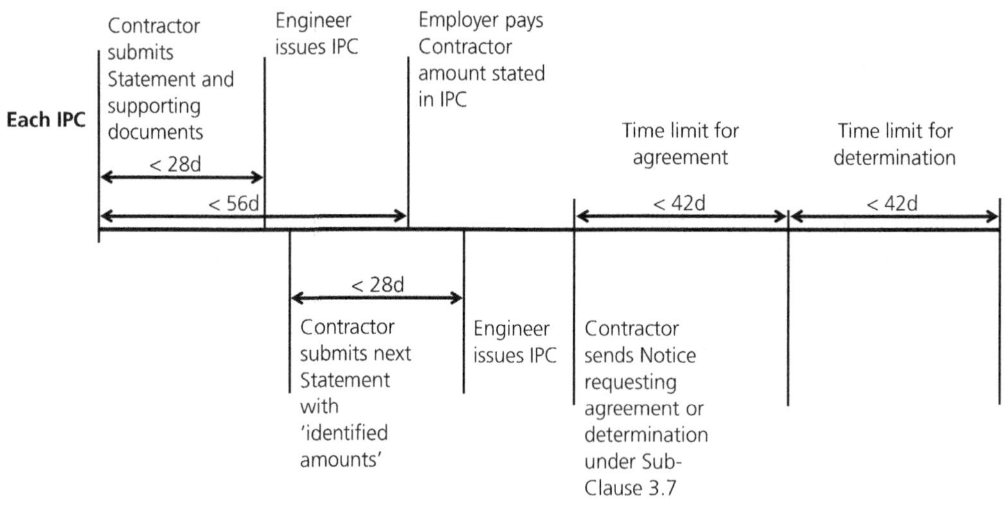

Example of IPC:

INTERIM PAYMENT CERTIFICATE No......							
PERIOD ENDING							
PROJECT:							
EMPLOYER:							
ENGINEER:							
CONTRACTOR:							
			CUMUL PREVIOUS PERIOD		**CUMUL THIS PERIOD**		**THIS PERIOD**
1	ESTIMATED VALUE OF WORKS EXECUTED INCL. VARIATIONS						
2	ADD/DEDUCT:						
		ADJUSTMENTS FOR CHANGES IN LAWS					
		ADJUSTMENTS FOR CHANGES IN COST					
3	DEDUCT:						
		RETENTION					
4	ADD:						
		ADVANCE PAYMENT					
	DEDUCT:						
		REIMBURSEMENT OF ADVANCE PAYMENT					
5	ADD:						
		PAYMENT FOR PLANT AND/OR MATERIALS SHIPPED OR DELIVERED					

	DEDUCT:						
		PAYMENT FOR PLANT AND/OR MATERIALS INCORPORATED IN WORKS					
6	ADD/DEDUCT:						
		OTHER AMOUNTS DUE INCLUDING UNDER SUB-CLAUSE 3.7					
7	ADD/DEDUCT:						
		AMOUNTS DUE UNDER PROVISIONAL SUMS					
8	ADD/DEDUCT:						
		RELEASE OF RETENTION					
9	DEDUCT:						
		AMOUNTS FOR USE OF UTLILTIES					
10	DEDUCT:						
		AMOUNT PREVIOUSLY CERTIFIED					
	NET AMOUNT CERTIFIED FOR PAYMENT THIS CERTIFICATE						
	Date:						
Engineer: Signature:							

12.6. Payments

Under Sub-Clause 14.7 [*Payment*] of RB 2017 and YB 2017, the Employer must pay the Contractor within the periods stated in the Contract Data. If no such periods are stated, the Employer must pay as follows:

(*a*) the advance payment within 21 days of the Employer receiving the Advance Payment Certificate;
(*b*) the amount certified in each IPC issued under

(i) Sub-Clause 14.6 [*Issue of IPC*] within 56 days of the Engineer receiving the Statement and supporting documents (regardless of the date when the IPC is issued) or
(ii) Sub-Clause 14.13 [*Issue of FPC*] within 28 days of the Employer receiving the IPC; and

(*c*) the amount certified in the FPC within 56 days of the Employer receiving the FPC.

Payment of the amount due in each currency must be made into the bank account for this currency, nominated by the Contractor, in the payment country specified in the Contract. The payment must reach the said bank account within the relevant period stated above.

Under Sub-Clause 14.7 [*Payment*] of SB 2017, if no periods for payment are stated in the Contract Data, the Employer must pay:

(*a*) the advance payment within 14 days of receipt of the Performance Security, the Advance Payment Guarantee and the Contractor's Statement;
(*b*) the interim payment due under

(i) Sub-Clause 14.6 [*Interim Payment*] within 56 days of the Employer receiving the Statement and supporting documents or
(ii) Sub-Clause 14.13 [*Final Payment*] within 42 days of the Employer receiving the Partially Agreed Final Statement (or, in the case of a deemed Partially Agreed Final Statement, within 84 days of the Employer receiving the draft final Statement); and

(*c*) the Final Payment within 56 days of the Employer

(i) receiving the Final Statement (or if the second paragraph of Sub-Clause 14.13 applies, after the expiry of 14 days from the Employer issuing the Notice stating the Final Payment) and
(ii) receiving (or the Contractor being deemed to have issued) the discharge under Sub-Clause 14.12 [*Discharge*].

12.7. Currencies of Payment

Under Sub-Clause 14.15 [*Currencies of Payment*], payments are to be in the currency or currencies named in the Contract Data. If more than one currency is named, payments are to be made as set out below.

If the Accepted Contract Amount was expressed in only one currency:

(*a*) the proportions or amounts of the Local and Foreign Currencies, and the fixed rates of exchange to be used for calculating the payments, are to be as stated in the Contract Data, (unless otherwise agreed by the Parties); if no rates of exchange are stated in the Contract Data, they are to be those prevailing on the Base Date and published by the central bank of the Country;
(*b*) payments and deductions under Sub-Clause 13.4 [*Provisional Sums*] and Sub-Clause 13.6 [*Adjustments for Changes in Laws*] shall be made in the applicable currencies and proportions; and
(*c*) other payments and deductions under sub-paragraphs (i) to (iv) of Sub-Clause 14.3 [*Application for Interim Payment*] are to be made in the currencies and proportions stated in the Contract Data.

Whenever additional payment (or a reduction) is agreed or determined under Sub-Clause 13.2 [*Value Engineering*] or Sub-Clause 13.3 [*Variation Procedure*], the amount payable in each of the applicable currencies must be specified. For this purpose, reference is to be made to the actual or expected currency proportions of the Cost of the varied work, and to the proportions of various currencies specified in sub-paragraph (a) above.

Payment of Delay Damages is to be made in the currencies and proportions specified in the Contract Data.

Payment of Performance Damages under YB 2017 and SB 2017 is to be made in the currencies and proportions specified in the Schedule of Performance Guarantees.

Other payments by the Contractor to the Employer are to be made in the currency in which the sum was expended by the Employer, or as otherwise agreed by the Parties.

If any amount payable by the Contractor to the Employer in a particular currency exceeds the sum payable by the Employer to the Contractor in that currency, the Employer may recover the balance from the sums otherwise payable to the Contractor in other currencies.

12.8. Late payment

If the Contractor does not receive payment in accordance with Sub-Clause 14.7 [*Payment*] the Contractor is entitled under Sub-Clause 14.8 [*Delayed Payment*] to receive financing charges compounded monthly on the unpaid amount, for the duration of the delay. This period of delay commences on the expiry of the time for payment specified in Sub-Clause 14.7 [*Payment*] irrespective of the date when the relevant IPC (or Notice of payment under SB 2017) was issued.

Unless otherwise stated in the Contract Data, these financing charges are to be calculated at the annual rate of 3% above:

(*a*) The average bank short-term lending rate to prime borrowers for the currency of payment at the place of payment; or
(*b*) where no such rate exists at that place, the same rate in the country of the currency of payment; or
(*c*) in the absence of such a rate at either place, the appropriate rate fixed by the law of the country of the currency of payment.

The Contractor will be entitled to payment of these financing charges by the Employer, provided he requests them, but without the need for the Contractor to submit a Statement or any formal Notice (including any requirement to comply with Sub-Clause 20.2 [*Claims for Payment and/or EOT*]) or certification; and without prejudice to any other right or remedy.

The reference to any other right or remedy relates in particular to the Contractor's right to suspend work under sub-paragraph (c) of Sub-Clause 16.1 [*Suspension by Contractor*] and the right to terminate the Contract under sub-paragraph 16.2.1 (c) of Sub-Clause 16.2 [*Termination by Contractor*]. Thus the right to financing charges is in addition to the right to suspend and the right to terminate.

The FIDIC 2017 Contracts Guide [2022] recommends that in such a situation of late payment, the Contractor should compile evidence of the late payment and make it available to the Engineer if requested. The Guide also recommends that the Contractor calculate the financing charges and attach the calculation to the request for financing charges in the form of an invoice. The Guide stresses that there is no need to include the amount in a Statement or for the Engineer to issue an IPC for the financing charges.

Hint:

If the Contractor does not include the financing charges in a Statement, but only issues an invoice as suggested by the FIDIC 2017 Contracts Guide [2022], the Contractor might never be paid the financing charges. The Contract does not fix a period for payment of the financing charges, and the remedies of suspension or termination are not available unless there has been a failure to pay an amount which was included in a Payment Certificate. Under such circumstances, the Contractor would be obliged to begin proceedings in a local court to enforce payment of the invoice.

However, if the Contractor includes the financing charges in a Statement under item (vi) of Sub-Clause 14.3 [*Application for Interim Payment*] and provides the necessary evidence (which should include evidence of the applicable rate as well as evidence of the late payment) the Engineer must include the financing charges in the IPC. Thereafter, the Employer must pay the financing charges within the periods set out under Sub-Clause 14.7 [*Payment*] and if the Employer fails to do so, the Contractor will have the right to suspend work or terminate the Contract, even if the Employer eventually pays the initial amount, in relation to which unpaid financing charges have arisen.

12.9. Statement at Completion

Within 84 days of the Date of Completion of the whole of the Works, Sub-Clause 14.10 [*Statement at Completion*] requires the Contractor to submit to the Engineer (the Employer under SB 2017) a Statement at completion with supporting documents, set out in accordance with Sub-Clause 14.3 [*Application for Interim Payment*] and showing:

(*a*) the value of all work done in accordance with the Contract up to the Date of Completion of the Works;
(*b*) any further sums which the Contractor considers to be due at the Date of Completion of the Works; and
(*c*) an estimate of any other amounts which the Contractor considers have or will become due after the Date of Completion of the Works, under the Contract or otherwise. These estimated amounts must be shown separately (from those of sub-paragraphs (a) and (b) above) and must include estimated amounts for

  (i) Claims for which the Contractor has submitted a Notice under Sub-Clause 20.2 [*Claims for Payment and/or EOT*];
  (ii) any matter referred to the DAAB under Sub-Clause 21.4 [*Obtaining DAAB's Decision*]; and
  (iii) any matter for which a NOD has been given under Sub-Clause 21.4 [*Obtaining DAAB's Decision*].

The '*value of all work done in accordance with the Contract*' excludes the estimated cost of rectification of any defects or the value of work to be completed, as listed in the TOC.

As well as the items listed under sub-paragraph (c), (i) to (iii), the estimate of any amounts that have or will become due after the Date of Completion must include the release of the Retention Money and any amounts retained with respect to defects and incomplete work listed in the TOC.

Following receipt of the Statement at completion, the Engineer must issue an IPC in accordance with Sub-Clause 14.6 [*Issue of IPC*].

Sub-Clause 14.10 [*Statement at Completion*] does not refer to Sub-Clause 14.14 [*Cessation of Employer's Liability*]. However, sub-paragraph (b) of Sub-Clause 14.14 [*Cessation of Employer's Liability*] states that the Employer shall have no liability towards the Contractor for any matter or thing under or in connection with the Contract or execution of the Works, except to the extent that the Contractor shall have included an amount expressly for it in the Statement of completion issued under Sub-Clause 14.10 [*Statement at Completion*]. The only exception to this cessation of the Employer's liability is if the matter or thing for which no amount was included in the Statement at completion only arose after the issue of the TOC for the Works.

In other words, if the Contractor forgets to include any item and the corresponding amount in the Statement at completion, he loses the right to claim payment for that item.

12.10. Final Statement

12.10.1 Draft Final Statement

Within 56 days of the issue (or deemed issue) of the Performance Certificate, the Contractor is required by Sub-Clause 14.11.1 [*Draft Final Statement*] to submit to the Engineer (the Employer under SB 2017), a draft final Statement which must:

(*a*) be in the same form as Statements previously submitted under Sub-Clause 14.3 [*Application for Interim Payment*];
(*b*) be submitted in one paper original, one electronic copy and additional paper copies (if any) as stated in the Contract Data; and
(*c*) show in detail, with supporting documents:

- (i) the value of all work done in accordance with the Contract;
- (ii) any further sums which the Contractor considers to be due at the date of the issue of the Performance Certificate, under the Contract or otherwise; and
- (iii) an estimate of any other amounts which the Contractor considers have or will become due after the issue of the Performance Certificate, under the Contract or otherwise, including estimated amounts, by reference to the matters described in sub-paragraphs (i) to (iii) of Sub-Clause 14.10 [*Statement at Completion*]. These estimated amounts must be shown separately.

With respect to sub-paragraph (iii) above, it should be noted that submission of the draft final Statement must not be delayed because of any referral under Sub-Clause 21.4 [*Obtaining DAAB's Decision*] or any arbitration under Sub-Clause 21.6 [*Arbitration*].

It should also be noted that at the date of the Performance Certificate, the second half of the Retention may not have been included in a Statement, and if this is the case, it should be included under sub-paragraph (iii).

Except for any amount under sub-paragraph (iii) above, if the Engineer (Employer under SB 2017) disagrees with or cannot verify any part of the draft final Statement, the Engineer (Employer under SB 2017) must promptly give a Notice to the Contractor. The Contractor must then submit any further information which the Engineer (Employer under SB 2017) may reasonably require within the time stated in this Notice.

After review by the Engineer (Employer under SB 2017) and discussions with the Contractor, the Contractor must modify the draft final Statement to reflect any agreements reached.

12.10.2 Agreed Final Statement

If there are no amounts under sub-paragraph (iii) of Sub-Clause 14.11.1 [*Draft Final Statement*] (that is, there are no outstanding Claims or disputes), the Contractor must then prepare and submit to the Engineer (Employer under SB 2017) the final Statement as agreed (the '*Final Statement*') under Sub-Clause 14.11.2 [*Agreed Final Statement*].

However, Sub-Clause 14.11.2 [*Agreed Final Statement*] also addresses the possibility that the Engineer (Employer under SB 2017) and Contractor are unable to agree on the amounts under sub-paragraphs (i) or (ii) of Sub-Clause 14.11.1 [*Draft Final Statement*]. In this case, or if there are amounts under sub-paragraph (iii) of Sub-Clause 14.11.1 [*Draft Final Statement*], the Contractor must then prepare and submit to the Engineer (Employer under SB 2017) a Statement, identifying separately the agreed amounts, the estimated amounts and the disagreed amount(s) (the '*Partially Agreed Final Statement*').

No time limit is imposed for the discussions between the Engineer and the Contractor at the end of which the Contractor must prepare and submit the Partially Agreed Final Statement. The Contractor is free to decide when to do so.

As mentioned above in relation to Sub-Clause 14.10 [*Statement at Completion*], there is no reference under Sub-Clause 14.11.2 [*Agreed Final Statement*] to Sub-Clause 14.14 [*Cessation of Employer's Liability*]. However, sub-paragraph (a) of Sub-Clause 14.14 [*Cessation of Employer's Liability*] states that the Employer shall have no liability towards the Contractor for any matter or thing under or in connection with the Contract or execution of the Works, except to the extent that the Contractor shall have included an amount expressly for it in the Final Statement or Partially Agreed Final Statement.

In other words, if the Contractor forgets to include any item and the corresponding amount in the Final Statement or Partially Agreed Final Statement, they lose the right to claim payment for that item.

12.11. Discharge, Final Certificate and Final Payment

When submitting the Final Statement or the Partially Agreed Final Statement, the Contractor is required by Sub-Clause 14.12 [*Discharge*] to submit a discharge which confirms that the total of said Statement represents full and final settlement of all amounts due to the Contractor under or in connection with the Contract.

Sub-Clause 14.12 [*Discharge*] also states that the discharge may specify that the total of the Statement is subject to any payment that may become due in respect of any Dispute for which a DAAB proceeding or arbitration is in progress under Sub-Clause 21.6 [*Arbitration*] and/or that it becomes effective after the Contractor has received full payment of the amount certified in the FPC, and the Performance Security.

In these respects, it should be remembered that under Sub-Clause 4.2.3 [*Return of the Performance Security*] the Employer should have released the Performance Security within 21 days of the issue of the Performance Certificate subject to the Contractor having complied with Sub-Clause 11.11 [*Clearance of Site*]. Under Sub-Clause 11.11 [*Clearance of Site*], the Contractor was to clear the Site promptly after the issue of the Performance Certificate and if the Contractor failed to comply within 28 days, the Employer was entitled to clear the Site at the Contractor's cost. Thus, in almost all cases, the Performance Security should already have been released by the Employer before the Contractor submitted its Final Statement or Partially Agreed Final Statement.

It should also be remembered that the Partially Agreed Final Statement must include all disagreed amounts and not only those which had already been referred to the DAAB or arbitration prior to submission of the draft final Statement. Therefore, before submitting the discharge, the Contractor should refer the disagreed amounts to the DAAB. The Contractor should then issue the discharge as suggested by FIDIC, indicating that the total of the Statement is subject to any payment that may become due in respect of any Dispute for which a DAAB proceeding or arbitration is in progress, and that it will become effective after the Contractor has received full payment of the amount certified in the FPC, and the Performance Security (if not already released).

The discharge issued under Sub-Clause 14.12 [*Discharge*] does not affect either Party's liability or entitlement in respect of any Dispute for which a DAAB proceeding or arbitration is in progress under Clause 21 [*Disputes and Arbitration*].

If the Contractor fails to submit the discharge, it is deemed to have been submitted and to have become effective when the Contractor has received full payment of the amount certified in the Final Payment Certificate (FPC) and the Performance Security has been released.

Note:

Nothing is stated about deemed submission of a discharge with respect to payment of a Partially Agreed Final Statement. This is entirely logical given that a Partially Agreed Final Statement leads to an IPC (see below) and therefore does not represent the full amount finally expected to be paid by the Employer.

Within 28 days of receiving the Final Statement or the Partially Agreed Final Statement (as the case may be) and the discharge under Sub-Clause 14.12 [*Discharge*] of RB 2017 and YB 2017, the Engineer is required by Sub-Clause 14.13 [*Issue of FPC*] to issue the FPC to the Employer (with a copy to the Contractor). This FPC must state

- the amount which the Engineer fairly considers is finally due, including any additions and/or deductions which have become due under Sub-Clause 3.7 [*Agreement or Determination*] or under the Contract or otherwise

- all amounts previously paid by the Employer and for all sums to which the Employer is entitled
- all amounts (if any) previously paid by the Contractor and/or received by the Employer under the Performance Security and
- the balance (if any) due from the Employer to the Contractor or from the Contractor to the Employer, as the case may be.

The position under SB 2017 is similar, except that the FPC is replaced by a Notice from the Employer.

> **Note:**
>
> The reference here to the issue of an FPC in response to a Partially Agreed Final Statement is confusing as the sub-clause goes on to say that the Engineer will issue an IPC (see below).

If the Contractor has not submitted a draft final Statement within 56 days of the issue of the Performance Certificate, the Engineer (Employer under SB 2017) must request the Contractor to do so. If the Contractor fails to submit a draft final Statement within a period of 28 days of receipt of the request, the Engineer must issue the FPC for the amount which the Engineer fairly considers to be due (under SB 2017, the Employer must issue a Notice stating the Final Payment).

If the Contractor has submitted a Partially Agreed Final Statement or if no Partially Agreed Final Statement has been submitted (but, to the extent that a draft final Statement submitted by the Contractor is deemed to be a Partially Agreed Final Statement), the Engineer (Employer under SB 2017) must proceed in accordance with Sub-Clause 14.6 [*Issue of IPC*] to issue an IPC (make an interim payment under SB 2017).

Figure 12.3 Final payment (RB 2017 & YB 2017)

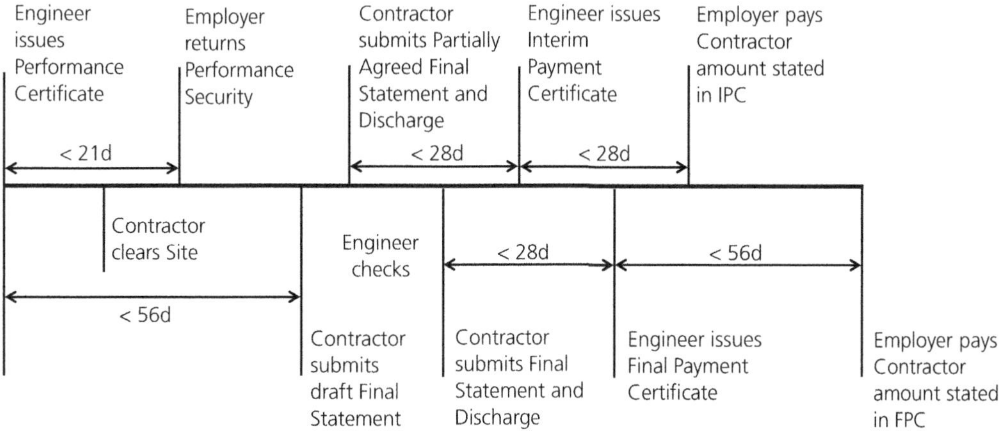

Example of Contractor's Discharge:

BETTA CONTRACTORS LTD.

A.N. OTHER & Partners
Consulting Engineers

15 September 2022

For the attention of Mr. Other, the Engineer

Subject: **Discharge under Sub-Clause 14.12**

Dear Sir,

In accordance with Sub-Clause 14.12 [*Discharge*] we hereby notify you that the total amount stated as due to us in our Final Statement/Partially Agreed Final Statement [*delete as appropriate*] represents full and final settlement of all moneys due to the Contractor under or in connection with the Contract, subject to the following limitations

(*a*) the total of the Statement is subject to any payment that may become due in respect of any Dispute for which a DAAB proceeding or arbitration is in progress

(*b*) this discharge is conditional upon

(i) full payment of the amount due to us under the Final Statement and

(ii) the return of our Performance Security [and our ES Performance Security].

Yours faithfully,

Contractor's Representative

cc The Employer

Example of Final Payment Certificate:

A.N. OTHER & Partners
Consulting Engineers

To:

THE EMPLOYER

03 February 2023

For the attention of: Mr. Okay, Employer's Representative

Subject: **Final Payment Certificate**

Dear Sir,

In accordance with Sub-Clause 14.13 [*Issue of FPC*] of the Contract, we hereby certify that the amount finally due to the Contractor under the Contract is as follows:

Final value of the Works	....
Less:	....
amounts previously paid by the Employer	....
sums to which the Employer is entitled	....
Sub-total	....
Plus:	....
amounts previously paid by the Contractor	....
amounts received by the Employer under the Performance Security	....
Balance due to/from the Contractor	....

Yours faithfully,

The Engineer

cc The Contractor

12.12. Cessation of Employer's Liability

As mentioned in Chapter 12.9 and 12.10 above, the Employer will not be liable to the Contractor for any matter or thing under or in connection with the Contract or execution of the Works, except to the extent that the Contractor shall have included an amount expressly for it in:

(*a*) the Final Statement or Partially Agreed Final Statement; and
(*b*) the Statement at completion (except for matters or things arising after the issue of the Taking-Over Certificate for the Works).

Moreover, under Sub-Clause 14.14 [*Cessation of Employer's Liability*] unless the Contractor makes or has made a Claim under Sub-Clause 20.2 [*Claims for Payment and/or EOT*] in respect of an amount or amounts under the FPC within 56 days of receiving a copy of the FPC, the Contractor shall be deemed to have accepted the amounts so certified. The Employer shall then have no further liability to the Contractor, other than to pay the amount due under the FPC and return the Performance Security to the Contractor.

It appears, therefore, that FIDIC contemplates the possibility of the Contractor making a Claim with respect to the agreed Final Payment Certificate despite having issued a discharge and possibly having received the final payment.

As indicated under Chapter 12.11 above, the Performance Security should have been returned to the Contractor well before the date for cessation of the Employer's liability.

Sub-Clause 14.14 [*Cessation of Employer's Liability*] does not limit the Employer's liability under the Employer's indemnification obligations, or the Employer's liability in any case of fraud, gross negligence, deliberate default or reckless misconduct by the Employer.

Smith G
ISBN 978-0-7277-6652-6
https://doi.org/10.1680/fcmh.66526.189

Chapter 13
Variations and adjustments

13.1. Right to Vary

Sub-Clause 1.1.86 **RB 2017** **Sub-Clause 1.1.88** **YB 2017** **Sub-Clause 1.1.78** **SB 2017**	***'Variation'*** *means any change to the Works, which is instructed as a variation under Clause 13 [Variations and Adjustments].*

Under Sub-Clause 13.1 [*Right to Vary*], the Engineer may initiate a Variation under Sub-Clause 13.3 [*Variation Procedure*] at any time prior to the issue of the Taking-Over Certificate for the Works. To do so, the Engineer may choose between the two procedures described in Chapters 13.2 and 13.3 below: a Variation instruction under Sub-Clause 13.3.1 [*Variation by Instruction*] or a request for a proposal under Sub-Clause 13.3.2 [*Variation by Request for Proposal*].

The Contractor must not make any alteration to and/or modification of the Permanent Works, unless and until the Engineer instructs a Variation under Sub-Clause 13.3.1 [*Variation by Instruction*].

The Contractor is bound by any such Variation instruction and must execute the Variation with due expedition and without delay, unless the Contractor promptly gives a Notice to the Engineer under Sub-Clause 13.1 [*Right to Vary*] stating (with detailed supporting particulars) that:

(*a*) the varied work was Unforeseeable having regard to the scope and nature of the Works described in the Specification;
(*b*) the Contractor cannot readily obtain the Goods required for the Variation; or
(*c*) the instruction will adversely affect the Contractor's ability to comply with Sub-Clause 4.8 [*Health and Safety Obligations*] and/or Sub-Clause 4.18 [*Protection of the Environment*].

The word '*unless*' is important in this paragraph. It means that if the Contractor gives such a Notice promptly, it is not bound to execute the Variation while awaiting the response from the Engineer. In the meantime, it must continue the execution of the Works without taking account of the contested instruction, unless the Engineer issues an instruction to suspend execution under Sub-Clause 8.9 [*Employer's Suspension*].

The above listed grounds for objecting to a Variation instruction are taken from RB 2017. The possible grounds for objecting under YB 2017 and SB 2017 are more extensive. Under YB 2017 and SB 2017, the Contractor may refer to two additional grounds:

(*d*) the instruction will have an adverse impact on the achievement of the Schedule of Performance Guarantees; or
(*e*) the instruction may adversely affect the Contractor's obligation to complete the Works so that they are fit for the purpose(s) for which they are intended under Sub-Clause 4.1 [*Contractor's General Obligations*].

Promptly after receiving such Contractor's Notice, the Engineer (Employer under SB 2017) must respond by giving a Notice to the Contractor which either cancels, confirms or varies the instruction. Any instruction so confirmed or varied is to be taken as an instruction under Sub-Clause 13.3.1 [*Variation by Instruction*].

If an instruction does not state that it is a Variation and the Contractor considers that the instruction does constitute a Variation (or involves work that is already part of an existing Variation) or does not comply with applicable Laws or will reduce the safety of the Works or is technically impossible, the Contractor is required by Sub-Clause 3.5 [*Engineer's Instructions*] of RB 2017 and YB 2017 (Sub-Clause 3.4 [*Instructions*] under SB 2017) to immediately give a Notice to the Engineer (Employer under SB 2017) with reasons. The Contractor's Notice must be sent before commencing any work related to the instruction. If the Engineer does not respond within 7 days of receiving this Notice, by giving a Notice confirming, reversing or varying the instruction, the Engineer (Employer under SB 2017) is deemed to have revoked the instruction. Otherwise the Contractor must comply with and will be bound by the terms of the Engineer's response (Employer's response under SB 2017).

It should be noted that two different situations are involved here, and that two separate Notices from the Contractor may be required (under two different sub-clauses). In one case, a failure to respond by the Engineer is deemed to be a cancellation of the instruction. In the other case, the outcome of a failure to respond is not expressly mentioned. Having two provisions which apply to broadly similar but slightly different situations could lead to complications. For example, the Contractor may contest an instruction that is not stated by the Engineer to be a Variation on the basis that the instruction does not comply with applicable Laws but may not contest the same instruction on the same grounds if the Engineer states that it is a Variation.

The situation is also likely to become complicated if the Contractor validly contests the instruction for safety reasons, but the Engineer confirms the instruction, the Contractor complies and an accident arises. What is the liability of the Engineer?

Under Sub-Clause 13.1 [*Right to Vary*] of RB 2017, a Variation may include:

(i) changes to the quantities of any item of work included in the Contract (but such changes do not constitute a Variation unless they result from an Engineer's instruction);
(ii) changes to the quality and other characteristics of any item of work;
(iii) changes to the levels, positions and/or dimensions of any part of the Works;
(iv) the omission of any work, unless it is to be carried out by others without the agreement of the Parties (see Chapter 13.6 below);

(v) any additional work, Plant, Materials or services necessary for the Permanent Works, including any associated Tests on Completion, boreholes and other testing and exploratory work; or
(vi) changes to the sequence or timing of the execution of the Works.

Sub-Clause 13.1 [*Right to Vary*] of YB 2017 and SB 2017 does not fix the possible scope of a Variation, as does RB 2017.

The only limitation is that, other than as stated under Sub-Clause 11.4 [*Failure to Remedy Defects*], a Variation may not comprise the omission of any work which is to be carried out by the Employer or by others, unless otherwise agreed by the Parties.

13.2. Variation by Instruction

The Engineer (Employer under SB 2017) may instruct a Variation by giving a Notice to the Contractor in accordance with Sub-Clause 3.5 [*Engineer's Instructions*] (Sub-Clause 3.4 [*Instructions*] under SB 2017). The Notice can specify any requirements for the recording of Costs, such as by means of Dayworks records (see below).

The Contractor must then proceed in accordance with Sub-Clause 13.3.1 [*Variation by Instruction*] with the execution of the Variation and, within 28 days of receiving the instruction (or other period proposed by the Contractor and agreed by the Engineer (Employer under SB 2017)), must submit detailed particulars including:

- a description of the varied work performed or to be performed, including details of the resources and methods adopted or to be adopted by the Contractor;
- a programme for its execution and the Contractor's proposal for any necessary modifications (if any) to the Programme according to Sub-Clause 8.3 [*Programme*] and any extension of the Time for Completion which is considered to become due; and
- the Contractor's proposal for adjustment of the Contract Price by valuing the Variation in accordance with Clause 12 [*Measurement and Valuation*], with supporting particulars (which shall include identification of any estimated quantities) and, if the Contractor incurs or will incur Cost as a result of any necessary modification to the Time for Completion, shall show the additional payment (if any) to which the Contractor considers themselves to be entitled). If the Parties have agreed to the omission of any work which is to be carried out by others, the Contractor's proposal may also include the amount of any loss of profit and other losses and damages suffered (or to be suffered) by the Contractor as a result of the omission (see Chapter 13.6 below).

The corresponding provisions under YB 2017 and SB 2017 do not mention Clause 12 or estimated quantities.

Note:

Unlike RB 2017, Sub-Clause 13.3.1 of YB 2017 and SB 2017 do not mention the inclusion in the evaluation of the Variation of the additional Costs arising from any necessary modification of the Time for Completion. This does not mean that the Contractor is not entitled to have such additional Costs included in the adjustment to the Contract Price, but the absence of the express mention could lead to disagreements.

Thereafter, the Contractor must submit any further particulars that the Engineer (Employer's Representative under SB 2017) may reasonably require.

The Engineer (Employer's Representative under SB 2017) must then proceed under Sub-Clause 3.7 [*Agreement or Determination*] to agree or determine:

(i) the EOT to which the Contractor is entitled, if any; and/or
(ii) the adjustment to the Contract Price (including valuation of the Variation in accordance with Clause 12 [*Measurement and Valuation*] using measured quantities of the varied work).

Again, YB 2017 and SB 2017 do not mention Clause 12 or quantities.

For Sub-Clause 3.7.3 [*Time limits*], the date at which the Engineer (Employer's Representative under SB 2017) receives the Contractor's submission (including any requested further particulars) will be the date of commencement of the time limit for agreement.

Note:

The Contractor's entitlement to an EOT and/or adjustment to the Contract Price in relation to a Variation, is not subject to compliance by the Contractor with Sub-Clause 20.2 [*Claims for Payment and/or EOT*], that is, there is no requirement for the Contractor to submit a Notice of Claim or a detailed Claim.

13.3. Variation proposals

Before instructing a Variation, by giving a Notice under Sub-Clause 13.3.2 [*Variation by Request for Proposal*] which describes the proposed change, the Engineer may request the Contractor to submit a proposal.

In such a case, the Contractor must respond as soon as practicable, by either:

- submitting a proposal, which must include all the matters described in sub-paragraphs (a) to (c) of Sub-Clause 13.3.1 [*Variation by Instruction*]; or
- giving reasons why the Contractor cannot comply, by reference to the matters described in sub-paragraphs (a) to (c) of Sub-Clause 13.1 [*Right to Vary*].

As soon as practicable after receiving the Contractor's proposal, the Engineer (Employer under SB 2017) must respond by giving a Notice to the Contractor giving their consent or otherwise. While awaiting the response to the proposal, the Contractor must not delay any work and must continue as though no proposal had been requested, unless instructed by the Engineer (Employer under SB 2017) to suspend work under Sub-Clause 8.9 [*Employer's Suspension*].

If the Engineer (Employer under SB 2017) gives consent to the proposal, with or without comments, the Engineer (Employer under SB 2017) must then instruct the Contractor to execute the Variation. Thereafter, the Contractor must submit any further particulars that the Engineer may reasonably require, and the Engineer (Employer's Representative under SB 2017) must seek to agree or determine the matter under Sub-Clause 3.7 [*Agreement or Determination*].

If the Engineer (Employer's Representative under SB 2017) does not consent to the proposal and if the Contractor has incurred Cost as a result of submitting it, the Contractor will be entitled, subject to Sub-Clause 20.2 [*Claims for Payment and/or EOT*], to payment of such Cost. In this respect, it will be necessary for the Contractor to demonstrate that because of the request for a proposal, the Contractor incurred Cost that would not otherwise have been incurred.

13.4. Value Engineering

At any time, the Contractor may submit to the Engineer (Employer under SB 2017) a written proposal which (in the Contractor's opinion) will, if adopted:

- accelerate completion;
- reduce the cost to the Employer of executing, maintaining or operating the Works;
- improve the efficiency or value to the Employer of the completed Works; or
- otherwise be of benefit to the Employer.

The proposal is to be prepared at the cost of the Contractor and must include the details as stated in sub-paragraphs (a) to (c) of Sub-Clause 13.3.1 [*Variation by Instruction*], that is, a description of the work, a programme for the work, an analysis of the impact on the Programme and a price proposal.

As soon as practicable after receiving such a proposal, the Engineer (Employer under SB 2017) must respond by a Notice to the Contractor stating their consent or otherwise. Such consent or otherwise shall be at the sole discretion of the Employer. The Contractor must not delay any work while awaiting a response and must continue as though no proposal had been submitted.

If the Engineer gives their consent to the proposal, with or without comments, the Engineer (Employer under SB 2017) must then instruct a Variation.

Thereafter, the last paragraph of Sub-Clause 13.3.1 [*Variation by Instruction*] applies and must include consideration by the Engineer (Employer's Representative under SB 2017) of the sharing (if any) of the benefit, costs and/or delay between the Parties stated in the Particular Conditions. Unlike RB 1999, the FIDIC 2017 Contracts give no guidance on how this sharing should be taken into account.

If a such a proposal from the Contractor which receives consent, includes a change in the design of part of the Permanent Works, then unless otherwise agreed by the Parties, the Contractor must design this part at its cost. Under RB 2017, sub-paragraphs (a) to (h) of Sub-Clause 4.1 [*Contractor's General Obligations*] apply in relation to submission, review of the design and so forth.

13.5. Valuation of the Variation

If the Variation was the subject of a proposal from the Contractor under Sub-Clause 13.2 [*Value Engineering*] or Sub-Clause 13.3.2 [*Variation by Request for Proposal*] which has received the consent of the Engineer (Employer under SB 2017), the amount of the Variation or at least the method of calculating the amount will have been agreed. However, if the Variation was instructed under Sub-Clause 13.3.1 [*Variation by Instruction*] the Engineer is required to seek to agree or to determine the valuation of the Variation and the adjustment to the Contract Price. In the case of RB 2017, the valuation is to be in accordance with Clause 12 [*Measurement and Valuation*] using measured quantities of the varied work.

Under Sub-Clause 12.3 [*Valuation of the Works*] of RB 2017, a step-by-step approach is set out with respect to the valuation of Variations.

(i) The appropriate rate or price for each item of work shall be the rate or price specified for such an item in the Bill of Quantities (or other Schedule).
(ii) If there is no such item, the rate or price is that specified for similar work.
(iii) If no specified rate or price is appropriate because the item of work is not of similar character, or is not executed under similar conditions, to any item in the Contract, a new rate or price is to be derived from any relevant rates or prices specified in the Bill of Quantities (or other Schedule), with reasonable adjustments to take account of the different character or different conditions.
(iv) If no specified rates or prices are relevant for the derivation of a new rate or price, the rate or price is to be derived from the reasonable Cost of executing the work, together with the applicable percentage for profit stated in the Contract Data (if not stated, at the rate of 5%), taking account of any other relevant matters.

It is important to note that with respect to step (iv), the Engineer must establish a new unit rate or a lump sum price. A lump sum price could be established by adding the applicable percentage for profit to the actual cost of executing the work after excluding any amount which is considered to be unreasonably incurred. However, the new unit rate cannot be established on the same basis because it would serve no useful purpose to convert actual costs reasonably incurred, plus profit, to a unit rate.

Instead, the new unit rate must be established based on an estimate of reasonable cost plus the profit percentage, assessed before the work is executed, or by an evaluation of the actual cost reasonably incurred for a portion of the work which is then converted to a unit rate applicable to the remainder of the work.

Under YB 2017 and SB 2017, if the Variation was instructed under Sub-Clause 13.3.1 [*Variation by Instruction*], the Engineer (Employer's Representative under SB 2017) is required to seek to agree or to determine the adjustments to the Contract Price and the Schedule of Payments, if any.

In doing so, the Engineer (Employer's Representative under SB 2017) is to follow the same step-by-step approach as under RB 2017, provided the Contract includes a Schedule of Rates and Prices which can serve as a baseline.

If the Contract does not include a Schedule of Rates and Prices, the adjustments are to be derived from the Cost Plus Profit of executing the work.

Note:

Whereas RB 2017 refers to new rates and prices being derived from the reasonable Cost Plus Profit, YB 2017 and SB 2017 do not mention 'reasonable'. It seems therefore that the Contractor is entitled to be reimbursed the actual Cost incurred plus profit at the rate stated in the Contract Data (if not stated, at the rate of 5%).

13.6. Omissions

Whenever a Variation includes the omission of any work for which:

- the value has not been agreed;
- the Contractor will incur (or has incurred) cost which, if the work had not been omitted, would have been deemed to be covered by a sum forming part of the Accepted Contract Amount;
- the omission of the work will result (or has resulted) in this sum not forming part of the Contract Price; and
- this cost is not deemed to be included in the valuation of any substituted work

then under Sub-Clause 12.4 [*Omissions*], the Contractor must include details, with detailed supporting particulars, in the Contractor's proposal under sub-paragraph (c) of Sub-Clause 13.3.1 [*Variation by Instruction*] which are then to be taken into account by the Engineer when valuing the Variation under Sub-Clause 3.7 [*Agreement or Determination*].

It should be remembered that under Sub-Clause 13.1 [*Right to Vary*], the Engineer may only omit work if no-one is to execute the omitted work, unless the Contractor agrees that it may be executed by others. If the Contractor does so agree, he is entitled to be paid the loss of profit related to the omitted work under sub-paragraph (c) of Sub-Clause 13.3.1 [*Variation by Instruction*].

13.7. Dayworks

For work of a minor or incidental nature, the Engineer may instruct that a Variation be executed on a daywork basis under Sub-Clause 13.5 [*Daywork*] (provided that the Contract includes a Dayworks Schedule). The work will then be valued based on records of the time spent by Equipment and labour, and of materials used, valued at the rates and prices listed in the Dayworks Schedule. If the Contract does not include a Dayworks Schedule, the Engineer may not instruct the Contractor to execute a Variation on this basis.

However, if the Dayworks Schedule does not contain rates for the necessary Contractor's Equipment, Temporary Works, Plant or Materials, the Contractor must submit to the Engineer one or more quotations from suppliers and/or subcontractors. Thereafter, the Engineer may instruct the Contractor to accept one of these quotations (but such an instruction is not to be taken as an instruction under Sub-Clause 5.2 [*Nominated Subcontractors*]).

Sub-Clause 13.5 [*Daywork*] states that, if the Engineer does not so instruct the Contractor within 7 days of receiving the quotations, the Contractor may accept any of these quotations at the Contractor's discretion. Thus, if the Engineer rejects all the quotations submitted by the Contractor, the Contractor is free to choose which of the submitted quotations to accept.

As soon as the instructed work begins, the Contractor must compile and submit to the Engineer records (in duplicate and with one electronic copy), in accordance with Sub-Clause 6.10 [*Contractor's Records*]. The records must be submitted each day and provide details of the resources used in executing the instructed work on the previous day.

If the record is correct and agreed, the Engineer must sign one copy and return it to the Contractor. If the record is not correct or agreed, the Engineer must proceed under Sub-Clause 3.7 [*Agreement or*

Determination] to agree or determine the resources. For Sub-Clause 3.7.3 [*Time limits*], the date at which the Contractor completes the Dayworks under the Variation shall be the date of commencement of the time limit for agreement under Sub-Clause 3.7.3 [*Time limits*].

It could be difficult for the Engineer to decide upon the value of the Dayworks, 84 days after the resources are used, unless they produced their own records at the time. If the instructed Dayworks are executed over a prolonged period, the task of the Engineer will be even more difficult especially if they wait until the Dayworks are completed before commencing the process under Sub-Clause 3.7 [*Agreement or Determination*].

After receipt of the Engineer's agreement to the records or the Engineer's determination, the Contractor must submit priced details in the next Statement, together with all applicable invoices, vouchers and accounts or receipts in substantiation of any Goods (other than Goods priced in the Daywork Schedule).

This means that the Contractor may not be able to include Dayworks in its Statements until three or four months after the resources are used, with corresponding payment being received approximately 2 months later. From a Contractor's viewpoint, this is unreasonable: why should the Engineer require 84 days from completion of the relevant Dayworks to decide upon the Contractor's entitlement, knowing that this period will be followed by a further period of 28 days to process the relevant IPC?

Unless otherwise stated in the Daywork Schedule, the rates and prices in the Daywork Schedule are deemed to include taxes, overheads and profit.

13.8. Provisional Sums

If the Contract includes a Provisional Sum, it is only to be used, and the Contractor is only to be paid from the Provisional Sum, if and to the extent that the Engineer instructs the Contractor under Sub-Clause 13.4 [*Provisional Sums*] to execute the work, supplies or services to which the Provisional Sum relates.

For each Provisional Sum, the Engineer may instruct:

(*a*) work to be executed (including Plant, Materials or services to be supplied) by the Contractor, and for which adjustments to the Contract Price shall be agreed or determined under Sub-Clause 13.3.1 [*Variation by Instruction*]; and/or

(*b*) Plant, Materials, works or services to be purchased by the Contractor from a nominated Subcontractor (as defined in Sub-Clause 5.2 [*Nominated Subcontractors*]) or otherwise; and for which there shall be included in the Contract Price:

  (i) the actual amounts paid (or due to be paid) by the Contractor; and

  (ii) a mark-up for overhead charges and profit, calculated as a percentage of these actual amounts by applying the relevant percentage rate (if any) stated in the applicable Schedule (if there is no such rate, the percentage rate stated in the Contract Data shall be applied).

The Engineer's instruction may include a requirement for the Contractor to submit quotations from suppliers and/or subcontractors for all (or some) of the items of the work to be executed or Plant,

Materials, works or services to be purchased. Thereafter, the Engineer may respond with a Notice either instructing the Contractor to accept one of these quotations or revoking the instruction. If the Engineer does not so respond within 7 days of receiving the quotations, the Contractor shall be entitled to accept any of these quotations at the Contractor's discretion. Thus, if the Engineer rejects all the quotations submitted by the Contractor, the Contractor is free to choose which of the submitted quotations to accept.

Strangely, it is stated in the third paragraph of Sub-Clause 13.4 [*Provisional Sums*], that an instruction from the Engineer instructing the Contractor to accept one of the submitted quotations is not to be taken as an instruction under Sub-Clause 5.2 [*Nominated Subcontractors*]), whereas sub-paragraph (b) refers expressly to Plant, Materials, works or services to be purchased by the Contractor from a nominated Subcontractor.

Each Statement that includes an amount forming part of a Provisional Sum must also include all necessary substantiation: applicable invoices, vouchers and accounts or receipts.

13.9. Adjustments for major changes in quantity

Under sub-paragraph (b) of Sub-Clause 12.3 [*Valuation of the Works*] of RB 2017, if the quantity of an item measured for payment differs significantly from the quantity stated in the Bill of Quantities (or other Schedule) a new rate or price is to be derived from the existing rate or price but only if four conditions are satisfied:

(i) the measured quantity of the item differs from the quantity stated in the Bill of Quantities (or other Schedule) by more than 10%;
(ii) the change in quantity multiplied by the rate or price specified in the Bill of Quantities (or other Schedule) for the item exceeds 0.01% of the Accepted Contract Amount;
(iii) the change in quantity directly changes the Cost per unit quantity of this item by more than 1%; and
(iv) the item is not described in the Bill of Quantities (or other Schedule) as a 'fixed rate item', 'fixed charge' or similar expression which indicates that the rate or price will not be subject to adjustment for any change in quantity.

It is easy to assess whether conditions (i), (ii) and (iv) are satisfied but to check whether condition (iii) is satisfied, the Engineer will require details of the Contractor's actual Costs for the measured quantity or a breakdown of the unit rate or price from the Contractor. If the Contractor refuses to provide this information, the Engineer will be obliged to make their own breakdown. Using this information, the Engineer will be able to assess the impact of the change in quantity on the Cost per unit. This can be done in two ways: either by comparing the actual Cost for the measured quantity compared to what the actual Cost would have been for the quantity indicated in the BOQ; or by assessing the impact on the price breakdown of the changed quantity.

In both cases, it is important to eliminate factors which have an impact on Costs but are unrelated to the change in quantity, such as increases in the cost of fuel between the Base Date and the time of executing the modified quantities. The FIDIC 2017 Contracts Guide [2022] recommends the first method but takes no account of the fact that it eliminates mistakes in the Contractor's bid. The second method may magnify the effect of such mistakes.

In most cases, a significant increase in quantity will reduce the Cost per unit because the fixed costs will be spread over a larger quantity. Conversely, a significant reduction in the quantity will cause the Cost per unit to rise because the fixed costs must be recovered from the smaller quantity. However, this is not always the case, and the Engineer must consider all the consequences, particularly if they use the second approach mentioned above when making the analysis. For example, if the Contractor mobilises additional or larger Equipment to deal with the additional quantities, the unit Cost may rise in comparison to the bid, because of the additional mobilisation and demobilisation.

13.10. Adjustments for changes in legislation

Sub-Clause 13.6 [*Adjustments for Changes in Laws*] foresees two different types of adjustment due to changes in legislation. The first deals with changes to the Contractor's Costs and/or delays caused by the changes in legislation, whereas the second deals with changes to the execution of the Works as a result of the changes in legislation.

If the Contractor suffers delay and/or incurs an increase in Cost as a result of any change in legislation of the kind listed below, the Contractor will be entitled, subject to Sub-Clause 20.2 [*Claims for Payment and/or EOT*], to EOT and/or payment of such Cost.

If there is a decrease in Cost as a result of any change in legislation of the kind listed below, the Employer will be entitled, subject to Sub-Clause 20.2 [*Claims for Payment and/or EOT*], to a reduction in the Contract Price.

If any adjustment to the execution of the Works becomes necessary as a result of any such change in legislation, the Contractor must promptly give a Notice to the Engineer (Employer under SB 2017), or the Engineer (Employer under SB 2017) shall promptly give a Notice to the Contractor (with detailed supporting particulars).

Thereafter, the Engineer (Employer under SB 2017) must either instruct a Variation under Sub-Clause 13.3.1 [*Variation by Instruction*] or request a proposal under Sub-Clause 13.3.2 [*Variation by Request for Proposal*].

The type of changes in legislation which give rise to such entitlements are as follow

(*a*) changes in the Laws of the Country (including the introduction of new Laws and the repeal or modification of existing Laws)
(*b*) changes in the judicial or official governmental interpretation or implementation of the Laws referred to in sub-paragraph (a) above
(*c*) changes to any permit, permission, license or approval obtained by the Employer or the Contractor under sub-paragraph (a) or (b), respectively, of Sub-Clause 1.13 [*Compliance with Laws*] or
(*d*) changes to the requirements for any permit, permission, licence and/or approval to be obtained by the Contractor under sub-paragraph (b) of Sub-Clause 1.13 [*Compliance with Laws*], made and/or officially published after the Base Date, which affect the Contractor in the performance of obligations under the Contract.

> **Note:**
>
> Sub-Clause 13.6 [*Adjustments for Changes in Laws*] applies only to changes in legislation in the Country where the Works are located.

13.11. Adjustments for Changes in Cost

It is standard practice within FIDIC Contracts to adjust the Contract Price to take account of increases (or decreases) in the cost of labour, Plant, Materials and consumables to be used in the Works. Sub-Clause 13.7 [*Adjustments for Changes in Cost*] sets out the basis for doing so. However, the sub-clause only applies if the Contract includes a Schedule of cost indexation, which contains the information necessary for the calculation of the adjustments.

It is not the intention of the sub-clause to fully compensate the Contractor for increases in labour and other costs (or the Employer for decreases in the costs) but to share the risk of such fluctuations which are almost inevitable over the lifespan of a normal construction project.

Therefore, to the extent that full compensation for any rise or fall in Costs is not covered by the sub-clause, the Accepted Contract Amount is deemed to have included amounts to cover the risk of other rises and falls in Costs. In other words, bidders must assess the risk when preparing their bids and include a provision if they consider that the sub-clause does not provide adequate protection.

The adjustment is to be calculated in accordance with the Schedule(s) of cost indexation and applied to the amount otherwise payable to the Contractor, as certified in Payment Certificates, for each of the currencies in which the Contract Price is payable. However, no adjustment is to be applied to work valued on the basis of Cost or current prices.

If the current costs index is not available at the time of calculating the adjustment, the Engineer (Employer under SB 2017) must use a provisional index for the issue of Interim Payment Certificates. When a current cost index is available, the adjustment must be recalculated accordingly.

Following the expiry of the Time for Completion, if the Contractor has not yet completed the Works, adjustment of prices thereafter is to be made using either

- each index or price applicable on the date 49 days before the expiry of the Time for Completion of the Works or
- the current index or price

whichever is more favourable to the Employer.

In other words, if the Contractor does not complete the Works on time (after taking account of any entitlement to EOT) the Contractor is not compensated for increases in Cost which occur after the end of the Time for Completion but continues to be compensated for increases in Cost that arose during the Time for Completion. However, if the Cost decreases after the Time for Completion, the Employer is entitled to benefit from the decrease.

Under earlier FIDIC contracts (except SB 1999), Sub-Clause 13.7 [*Adjustments for Changes in Cost*] included a formula to be used for the calculation of the adjustment. The standard wording of the FIDIC 2017 contracts does not provide the formula but it is proposed within the Notes on the Preparation of Special Provisions:

$$Pn = a + b\frac{Ln}{Lo} + c\frac{En}{Eo} + d\frac{Mn}{Mo} + \ldots$$

where

'Pn' is the adjustment multiplier to be applied to the estimated contract value, in the relevant currency, of the work carried out in period 'n', which is normally a month

'a' is a fixed coefficient, representing the non-adjustable portion (suggested by FIDIC to be 0.10)

'b', 'c', 'd' … are coefficients representing the estimated proportion of each cost element related to the execution of the Works, such as labour, equipment and materials

'Ln', 'En', 'Mn' … are the current cost indices or reference prices (stated in the Schedule of cost indexation) for period 'n', expressed in the relevant currency of payment, applicable to the relevant tabulated cost element on the date 49 days prior to the last day of the period (to which the particular Payment Certificate relates) and

'Lo', 'Eo', 'Mo' … are the base cost indices or reference prices, expressed in the relevant currency of payment, each of which is applicable to the relevant tabulated cost element on the Base Date.

Other cost elements such as diesel, cement, steel, bitumen and so forth can be added depending upon the nature of the Works. The total of the weightings (coefficients) of the various cost elements (a, b, c, d, etc.) must not exceed 1. These weightings are to be adjusted if they are rendered unreasonable, unbalanced or inapplicable, as a result of Variations.

Sometimes, the Employer may decide when preparing the bidding documents which cost elements to include, which has the benefit of making it easier to compare bids. Sometimes the Employer also fixes the weightings or fixes a range for each rating. This has the advantage of limiting the possibility for bidders to exaggerate the effect of cost increases by quoting very high weightings for materials which are expected to incur significant cost increases.

If not, the choice of cost elements, weightings and sources of indices can be left to bidders.

In these respects, several MDBs include the following sample schedules in their Standard Bidding Documents.

It should be noted that several MDBs do not provide finance to cover the adjustment for changes in Cost.

The calculation of the adjustment is one of the biggest causes of disputes. Disagreements arise with respect to the source of indices and/or their values, with respect to the coefficients (a), (b), (c) and so forth, and with respect to the amount to which the adjustment is to be applied. Disagreements also arise when a modified version of the standard formula is used or restrictions are placed upon when it is to come into operation.

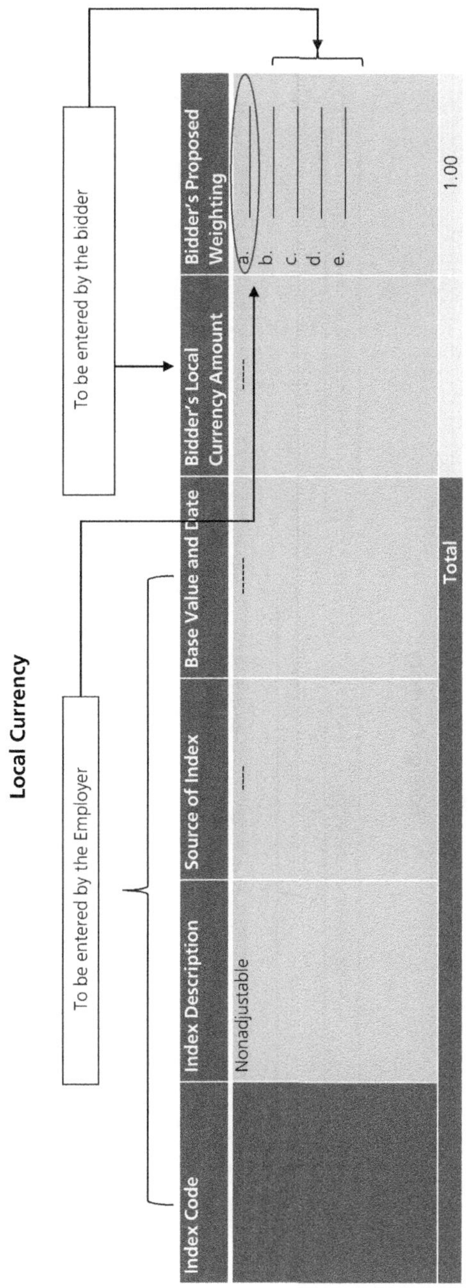

Figure 13.1 Schedule of cost indexation

Figure 13.2 Schedule of cost indexation

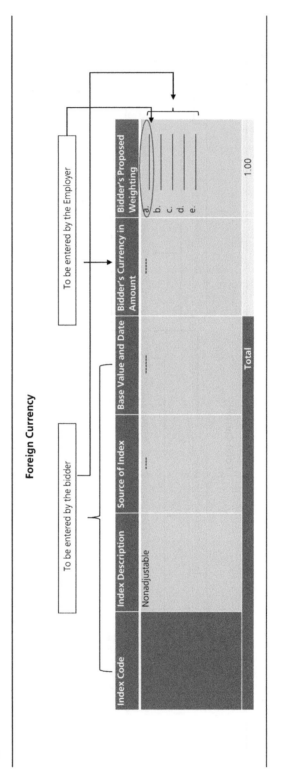

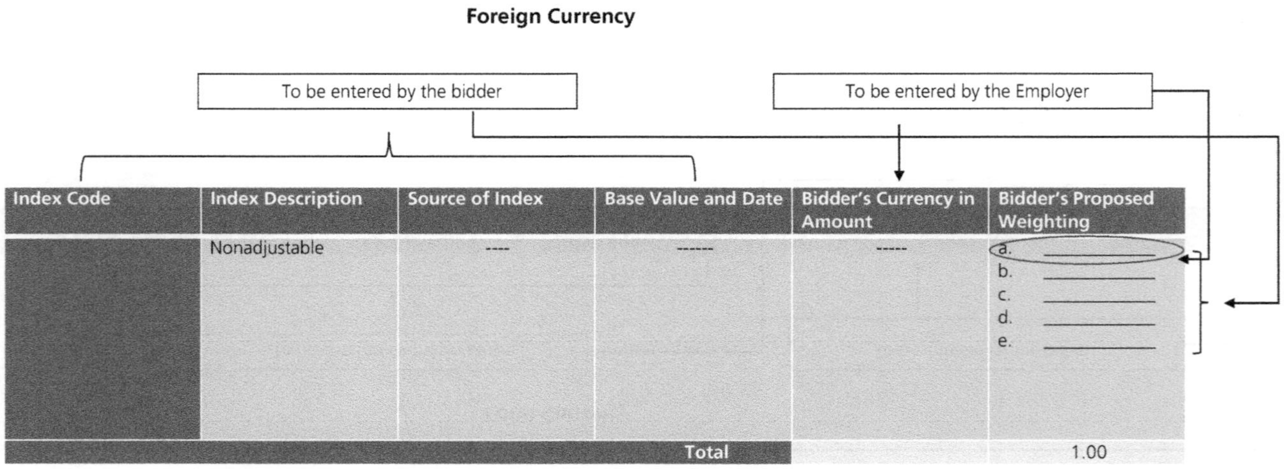

Example from a contract for supply and installation of electrical equipment:

'The date of adjustment shall be defined as the last day of the 18th month after the effective date of the contract, but not later than the last day of delivery time as indicated in the Bid, if the last day of delivery period occurs earlier.

$$\frac{PC1}{PC0} = \left(a + b\frac{MC1}{MC0} + c\frac{MS1}{MS0} + d\frac{MO1}{MO0} + e\frac{CPI1}{CPI0}\right) - 1$$

where

MC is the index for the price of copper published by the London Metals Exchange
MS is the index for the price of steel HRC FOB China (Argus) published by the London Metals Exchange
MO is the index for the price of oil (WTI Crude) published by Oilprice.com, and
CPI is the China Consumer Price Index published by tradingeconomics.com.

Example: In a contract for supply and installation of [illegible] equipment

[illegible]

Smith G
ISBN 978-0-7277-6652-6
https://doi.org/10.1680/fcmh.66526.205

Chapter 14
Suspension and termination

14.1. Termination by the Employer due to Contractor's default

If the Contractor breaches the Contract, the ultimate sanction is for the Employer to terminate the Contract under Sub-Clause 15.2 [*Termination for Contractor's Default*]. However, as a preliminary step prior to such a serious action, the Contractor may be given a Notice to Correct the situation under Sub-Clause 15.1 [*Notice to Correct*].

14.1.1 Notice to Correct

By giving the Contractor a Notice to Correct, the Engineer (the Employer under SB 2017) may require the Contractor to remedy within a specified time any failure to fulfil any obligation under the Contract.

The Notice to Correct must describe the Contractor's failure, state the sub-clause and/or provisions of the Contract under which the Contractor has the obligation, and fix a reasonable period within which the Contractor must remedy the failure. The reasonableness of the period is to be ascertained with due regard to the nature of the failure as well as the work and/or other action required to remedy it. The period stated in the Notice to Correct is not to be taken as an extension of the Time for Completion.

After receiving a Notice to Correct, the Contractor must immediately respond by giving a Notice to the Engineer (the Employer under SB 2017) describing the measures the Contractor will take to remedy the failure, and stating the date when the measures will be commenced to ensure compliance within the time specified in the Notice to Correct.

Any Contractor receiving such a Notice to Correct should understand that the Employer is taking a step along the path to termination and react accordingly. On the other hand, no Engineer should issue a Notice to Correct without first consulting with the Employer, and receiving confirmation that the Employer is prepared to issue a Notice of intention to terminate the Contract under Sub-Clause 15.2.1 [*Notice*] (see below) if the Contractor does not comply. If the Employer does not react to a Contractor's failure to comply with a Notice to Correct, by issuing the Notice of intention to terminate, the Engineer will lose credibility in the eyes of the Contractor.

14.1.2 Notice of intention to terminate/Notice of termination

The Employer is entitled under Sub-Clause 15.2 [*Termination for Contractor's Default*] to follow the termination procedure set out under Sub-Clause 15.2.1 [*Notice*] and as described below, if the Contractor:

(*a*) fails to comply with:

- (i) a Notice to Correct;
- (ii) a binding agreement, or final and binding determination, under Sub-Clause 3.7 [*Agreement or Determination*]; or

(iii) a decision of the DAAB under 21.4 [*Obtaining DAAB's Decision*] (whether binding or final and binding)

and such failure constitutes a material breach of the Contractor's obligations under the Contract;

(*b*) abandons the Works or otherwise plainly demonstrates an intention not to continue performance of the Contractor's obligations under the Contract;
(*c*) without reasonable excuse fails to proceed with the Works in accordance with Clause 8 [*Commencement, Delays and Suspension*] or, if there is a maximum amount of Delay Damages stated in the Contract Data, its failure to comply with Sub-Clause 8.2 [*Time for Completion*] is such that the Employer would be entitled to Delay Damages that exceed this maximum amount;
(*d*) without reasonable excuse fails to comply with a Notice of rejection given by the Engineer under Sub-Clause 7.5 [*Defects and Rejection*] or an Engineer's instruction under Sub-Clause 7.6 [*Remedial Work*], within 28 days of receiving it;
(*e*) fails to comply with Sub-Clause 4.2 [*Performance Security*];
(*f*) subcontracts the whole, or any part of, the Works in breach of Sub-Clause 5.1 [*Subcontractors*], or assigns the Contract without the required agreement under Sub-Clause 1.7 [*Assignment*];
(*g*) becomes bankrupt or insolvent; goes into liquidation, administration, reorganisation, winding-up or dissolution; becomes subject to the appointment of a liquidator, receiver, administrator, manager or trustee; enters into a composition or arrangement with the Contractor's creditors; or any act is done or any event occurs which is analogous to or has a similar effect to any of these acts or events under applicable Laws;

or if the Contractor is a JV:

(i) any of these matters apply to a member of the JV, and
(ii) the other member(s) do not promptly confirm to the Employer that, in accordance with Sub-Clause 1.14(a) [*Joint and Several Liability*], such member's obligations under the Contract shall be fulfilled in accordance with the Contract; or

(*h*) is found, based on reasonable evidence, to have engaged in corrupt, fraudulent, collusive or coercive practice at any time in relation to the Works or to the Contract.

The termination of the Contract does not prejudice any other rights of the Employer under the Contract or otherwise.

In this respect, it should be remembered that under sub-paragraph (d) of Sub-Clause 4.2.2. [*Claims under the Performance Security*] the Employer is entitled to make a claim under the Performance Security in the event of circumstances which entitle the Employer to terminate the Contract under Sub-Clause 15.2 [*Termination for Contractor's Default*], irrespective of whether a Notice of termination has been given. Thus, the Employer can make a claim under the Performance Security due to the termination, before triggering the termination procedure as described below.

It should be noted that sub-paragraph (a) requires the Contractor's failure to be a '*material breach*' of the Contractor's obligations and sub-paragraphs (c) and (d) require the failure to be '*without reasonable excuse*'. These requirements introduce some subjectivity, and before deciding whether to commence the termination procedure the Employer should consider carefully whether the Contractor's failure is sufficient to justify such a termination. The consequences of wrongful termination can be very serious. An unjustified Notice of termination would constitute a breach of Contract by the Employer which would justify termination by the Contractor.

In the case of sub-paragraphs (a) to (e), the Employer commences the procedure by a Notice to the Contractor under Sub-Clause 15.2.1 [*Notice*] of the Employer's intention to terminate the Contract.

There is no express requirement under Sub-Clause 15.2.1 [*Notice*] for the Employer to describe the failure which gives rise to the Notice. However, Sub-Clause 15.2.2 [*Termination*] goes on to state that unless the Contractor remedies the matter described in the Notice within 14 days of its receipt, the Employer may, by giving a second Notice to the Contractor, immediately terminate the Contract and the date of termination shall be the date the Contractor receives this second Notice. Therefore, there is a requirement for the Notice to describe the matter; that is: to specify the failure, ideally by reference to the applicable sub-paragraph, with details, when applicable, of why the failure is considered to be a '*material breach*' or '*without reasonable excuse*'.

Having sent the Notice of intention to terminate, the Employer is not obliged to send the Notice of termination at the end of the 14-day period. If the Contractor has remedied the failure, the Employer will no longer have grounds for a Notice of termination. If the Contractor has begun to remedy the failure, it may be better for the Employer to allow more time for the Contractor to complete the corrective measures, before deciding whether to proceed with the termination. However, the Employer should not wait too long before deciding, as the right to terminate might be taken as waived.

In the case of sub-paragraph (f), (g) or (h), the Employer is entitled to serve a Notice of termination under Sub-Clause 15.2.1 [*Notice*], without the need for a Notice of intention to terminate, and under Sub-Clause 15.2.2 [*Termination*], the date of termination shall be the date the Contractor receives this Notice of termination.

In all cases, the Notice(s) must refer to the relevant sub-clause(s).

14.1.3 After termination

After termination of the Contract under Sub-Clause 15.2.2 [*Termination*], the Contractor is required by Sub-Clause 15.2.3 [*After termination*] to:

(*a*) immediately comply with any reasonable instructions included in a Notice from the Employer:

- (i) for the assignment of any subcontract; and
- (ii) for the protection of life or property or for the safety of the Works;

(*b*) deliver to the Engineer (Employer under SB 2017):

- (i) any Goods required by the Employer (which may include the Plant and Materials for use in the Works as well as the Contractor's Equipment);
- (ii) all Contractor's Documents; and
- (iii) all other design documents made by or for the Contractor to the extent, if any, that the Contractor is responsible for the design of part of the Permanent Works under Sub-Clause 4.1 [*Contractor's General Obligations*]; and

(*c*) leave the Site and, if the Contractor does not do so, the Employer will have the right to expel the Contractor from the Site.

After such a termination, the Employer may complete the Works and/or arrange for others to do so. In doing so, the Employer and/or these others may use any Goods and Contractor's Documents (and any other design documents) made by or on behalf of the Contractor. As well as Plant and Materials to be used in the Works, 'Goods' includes the Contractor's Equipment (cranes, excavators, trucks, cars, offices, stores, workshops, computers, etc.). Thus, when the Contractor leaves the Site, it should take nothing with it except for personal items.

14.1.4 Valuation after termination

After termination of the Contract under Sub-Clause 15.2 [*Termination for Contractor's Default*], the Engineer (the Employer's Representative under SB 2017) is required by Sub-Clause 15.3 [*Valuation after Termination for Contractor's Default*] to proceed under Sub-Clause 3.7 [*Agreement or Determination*] to agree or determine the value of the Permanent Works, Goods and Contractor's Documents, and any other sums due to the Contractor for work executed in accordance with the Contract. The date of commencement of the time limit for agreement under Sub-Clause 3.7.3 [*Time limits*], is the date of termination.

This valuation must include any additions and/or deductions, and the balance due (if any), by reference to the matters described in sub-paragraphs (a) and (b) of Sub-Clause 14.13 [*Issue of FPC*] but must exclude the value of any defect or non-compliance in the Contractor's Documents, Materials, Plant or Permanent Works.

It should be remembered that under sub-paragraph (b) of Sub-Clause 14.13 [*Issue of FPC*], the Engineer (the Employer under SB 2017) must give credit for any amount received by the Employer under the Performance Security. Under Sub-Clause 14.2.3 [*Repayment of Advance Payment*], account must also be taken of any advance payment that has not yet been reimbursed. As such an amount is immediately due and payable to the Employer upon termination, the Employer will be entitled to claim payment under the Advance Payment Guarantee.

14.1.5 Payment after termination

Under Sub-Clause 15.4 [*Payment after Termination for Contractor's Default*], the Employer may withhold payment to the Contractor of the amounts agreed or determined under Sub-Clause 15.3 [*Valuation after Termination for Contractor's Default*] until all amounts due from the Contractor have been established.

In this respect, after termination of the Contract under Sub-Clause 15.2 [*Termination for Contractor's Default*], the Employer will be entitled, subject to Sub-Clause 20.2 [*Claims for Payment and/or EOT*], to payment by the Contractor of:

(*a*) the additional costs of execution of the Works which the Employer would not have incurred had it not been necessary to terminate the Contract, and all other costs reasonably incurred by the Employer, such as any additional fees charged by the Engineer;
(*b*) any losses and damages suffered by the Employer in completing the Works within the limitations set out under Sub-Clause 1.15 [*Limitation of Liability*] of RB 2017 and YB 2017 (Sub-Clause 1.14 of SB 2017); and

(*c*) Delay Damages, if the Works or a Section had not been taken over under Sub-Clause 10.1 [*Taking Over the Works and Sections*] at the date of termination under Sub-Clause 15.2 [*Termination for Contractor's Default*] and the date of termination occurred after the expiry of the Time for Completion of the Works (or Section). If the date of termination was prior to the expiry of the Time for Completion of the Works (or Section), the Employer will not be entitled to the Delay Damages.

Clearly, the amounts to which the Employer may be entitled under sub-paragraphs (a) and (b) cannot be established before completion of the Works (by the Employer or another contractor on behalf of the Employer) and the remedying of any defects notified during the DNP. This could be several years after the termination. Nevertheless, the Employer must submit a Notice of Claim under Sub-Clause 20.2 [*Claims for Payment and/or EOT*] within 28 days of when the Employer should have become aware of the event which gave rise to the claim, which was the termination.

After such a completion of the Works, the Employer must give another Notice to the Contractor under Sub-Clause 15.2.4 [*Completion of the Works*] confirming that the Contractor's Equipment and Temporary Works will be released to the Contractor at or near the Site. The Contractor must then promptly arrange their removal, at its own risk and cost. However, if the Contractor owes any amount to the Employer, these items may be sold by the Employer (subject to the applicable Laws) in order to recover this amount. Only then, is any balance of the proceeds paid to the Contractor.

14.2. Termination by the Employer 'for convenience'

At any time, the Employer may terminate the Contract for the Employer's convenience, by giving a Notice of such termination to the Contractor. The Notice must state that it is given under Sub-Clause 15.5 [*Termination for Employer's Convenience*].

After giving such a Notice, the Employer must immediately:

(*a*) cease using any of the Contractor's Documents, which must be returned to the Contractor, except those for which the Contractor has received payment or for which payment is due under a Payment Certificate;
(*b*) if Sub-Clause 4.6 [*Co-operation*] applies, cease using any Contractor's Equipment, Temporary Works, access arrangements and/or other facilities or services of the Contractor; and
(*c*) make arrangements to return the Performance Security to the Contractor.

Termination under this sub-clause takes effect 28 days after the later of the dates on which the Contractor receives this Notice or the Employer returns the Performance Security.

The Contractor's obligations after such termination are set out under Sub-Clause 16.3 [*Contractor's Obligations after Termination*]. The Contractor must promptly:

(*a*) cease all further work, except for such work as may have been instructed by the Engineer (the Employer under SB 2017) for the protection of life or property or for the safety of the Works;
(*b*) hand over to the Engineer all Contractor's Documents, Plant, Materials and other work for which the Contractor has been paid; and
(*c*) remove all other Goods from the Site, except as necessary for safety, and leave the Site.

If the Contractor incurs Cost as a result of carrying out the work instructed under sub-paragraph (a), it will be entitled, subject to Sub-Clause 20.2 [*Claims for Payment and/or EOT*], to be paid such Cost Plus Profit.

Unless and until the Contractor has received payment of the amount due under Sub-Clause 15.6 [*Valuation after Termination for Employer's Convenience*], the Employer must not continue the execution of any of the Works or arrange for others to do so.

After this termination, the Contractor is required by Sub-Clause 16.3 [*Contractor's Obligations after Termination*] to submit as soon as practicable, detailed supporting particulars (as reasonably required by the Engineer (Employer under SB 2017)), of the value of work done and the amount of any loss of profit or other losses and damages suffered by the Contractor as a result of the termination.

The value of work done must include the matters described in sub-paragraphs (a) to (e) of Sub-Clause 18.5 [*Optional Termination*], and any additions and/or deductions, and the balance due (if any), by reference to the matters described in sub-paragraphs (a) and (b) of Sub-Clause 14.13 [*Issue of FPC*]. It must also include the amount of any loss of profit or other losses and damages suffered by the Contractor because of this termination.

The matters described in sub-paragraphs (a) to (e) of Sub-Clause 18.5 [*Optional Termination*] are:

(*a*) the amounts payable for any work carried out for which a price is stated in the Contract;
(*b*) the Cost of Plant and Materials ordered for the Works which have been delivered to the Contractor, or of which the Contractor is liable to accept delivery. This Plant and these Materials become the property of (and will be at the risk of) the Employer when paid for by the Employer, and the Contractor must make the items available to the Employer;
(*c*) any other Cost or liability which in the circumstances was reasonably incurred by the Contractor in the expectation of completing the Works;
(*d*) the Cost of removal of Temporary Works and Contractor's Equipment from the Site and the return of these items to the Contractor's place of business in the Contractor's country (or to any other destination(s) at no greater cost); and
(*e*) the Cost of repatriation of the Contractor's staff and labour employed wholly in connection with the Works at the date of termination.

Under Sub-Clause 15.6 [*Valuation after Termination for Employer's Convenience*], the Engineer (the Employer's Representative under SB 2017) must then proceed under Sub-Clause 3.7 [*Agreement or Determination*] to agree or determine the value of work done and the loss of profit or other losses and damages. The date at which the Engineer (the Employer's Representative under SB 2017) receives the Contractor's particulars is the date of commencement of the time limit for agreement under Sub-Clause 3.7.3 [*Time limits*].

After an agreement has been reached or a determination has been issued, the Engineer must issue a Payment Certificate under Sub-Clause 15.6 [*Valuation after Termination for Employer's Convenience*] for the amount so agreed or determined, without the need for the Contractor to submit a Statement (although it will already have submitted supporting particulars). Under SB 2017, the Employer must pay the amount agreed or determined, also without the need for the Contractor to submit a Statement.

Under Sub-Clause 15.7 [*Payment after Termination for Employer's Convenience*] of RB 2017 and YB 2017, the Employer must pay the Contractor the amount certified in the Payment Certificate.

Under all three contracts, the payment must be made within 112 days after the Engineer (the Employer under SB 2017) receives the Contractor's submission under Sub-Clause 15.6 [*Valuation after Termination for Employer's Convenience*].

From the Contractor's viewpoint, this may seem long, but given that the Engineer (the Employer's Representative under SB 2017) may take 84 days to issue a determination under Sub-Clause 3.7 [*Agreement or Determination*], the period only allows 28 days more for the Engineer to issue the Payment Certificate and for the Employer to pay. If the Employer wishes to continue the Works by itself or with others, it is in the interest of the Employer to pay as quickly as possible because work cannot resume until the Contractor has received the payment.

14.3. Suspension by the Contractor

Under Sub-Clause 16.2 [*Termination by Contractor*], the Contractor has similar rights to terminate the Contract for default by the Employer as the Employer has under Sub-Clause 15.2 [*Termination for Contractor's Default*] for default by the Contractor. In addition, the Contractor has the right to suspend work under Sub-Clause 16.1 [*Suspension by Contractor*] in relation to a small number of failures by the Employer.

Under Sub-Clause 16.1 [*Suspension by Contractor*], the Contractor may suspend work (or reduce the rate of work) due to any of the defaults listed under the sub-clause unless and until the Employer has remedied its default. Prior to such suspension, the Contractor must give not less than 21 days' Notice to the Employer (not the Engineer). The Notice must state that it is given under Sub-Clause 16.1 [*Suspension by Contractor*], and although there is no express requirement for the Notice to specify the default, this is clearly the case, as the Contractor may only suspend work or slow down if the Employer has failed to remedy the default as described in the Notice, more than 21 days after receiving the Notice.

The defaults which give rise to the right to suspend or slow down the rate of work are as follow.

(*a*) The Engineer fails to certify in accordance with Sub-Clause 14.6 [*Issue of IPC*] (this ground does not exist in SB 2017).
(*b*) The Employer fails to provide reasonable evidence of its financial arrangements within 28 days of a request from the Contractor in accordance with Sub-Clause 2.4 [*Employer's Financial Arrangements*].
(*c*) The Employer fails to comply with the payment periods under Sub-Clause 14.7 [*Payment*].
(*d*) The Employer fails to comply with:

  (i) a binding agreement, or final and binding determination under Sub-Clause 3.7 [*Agreement or Determination*]; or
  (ii) a decision of the DAAB under 21.4 [*Obtaining DAAB's Decision*] (whether binding or final and binding)

  and such failure constitutes a material breach of the Employer's obligations under the Contract.

The arguments which arise most frequently with respect to the Contractor's right to suspend or slow down the work relate to the start date for calculating the period for issuing the IPC and for making payment. The Engineer and the Employer often argue that the periods did not begin on the date of submission of the Contractor's Statement, because not all necessary supporting documents were attached. In such a situation, the FIDIC 2017 Contracts Guide [2022] suggests that the Engineer or Employer should immediately contest the validity of the Contractor's Notice of suspension. However, such delays to the start of the certification and payment periods should be minor in most cases, if the Engineer complies with its obligation under Sub-Clause 1.3 [*Notices and Other Communications*] not to unreasonably withhold or delay the certification (see Chapter 12.5 above).

Further sources of frequent argument are the estimate of the Contract Price for which the Employer is required to provide evidence of its financial arrangements and the sufficiency of the evidence provided to demonstrate that the funds will be available.

It should also be noted that, under sub-paragraph (d), the Employer's non-compliance must be so significant that it represents a material breach of the Employer's obligations. Thus, a failure to comply with an agreement, determination or DAAB decision which is relatively minor would not justify suspension or slowdown.

If the Contractor does suspend or slow down the rate of work, after the 21-day period, and the Employer subsequently remedies the default (before the Contractor gives a Notice of termination under Sub-Clause 16.2 [*Termination by Contractor*]), the Contractor must resume normal working as soon as is reasonably practicable. The reasonableness of the period for the return to normal working will depend on the circumstances, such as the duration of the suspension, the extent to which the Contractor had demobilised resources, the extent to which the Contractor could foresee the date at which the Employer would remedy the default and so forth.

If the Contractor suffers delay and/or incurs Cost as a result of suspending work (or reducing the rate of work), the Contractor will be entitled, subject to Sub-Clause 20.2 [*Claims for Payment and/or EOT*], to EOT and/or payment of such Cost Plus Profit.

Moreover, the suspension or slowdown will not prejudice the Contractor's entitlements to financing charges under Sub-Clause 14.8 [*Delayed Payment*] and to termination under Sub-Clause 16.2 [*Termination by Contractor*].

14.4. Termination by the Contractor due to Employer's default

The Contractor is entitled under Sub-Clause 16.2 [*Termination by Contractor*] to follow the termination procedure set out under Sub-Clause 16.2.1 [*Notice*] and as described below, if any of the following circumstances arise.

(*a*) The Contractor does not receive the reasonable evidence within 42 days of giving a Notice under Sub-Clause 16.1 [*Suspension by Contractor*] in respect of a failure to comply with Sub-Clause 2.4 [*Employer's Financial Arrangements*].

(*b*) The Engineer fails, within 56 days of receiving a Statement and supporting documents, to issue the relevant Payment Certificate.

(*c*) The Contractor does not receive the amount due under any Payment Certificate within 42 days of the expiry of the time stated in Sub-Clause 14.7 [*Payment*].
(*d*) The Employer fails to comply with:

(i) a binding agreement, or final and binding determination under Sub-Clause 3.7 [*Agreement or Determination*]; or
(ii) a decision of the DAAB under Sub-Clause 21.4 [*Obtaining DAAB's Decision*] (whether binding or final and binding)

and such failure constitutes a material breach of the Employer's obligations under the Contract.

(*e*) The Employer substantially fails to perform, and such failure constitutes a material breach of the Employer's obligations under the Contract.
(*f*) The Contractor does not receive a Notice of the Commencement Date under Sub-Clause 8.1 [*Commencement of Works*] within 84 days of receiving the Letter of Acceptance.
(*g*) The Employer:

(i) fails to comply with Sub-Clause 1.6 [*Contract Agreement*]; or
(ii) assigns the Contract without the required agreement under Sub-Clause 1.7 [*Assignment*].

(*h*) A prolonged suspension affects the whole of the Works as described in sub-paragraph (b) of Sub-Clause 8.12 [*Prolonged Suspension*].
(*i*) The Employer becomes bankrupt or insolvent; goes into liquidation, administration, reorganisation, winding-up or dissolution; becomes subject to the appointment of a liquidator, receiver, administrator, manager or trustee; enters into a composition or arrangement with the Employer's creditors; or any act is done, or any event occurs which is analogous to or has a similar effect to any of these acts or events under applicable Laws.
(*j*) The Employer is found, based on reasonable evidence, to have engaged in corrupt, fraudulent, collusive or coercive practice at any time in relation to the Works or to the Contract.

As was the case under Sub-Clause 16.1 [*Suspension by Contractor*], the Employer's failure to comply with an agreement, determination or DAAB decision must constitute a material breach of the Contract if it is to justify termination under sub-paragraph (d). This is also the case, under sub-paragraph (e), in relation to a failure by the Employer to perform any other obligation under the Contract.

Under SB 2017, as there is no Engineer and no requirement for Payment Certificates, the grounds set out under sub-paragraph (b) do not exist. Moreover, the Employer has no obligation under Sub-Clause 1.6 [*Contract Agreement*] other than to pay stamp duty. Therefore, it is extremely doubtful whether the Contractor could terminate the Contract under sub-paragraph (f) (i) which corresponds to sub-paragraph (g) (i) above.

Termination of the Contract does not prejudice any other rights of the Contractor, under the Contract or otherwise. Thus, if the termination arises from late payment, the Contractor is still entitled to financing charges calculated up to the date of receipt of the due amount, even after the date of termination.

In the case of sub-paragraphs (a) to (g) (i), the Contractor commences the procedure by giving a Notice to the Employer under Sub-Clause 16.2.1 [*Notice*] of the Contractor's intention to terminate the Contract.

Unless the Employer remedies the matter described in the Notice within 14 days of its receipt, the Contractor may, by giving a second Notice to the Employer (under Sub-Clause 16.2.2 [*Termination*]), immediately terminate the Contract. The date of termination shall be the date the Employer receives this second Notice.

In the case of sub-paragraphs (g) (ii) to (j), the Contractor is entitled to serve a Notice of termination under Sub-Clause 16.2.1 [*Notice*], without the need for a Notice of intention to terminate and, under Sub-Clause 16.2.2 [*Termination*], the date of termination shall be the date the Contractor receives this Notice of termination.

Under sub-paragraph (*a*), the procedure is particularly complicated. The process begins with a request from the Contractor under Sub-Clause 2.4 [*Employer's Financial Arrangements*] for reasonable evidence that adequate financial arrangements have been made and are being maintained. If the Employer does not provide the reasonable evidence within 28 days, the Contractor may serve Notice under Sub-Clause 16.1 [*Suspension by Contractor*] and, after 21 days from the date of serving the Notice, the Contractor may suspend or slow down all or part of the Works, unless the Employer has provided the evidence. If, 42 days after the Notice of suspension was received by the Employer, the evidence has still not been provided, the Contractor may serve a Notice of its intention to terminate under Sub-Clause 16.2.1 [*Notice*]. After receipt of this Notice, the Employer has a further period of 14 days in which to provide the evidence, failing which, the Contractor may send a Notice of termination under Sub-Clause 16.2.2 [*Termination*]. It should be noted that the Contractor is not obliged to suspend or slow down any work, but the Notice of suspension under Sub-Clause 16.1 [*Suspension by Contractor*] must be issued, otherwise the Contractor will have no right to issue a Notice of intention to terminate under Sub-Clause 16.2.1 [*Notice*].

In all cases, the Notice(s) must refer to the relevant sub-clause(s).

If the Contractor suffers delay and/or incurs Cost during the above period of 14 days, the Contractor shall be entitled, subject to Sub-Clause 20.2 [*Claims for Payment and/or EOT*], to EOT and/or payment of such Cost Plus Profit.

14.4.1 After Termination

As for termination of the Contract under Sub-Clause 15.5 [*Termination for Employer's Convenience*], following termination by the Contractor under Sub-Clause 16.2 [*Termination by Contractor*], the Contractor is required by Sub-Clause 16.3 [*Contractor's Obligations after Termination*] to promptly

(*a*) cease all further work, except for such work as may have been instructed by the Engineer (the Employer under SB 2017) for the protection of life or property or for the safety of the Works
(*b*) hand over to the Engineer all Contractor's Documents, Plant, Materials and other work for which the Contractor has been paid and
(*c*) remove all other Goods from the Site, except as necessary for safety, and leave the Site.

If the Contractor incurs Cost as a result of carrying out such instructed work the Contractor will be entitled, subject to Sub-Clause 20.2 [*Claims for Payment and/or EOT*], to be paid such Cost Plus Profit.

Figure 14.1 Termination under Sub-Clause 16.2 (a)

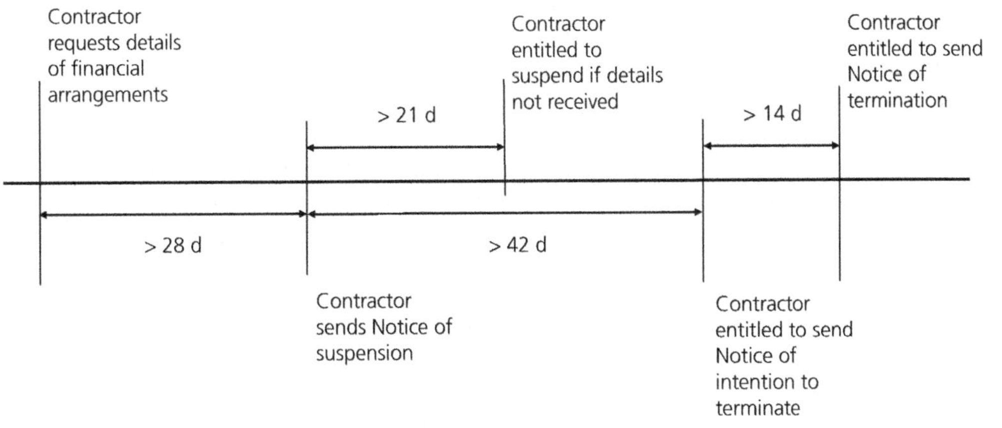

Figure 14.2 Suspension and termination other than under Sub-Clause 16.2 (a)

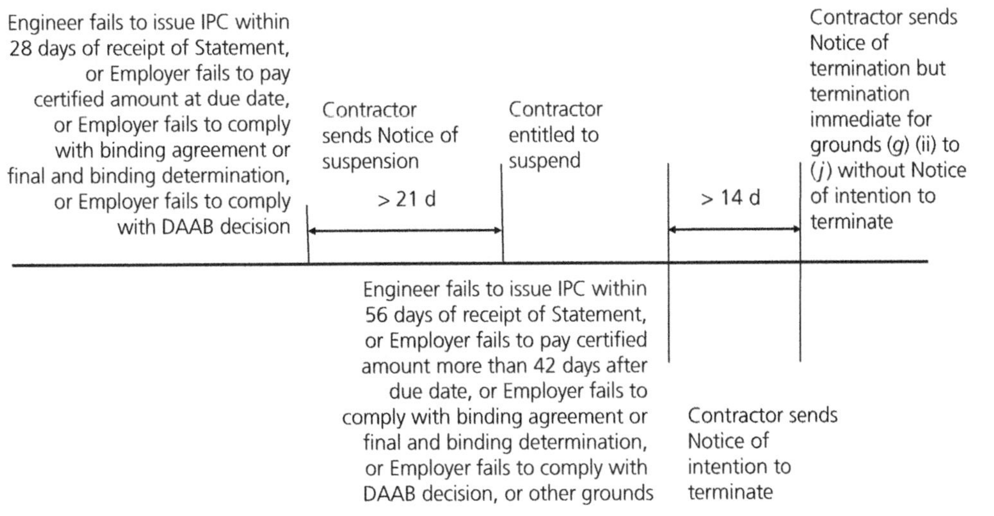

14.4.2 Payment after Termination by Contractor

After termination under Sub-Clause 16.2 [*Termination by Contractor*], the Employer must promptly pay the Contractor in accordance with Sub-Clause 18.5 [*Optional Termination*] and, subject to the Contractor's compliance with Sub-Clause 20.2 [*Claims for Payment and/or EOT*], pay the Contractor the amount of any loss of profit or other losses and damages suffered by the Contractor as a result of this termination (Sub-Clause 16.4 [*Payment after Termination by Contractor*]).

Under Sub-Clause 18.5 [*Optional Termination*], after the date of termination the Contractor must, as soon as practicable, submit detailed supporting particulars (as reasonably required by the Engineer (the Employer under SB 2017)) of the value of the work done, which must include:

(*a*) the amounts payable for any work carried out for which a price is stated in the Contract;
(*b*) the Cost of Plant and Materials ordered for the Works which have been delivered to the Contractor, or of which the Contractor is liable to accept delivery. This Plant and Materials shall become the property of (and be at the risk of) the Employer when paid for by the Employer, and the Contractor shall place the same at the Employer's disposal;
(*c*) any other Cost or liability which in the circumstances was reasonably incurred by the Contractor in the expectation of completing the Works;
(*d*) the Cost of removal of Temporary Works and Contractor's Equipment from the Site and the return of these items to the Contractor's place of business in the Contractor's country (or to any other destination(s) at no greater cost); and
(*e*) the Cost of repatriation of the Contractor's staff and labour employed wholly in connection with the Works at the date of termination.

The Engineer (the Employer's Representative under SB 2017) must then proceed under Sub-Clause 3.7 [*Agreement or Determination*] to agree or determine the value of work done. The date of receipt of the Contractor's particulars shall be the date of commencement of the time limit for agreement under Sub-Clause 3.7.3 [*Time limits*].

It should be remembered that under sub-paragraph (b) Sub-Clause 4.2.3 [*Return of the Performance Security*] the Employer must return the Performance Security to the Contractor promptly after the date of termination.

As was the case for termination by the Employer, the Contractor should not start the termination procedure without seeking legal advice as a wrongful termination may be treated as a serious breach of Contract by the Contractor which would justify the termination of the Contract by the Employer.

Contractor's Entitlement to Payment after Termination				
	Termination for Contractor's default	**Termination for Convenience**	**Termination for Employer's default**	**Optional Termination**
Value of the Permanent Works, Goods and Contractor's Documents, and any other sums due to the Contractor for work executed in accordance with the Contract	X			
Any additions and/or deductions which have become due under Sub-Clause 3.7 [*Agreement or Determination*] or under the Contract or otherwise	X			
Amounts payable for any work carried out for which a price is stated in the Contract		X	X	X
The Cost of Plant and Materials ordered for the Works which have been delivered to the Contractor, or of which the Contractor is liable to accept delivery		X	X	X
Any other Cost or liability which in the circumstances was reasonably incurred by the Contractor in the expectation of completing the Works		X	X	X
The Cost of removal of Temporary Works and Contractor's Equipment from the Site and the return of these items to the Contractor's place of business in the Contractor's country (or to any other destination(s) at no greater cost)		X	X	X
The Cost of repatriation of the Contractor's staff and labour employed wholly in connection with the Works at the date of termination		X	X	X
Any loss of profit or other losses and damages suffered by the Contractor as a result of this termination		X	X	

emerald PUBLISHING ice Publishing

Smith G
ISBN 978-0-7277-6652-6
https://doi.org/10.1680/fcmh.66526.219

Chapter 15
Contractor's care of the works, employer's risks, exceptional events

15.1. Contractor's care of the Works

In general, under Sub-Clause 17.1 [*Responsibility for Care of the Works*] the Contractor is fully responsible for the care of the Works, Goods and Contractor's Documents from the Commencement Date until the Date of Completion of the Works, when responsibility for the care of the Works passes to the Employer. That is to say that until the Works are taken over the Contractor must rectify the loss or damage at the Contractor's risk and cost, so that the Works, Goods or Contractor's Documents (as the case may be) comply with the Contract.

Similarly, if a Taking-Over Certificate is issued (or is deemed to be issued) for any Section or Part, responsibility for the care of the relevant Section or Part passes to the Employer at the date of taking over (or deemed taking over).

The only exceptions are if the Contract is terminated, or if the loss or damage results from an event which is the Employer's responsibility under Sub-Clause 17.2 [*Liability for Care of the Works*] (see below).

If the Contract is terminated in accordance with the Contract or otherwise, the Contractor ceases to be responsible for the care of the Works from the date of termination.

After the Date of Completion, when responsibility for care of the Works, Section or Part passes to the Employer, the Contractor is responsible only for the care of any work which was outstanding on the relevant Date of Completion, until this outstanding work has been completed.

Nevertheless, the Contractor remains liable for any loss or damage caused by the Contractor after the issue of a Taking-Over Certificate. The Contractor will also be liable for any loss or damage, which occurs after the issue of a Taking-Over Certificate and which arose from an event for which the Contractor was liable, which occurred before the issue of this Taking-Over Certificate.

15.2. Employer's Risks

The FIDIC 1999 Contracts contained a list of events, for which the associated risk of loss or damage to the Works was clearly stated to be borne by the Employer.

The FIDIC 2017 Contracts are similar although the events are not so clearly categorised as 'Employer's Risks'.

Under Sub-Clause 17.2 [*Liability for Care of the Works*], the Contractor will not be liable for loss or damage to the Works, Goods or Contractor's Documents caused by any of the listed events.

(However, the Contractor will remain liable for remedying defects in any Works, Goods or Contractor's Documents which had been rejected by the Engineer before the occurrence of any of the listed events).

The list under RB 2017 is as follows:

(*a*) *'interference, whether temporary or permanent, with any right of way, light, air, water or other easement (other than that resulting from the Contractor's method of construction) which is the unavoidable result of the execution of the Works in accordance with the Contract;*
(*b*) *use or occupation by the Employer of any part of the Permanent Works, except as may be specified in the Contract;*
(*c*) *fault, error, defect or omission in any element of the design of the Works by the Employer or which may be contained in the Specification and Drawings (and which an experienced contractor exercising due care would not have discovered when examining the Site and the Specification and Drawings before submitting the Tender) other than design carried out by the Contractor in accordance with the Contractor's obligations under the Contract;*
(*d*) *any operation of the forces of nature (other than those allocated to the Contractor in the Contract Data) which is Unforeseeable or against which an experienced contractor could not reasonably have been expected to have taken adequate preventative precautions;*
(*e*) *any of the events or circumstances listed under sub-paragraphs (a) to (f) of Sub-Clause 18.1 [Exceptional Events]; and/or*
(*f*) *any act or default of the Employer's Personnel or the Employer's other contractors.'*

Under YB 2017, the only difference with respect to the above list is that sub-paragraph (c) refers to the Employer's Requirements rather than the Specification and Drawings. Under sub-paragraph (c) of SB 2017, the event is limited to a fault, error, defect and so forth in any element of design by the Employer.

If any of the events listed above results in damage to the Works, Goods or Contractor's Documents, the Contractor must promptly give a Notice to the Engineer. Thereafter, the Contractor must rectify the loss and/or damage to the extent instructed by the Engineer. Such instruction is deemed to have been given under Sub-Clause 13.3.1 [*Variation by Instruction*]. As such, the Contractor will be entitled to an EOT if and to the extent that the completion of the Works (or Section) is delayed, without the need for a Notice of Claim under Sub-Clause 20.2 [*Claims for Payment and/or EOT*].

However, if the event falls within the definition of an Exceptional Event under Sub-Clause 18.1 [*Exceptional Events*] the Contractor will be entitled to recover the Cost, only if the Exceptional Event is of the kind described in sub-paragraphs (a) to (e) of Sub-Clause 18.1 [*Exceptional Events*] and, in the case of sub-paragraphs (b) to (e) of that sub-clause, occurs in the Country (see below).

If the loss or damage to the Works, Goods or Contractor's Documents results from a combination of any of the events listed above, and a cause for which the Contractor is liable and the Contractor suffers a delay and/or incurs Cost from rectifying the loss and/or damage, subject to Sub-Clause 20.2 [*Claims for Payment and/or EOT*], the Contractor will be entitled to a proportion of EOT and/or Cost Plus Profit to the extent that any of the above events has contributed to such delays and/or Costs. In other words, liability for the delay and additional Costs will be apportioned between the Contractor and the Employer.

15.3. Indemnities

Under Sub-Clause 17.4 [*Indemnities by Contractor*], the Contractor must indemnify and hold harmless the Employer, the Employer's Personnel, and their respective agents, against and from all third-party claims, damages, losses and expenses (including legal fees and expenses) in respect of:

(*a*) bodily injury, sickness, disease or death of any person arising out of, or in the course of, or by reason of the Contractor's execution of the Works, unless attributable to any negligence, wilful act or breach of the Contract by the Employer, the Employer's Personnel, or any of their respective agents; and

(*b*) damage to or loss of any property, real or personal (other than the Works), to the extent that such damage or loss:

- arises out of, or in the course of, or by reason of the Contractor's execution of the Works, and
- is attributable to any negligence, wilful act or breach of the Contract by the Contractor, the Contractor's Personnel, their respective agents, or anyone directly or indirectly employed by any of them.

Under RB 2017, to the extent that the Contractor is responsible for the design of part of the Permanent Works under Sub-Clause 4.1 [*Contractor's General Obligations*], and/or any other design under the Contract, the indemnity also covers all acts, errors or omissions in carrying out the Contractor's design obligations that result in the Works (or Section or Part or major item of Plant, if any), when completed, not being fit for the purpose(s) for which they are intended under Sub-Clause 4.1 [*Contractor's General Obligations*].

Under Sub-Clause 17.6 [*Shared Indemnities*], the Contractor's liability to indemnify the Employer will be reduced proportionately to the extent that any event listed under Sub-Clause 17.2 [*Liability for Care of the Works*] may have contributed to the said damage, loss or injury.

Under Sub-Clause 17.5 [*Indemnities by Employer*] the Employer must indemnify and hold harmless the Contractor, the Contractor's Personnel, and their respective agents, against and from all third-party claims, damages, losses and expenses (including legal fees and expenses) in respect of:

(*a*) bodily injury, sickness, disease or death, or loss of or damage to any property other than the Works, which is attributable to any negligence, wilful act or breach of the Contract by the Employer, the Employer's Personnel, or any of their respective agents; and

(*b*) damage to or loss of any property, real or personal (other than the Works), to the extent that such damage or loss arises out of any event described under sub-paragraphs (a) to (f) of Sub-Clause 17.2 [*Liability for Care of the Works*].

Under Sub-Clause 17.6 [*Shared Indemnities*], the Employer's liability to indemnify the Contractor, will be reduced proportionately to the extent that any event for which the Contractor is responsible may have contributed to the said damage, loss or injury.

15.4. Exceptional Events

Sub-Clause 18.1 [*Exceptional Events*] defines an event or circumstance as an 'Exceptional Event' if four conditions are satisfied:

(i) the event or circumstance must be beyond a Party's control;
(ii) that Party could not reasonably have provided against the event or circumstance before entering into the Contract;
(iii) having arisen, that Party could not reasonably have avoided or overcome the event or circumstance; and
(iv) it is not substantially attributable to the other Party.

> **Note:**
>
> The 2022 Reprint states that to qualify as an 'Exceptional Event' the four conditions must be satisfied, and the event must be exceptional.

Provided that the necessary conditions are all satisfied, examples of events or circumstances which often qualify as Exceptional Events include:

(*a*) war, hostilities (whether war be declared or not), invasion, act of foreign enemies;
(*b*) rebellion, terrorism, revolution, insurrection, military or usurped power, or civil war;
(*c*) riot, commotion or disorder by persons other than the Contractor's Personnel and other employees of the Contractor and Subcontractors;
(*d*) strike or lockout not solely involving the Contractor's Personnel and other employees of the Contractor and Subcontractors;
(*e*) encountering munitions of war, explosive materials, ionising radiation or contamination by radioactivity, except as may be attributable to the Contractor's use of such munitions, explosives, radiation or radioactivity; or
(*f*) natural catastrophes such as earthquake, tsunami, volcanic activity, hurricane or typhoon.

If such an Exceptional Event arises and it prevents or will prevent one of the Parties (the 'affected Party') from performing an obligation under the Contract, Sub-Clause 18.2 [*Notice of an Exceptional Event*] requires the affected Party to give a Notice to the other Party (copied to the Engineer) of the Exceptional Event and specifying the obligation(s) the performance of which is or will be prevented (the 'prevented obligations'). This requirement to specify the obligation(s) which cannot be performed is often overlooked. Moreover, it is not sufficient that the Exceptional Event makes it more difficult or more expensive to perform the obligation, performance must be prevented.

The Notice must be given within 14 days of the date on which the affected Party became aware, or should have become aware, of the Exceptional Event. Thereafter, the affected Party will be excused from performance of the prevented obligations from the date from which such performance was prevented until the date when performance can begin or resume. However, if this Notice is not received by the other Party until after this period of 14 days, the affected Party is excused performance of the prevented obligations only from the date on which the Notice is received by the other Party.

It should be noted that the affected Party will be excused from performance of the prevented obligations, but not others. The affected Party will remain bound to perform all obligations for which performance is not prevented. This includes the obligations of either Party to make payments due to the other Party under the Contract.

Under Sub-Clause 18.3 [*Duty to Minimise Delay*], each Party must use all reasonable endeavours to minimise any delay in the performance of the Contract due to an Exceptional Event. The expression '*all reasonable endeavours*' does not have same legal implications in all countries, and appropriate legal advice should be sought with respect to the meaning of the expression under the applicable Laws.

After giving the first Notice under Sub-Clause 18.2 [*Notice of an Exceptional Event*] the affected Party must give further Notices describing the effect of the Exceptional Event every 28 days until the effects end.

Immediately after the effect of the Exceptional Event ends (that is, as soon as the affected Party can commence or resume performance of the affected obligation(s)), the affected Party must give a Notice to the other Party. If the affected Party fails to do so, the other Party may give a Notice to the affected Party stating that the other Party considers that the affected Party's performance is no longer prevented by the Exceptional Event, with reasons.

If the Contractor is the affected Party and suffers delay and/or incurs Cost by reason of the Exceptional Event of which they gave a Notice under Sub-Clause 18.2 [*Notice of an Exceptional Event*], the Contractor is entitled to an EOT under Sub-Clause 18.4 [*Consequences of the Exceptional Event*] and subject to Sub-Clause 20.2 [*Claims for Payment and/or EOT*]. The Contractor is entitled to reimbursement of the Cost only if the Exceptional Event is:

- of the kind described in sub-paragraphs (a) to (e) of Sub-Clause 18.1 [*Exceptional Events*]; and,
- in the case of sub-paragraphs (b) to (e) of that Sub-Clause, occurs in the Country.

This means that for the Contractor to be entitled to payment of the Cost, the Exceptional Event must be man-made, and it must occur in the Country where the Works are located (except for war, hostilities, invasion, act of foreign enemies).

Either Party may give to the other Party a Notice of termination of the Contract under Sub-Clause 18.5 [*Optional Termination*], if the execution of substantially all the Works in progress is prevented for a continuous period of 84 days by an Exceptional Event of which Notice has been given under Sub-Clause 18.2 [*Notice of an Exceptional Event*], or for multiple periods due to the same Exceptional Event which total more than 140 days.

In this event, the date of termination will be the date 7 days after the Notice is received by the other Party, and the Contractor must then proceed in accordance with Sub-Clause 16.3 [*Contractor's Obligations after Termination*] to demobilise and so forth (see Chapter 14.4.1).

> **Note:**
>
> The period of Notice is different from that for other grounds for termination, for which the period is either 14 days from receipt of a Notice of intention to terminate under Sub-Clause 16.2.1 [*Notice*] or immediately upon receipt of a Notice of termination under Sub-Clause 16.2.2 [*Termination*].

As soon as practicable after the date of termination, the Contractor must submit detailed supporting particulars (as reasonably required by the Engineer) of the value of the work done, including:

(*a*) the amounts payable for any work carried out for which a price is stated in the Contract;
(*b*) the Cost of Plant and Materials ordered for the Works which have been delivered to the Contractor, or of which the Contractor is liable to accept delivery;
(*c*) any other Cost or liability reasonably incurred by the Contractor in the expectation of completing the Works;
(*d*) the Cost of removal of Temporary Works and Contractor's Equipment from the Site and the return of these items to the Contractor's place of business in the Contractor's country (or to any other destination(s) at no greater cost); and
(*e*) the Cost of repatriation of the Contractor's staff and labour employed wholly in connection with the Works at the date of termination.

Following receipt of these particulars and without the need for the Contractor to submit a Statement, the Engineer must proceed under Sub-Clause 3.7 [*Agreement or Determination]* to agree or determine the value of work done. The date of commencement of the time limit for agreement under Sub-Clause 3.7.3 [*Time limits*] shall be the date the Engineer receives the Contractor's particulars.

The Engineer must then issue a Payment Certificate, under Sub-Clause 14.6 [*Issue of IPC*], for the amount so agreed or determined.

> **Note:**
>
> Clause 18 [*Exceptional Events*] contains no express provision with respect to payment of the amount certified, whereas in the event of termination for Employer's convenience or termination due to Employer's default, payment must be made promptly (i.e. without delay).

15.5. Release from Performance under the Law

If any event arises which is outside the control of the Parties (including, but not limited to, an Exceptional Event) which:

(*a*) makes it impossible or unlawful for either Party or both Parties to fulfil their contractual obligations; or
(*b*) under the law governing the Contract, entitles the Parties to be released from further performance of the Contract,

and if the Parties are unable to agree on an amendment to the Contract that would permit the continued performance of the Contract, then either Party may give a Notice to the other Party of the event.

Following receipt of such a Notice, both Parties shall be discharged from further performance, but without prejudice to the rights of either Party in respect of any previous breach of the Contract.

The amount then payable by the Employer to the Contractor shall be the same as would have been payable under Sub-Clause 18.5 [*Optional Termination*] and that amount must be certified by the Engineer, as if the Contract had been terminated under that sub-clause. Again, no mention is made of the period for the Employer to pay the certified amount.

Smith G
ISBN 978-0-7277-6652-6
https://doi.org/10.1680/fcmh.66526.227

Chapter 16
Insurances

The Contractor must take out and maintain all the insurances for which the Contractor is responsible under the Contract. The insurances must be with insurers and in terms to which the Employer has given consent. These terms must be consistent with any terms agreed by the Parties before the date of the Letter of Acceptance (under SB 2017: before the signature of the Contract Agreement by both Parties).

The permitted deductible limits allowed in any policy must not exceed the amounts stated in the Contract Data (if not stated, the amounts agreed with the Employer).

If requested to do so by the Employer, the Contractor must provide copies of the policies for the insurances which the Contractor is required to take out.

Each time that a premium is paid, the Contractor must promptly submit to the Employer (with a copy to the Engineer) either a copy of the receipt of payment, or confirmation from the insurers that the premium has been paid.

The Contractor is also responsible for:

- notifying the insurers of any changes in the nature, extent or programme for the execution of the Works; and
- the adequacy and validity of the insurances in accordance with the Contract throughout the performance of the Contract.

If the Contractor fails to take out or maintain any of the required insurances the Employer may take out and/or maintain the insurances and recover the cost from the Contractor, by deducting the amount(s) paid from any moneys due to the Contractor or otherwise recover the same as a debt from the Contractor. There is no requirement for a Notice of Claim or a Claim under Clause 20 [*Employer's and Contractor's Claims*].

If either Party fails to comply with any condition of the insurances, the defaulting Party must indemnify the other Party against all direct losses and claims (including legal fees and expenses) arising from the failure.

Where there is a shared liability for the loss and/or damage suffered, the loss must be borne by each Party in proportion to each Party's liability, provided that the non-recovery from insurers has not been caused by a breach of Clause 19 by the Contractor or the Employer. If non-recovery from insurers has been caused by such a breach, the defaulting Party must bear the loss suffered.

The insurances required under Sub-Clause 19.2 [*Insurances to be provided by the Contractor*], and as described in the table below, are the minimum required by the Employer, and the Contractor may take out additional insurance cover at its own expense.

In addition, the Contractor must provide at its cost, all other insurances required by the Laws of the countries where (any part of) the Works are being carried out, at the Contractor's own cost.

Other insurances required by local practice (if any) must be detailed in the Contract Data.

	Works	Goods	Professional Liability	Third Parties	Employees
Coverage	Loss or damage to Works, Materials, Plant and Contractor's Documents due to defective design or construction	Loss or damage to Goods	Against liability arising out of any act, error or omission by the Contractor in carrying out the Contractor's design obligations and, if stated in the Contract Data, liability for failure to achieve fitness for purpose	Liability for death or injury to any person, or loss of or damage to any property (other than the Works)	Liability in respect of injury, sickness, disease or death of any person employed by the Contractor or any of the Contractor's other personnel arising out of the execution of the Works
Amount	Full replacement cost + 15%	As stated in the Contract Data (if not specified or stated, for their full replacement value including delivery to Site)	Not less than as stated in the Contract Data or, if not stated, as agreed with the Employer	Not less than as stated in the Contract Data	Not stated
Insured	Joint names of the Contractor and the Employer	Joint names of the Contractor and the Employer	Contractor	Joint names of the Contractor and the Employer with cross liability provision	The Contractor, Employer and the Engineer
Period	From Commencement Date to Taking Over of Works plus until issue of Performance Certificate with respect to completion of outstanding work and execution of DNP obligations	From the time the Goods are delivered to the Site until they are no longer required for the Works	As stated in Contract Data	Before the Contractor begins any work on the Site until the issue of the Performance Certificate	The whole time that the Contractor's Personnel are assisting in the execution of the Works

	Works	Goods	Professional Liability	Third Parties	Employees
Possible Exclusions	(i) the cost of making good any part of the Works which is defective or otherwise does not comply with the Contract, provided that it does not exclude the cost of making good any loss or damage to any other part of the Works attributable to such defect or non-compliance; (ii) indirect or consequential loss or damage including any reductions in the Contract Price due to delay; (iii) wear and tear, shortages and pilferages; and (iv) unless otherwise stated in the Contract Data, the risks arising from Exceptional Events			Loss or damage caused by an Exceptional Event.	Losses and claims to the extent that they arise from any act or neglect of the Employer or of the Employer's Personnel

Smith G
ISBN 978-0-7277-6652-6
https://doi.org/10.1680/fcmh.66526.231

Chapter 17
Claims

Sub-Clause 1.1.6	***'Claim'*** *means a request or assertion by one Party to the other Party for an entitlement or relief under any Clause of the Conditions or otherwise in connection with, or arising out of, the Contract or the execution of the Works.*

Sub-Clause 20.1 [*Claims*] states that such a Claim may arise:

(*a*) *if the Employer considers that it is entitled to any additional payment from the Contractor (or reduction in the Contract Price) and/or to an extension of the DNP; or*
(*b*) *if the Contractor considers that it is entitled to an additional payment from the Employer and/or to EOT; or*
(*c*) *if either Party considers that they are entitled to another entitlement or relief against the other Party.*

Such other entitlement or relief may be of any kind (including in connection with any certificate, determination, instruction, Notice, opinion or valuation of the Engineer) except to the extent that it involves any alleged entitlement of the kind referred to in sub-paragraphs (a) or (b) above. For example, such a claim might concern deductions by the Engineer from amounts mentioned in a Contractor's Statement.

It should be noted that the above definition has been slightly modified by the 2022 Reprint (see Chapter 19).

In the case of a Claim under sub-paragraph (c) above, where the other Party or the Engineer has disagreed with the requested entitlement or relief (or is deemed to have disagreed if they do not respond within a reasonable time), a Dispute will not be deemed to have arisen, but the claiming Party may, by giving a Notice, refer the Claim to the Engineer and Sub-Clause 3.7 [*Agreement or Determination*] will apply. This Notice must be given as soon as practicable after the claiming Party becomes aware of the disagreement (or deemed disagreement) and must include details of the claiming Party's case and the other Party's or the Engineer's disagreement (or deemed disagreement in the event of a lack of response).

If either Party considers that it is entitled to make a Claim under sub-paragraphs (a) or (b) above, the procedure set out under Sub-Clause 20.2 [*Claims for Payment and/or EOT*] applies. The sub-clauses which expressly require compliance with Sub-Clause 20.2 [*Claims for Payment and/or EOT*] for each of the FIDIC 2017 contracts are summarised in the tables below.

RB 2017 – SUB-CLAUSES GIVING EXPRESS ENTITLEMENT TO CLAIM

Clause	Topic	Claiming Party	Entitlement to claim
1.9	Late drawings or instructions	Contractor	EOT and/or Cost Plus Profit
1.13	Late provision of permits	Contractor	EOT and/or Cost Plus Profit
1.13	Contractor's failure to obtain or comply with permits, etc., or failure to assist Employer	Employer	Costs
2.1	Delay in providing right of access to and possession of Site	Contractor	EOT and/or Cost Plus Profit
4.6	Instructions to cooperate	Contractor	EOT and/or Cost Plus Profit
4.7	Errors in setting-out information	Contractor	EOT and/or Cost Plus Profit
4.12	Unforeseeable Physical Conditions	Contractor	EOT and/or Cost
4.15	Changes to access after Base Date	Contractor	EOT and/or Cost
4.23	Geological or archaeological findings	Contractor	EOT and/or Cost
7.4	Delay to testing caused by Engineer	Contractor	EOT and/or Cost Plus Profit
7.4	Delay to testing caused by Contractor	Employer	Costs
7.5	Rejection and retesting	Employer	Costs
7.6	Remedial work attributable to Employer	Contractor	EOT and/or Cost Plus Profit
7.6	Failure to execute remedial work	Employer	Costs
8.5	Extension of Time for Completion	Contractor	EOT
8.6	Delay caused by authorities	Contractor	EOT
8.7	Measures to improve rate of progress	Employer	Costs
8.8	Delay Damages	Employer	Delay Damages
8.10	Employer's suspension	Contractor	EOT and/or Cost Plus Profit
9.2	Delay by Contractor to Tests on Completion	Employer	Costs
9.4	Failure to pass Tests on Completion	Employer	Reduction in Contract Price
10.2	Taking Over parts of the Works	Contractor	Cost Plus Profit
10.3	Interference with Tests on Completion	Contractor	EOT and/or Cost Plus Profit
11.3	Extension of DNP	Employer	Extension of DNP
11.4	Failure to remedy defects	Employer	Performance Damages or reduction in Contract Price or reimbursement of sums paid plus demolition costs plus financing costs
11.7	Failure to give right of access after Taking Over	Contractor	Cost Plus Profit
11.8	Search by Contractor for cause of defect shows defect is not Contractor's responsibility	Contractor	Cost Plus Profit

RB 2017 – SUB-CLAUSES GIVING EXPRESS ENTITLEMENT TO CLAIM			
Clause	**Topic**	**Claiming Party**	**Entitlement to claim**
11.8	Search by Employer for cause of defect shows defect is Contractor's responsibility	Employer	Costs of search
11.11	Failure to clear Site	Employer	Costs
13.3	Cost of proposal requested by Engineer but not implemented	Contractor	Cost
13.6	Changes in legislation	Contractor	EOT and/or Cost
13.6	Changes in legislation	Employer	Reduction in Contract Price
14.14	Contractor contests FPC	Contractor	Adjustment to FPC
15.4	Termination for Contractor's default	Employer	Costs plus loss and damage*
16.1	Suspension by Contractor	Contractor	EOT and/or Cost Plus Profit
16.2	Delay and/or costs incurred after Notice of intention to terminate by Contractor	Contractor	EOT and/or Cost Plus Profit
16.3	Contractor's obligations after termination	Contractor	Cost Plus Profit
16.4	Payment after termination by Contractor	Contractor	Loss of profit or other losses and damages
17.2	Loss or damage to the Works, Goods or Contractor's Documents	Contractor	EOT and/or Cost Plus Profit
18.4	Consequences of an Exceptional Event	Contractor	EOT and/or Cost

*The FIDIC 2017 Contracts Guide [2022] at page 502 suggests that the Employer is also entitled to claim Delay Damages under Sub-Clause 15.4. This is not strictly correct. A claim for Delay Damages can be made under Sub-Clause 8.8 if the Time for Completion expired before the date of termination, subject to having issued a Notice of Claim within 28 days of the expiry of the Time for Completion. If termination occurs before the expiry of the Time for Completion, the Employer has no right to claim Delay Damages.

YB 2017 – SUB-CLAUSES GIVING EXPRESS ENTITLEMENT TO CLAIM			
Clause	**Topic**	**Claiming Party**	**Entitlement to claim**
1.9	Errors in the Employer's Requirements	Contractor	EOT and/or Cost Plus Profit
1.13	Late provision of permits	Contractor	EOT and/or Cost Plus Profit
1.13	Contractor's failure to obtain or comply with permits, etc., or failure to assist Employer	Employer	Costs
2.1	Delay in providing right of access to and possession of Site	Contractor	EOT and/or Cost Plus Profit
4.6	Instructions to cooperate	Contractor	EOT and/or Cost Plus Profit
4.7	Errors in setting-out information	Contractor	EOT and/or Cost Plus Profit
4.12	Unforeseeable Physical Conditions	Contractor	EOT and/or Cost

YB 2017 – SUB-CLAUSES GIVING EXPRESS ENTITLEMENT TO CLAIM			
Clause	**Topic**	**Claiming Party**	**Entitlement to claim**
4.15	Changes to access after Base Date	Contractor	EOT and/or Cost
4.23	Geological or archaeological findings	Contractor	EOT and/or Cost
5.2	Resubmission and review of Contractor's Documents	Employer	Costs
7.4	Delay to testing caused by Engineer	Contractor	EOT and/or Cost Plus Profit
7.4	Delay to testing caused by Contractor	Employer	Costs
7.5	Rejection and retesting	Employer	Costs
7.6	Remedial work attributable to Employer	Contractor	EOT and/or Cost Plus Profit
7.6	Failure to execute remedial work	Employer	Costs
8.5	Extension of Time for Completion	Contractor	EOT
8.6	Delay caused by authorities	Contractor	EOT
8.7	Measures to improve rate of progress	Employer	Costs
8.8	Delay Damages	Employer	Delay Damages
8.10	Employer's suspension	Contractor	EOT and/or Cost Plus Profit
9.2	Delay by Contractor to Tests on Completion	Employer	Costs
9.4	Failure to pass Tests on Completion	Employer	Reduction in Contract Price
10.2	Taking Over parts of the Works	Contractor	Cost Plus Profit
10.3	Interference with Tests on Completion	Contractor	EOT and/or Cost Plus Profit
11.3	Extension of DNP	Employer	Extension of DNP
11.4	Failure to remedy defects	Employer	Performance Damages or reduction in Contract Price or reimbursement of sums paid plus demolition costs plus financing costs
11.7	Failure to give right of access after Taking Over	Contractor	Cost Plus Profit
11.8	Search by Contractor for cause of defect shows defect is not Contractor's responsibility	Contractor	Cost Plus Profit
11.8	Search by Employer for cause of defect shows defect is Contractor's responsibility	Employer	Costs of search
11.11	Failure to clear Site	Employer	Costs
12.2	Delay to Tests after Completion attributable to Employer	Contractor	Cost Plus Profit
12.3	Retesting	Employer	Costs
12.4	Failure to pass Tests after Completion	Employer	Performance Damages
12.4	Failure to give right of access for adjustments	Contractor	Cost Plus Profit

YB 2017 – SUB-CLAUSES GIVING EXPRESS ENTITLEMENT TO CLAIM			
Clause	**Topic**	**Claiming Party**	**Entitlement to claim**
13.3	Cost of proposal requested by Engineer but not implemented	Contractor	Cost
13.6	Changes in legislation	Contractor	EOT and/or Cost
13.6	Changes in legislation	Employer	Reduction in Contract Price
14.14	Contractor contests FPC	Contractor	Adjustment to FPC
15.4	Termination for Contractor's default	Employer	Costs plus loss and damage*
16.1	Suspension by Contractor	Contractor	EOT and/or Cost Plus Profit
16.2	Delay and/or costs incurred after Notice of intention to terminate by Contractor	Contractor	EOT and/or Cost Plus Profit
16.3	Contractor's obligations after termination	Contractor	Cost Plus Profit
16.4	Payment after termination by Contractor	Contractor	Loss of profit or other losses and damages
17.2	Loss or damage to the Works, Goods or Contractor's Documents	Contractor	EOT and/or Cost Plus Profit
18.4	Consequences of an Exceptional Event	Contractor	EOT and/or Cost

*The FIDIC 2017 Contracts Guide [2022] at page 502 suggests that the Employer is also entitled to claim Delay Damages under Sub-Clause 15.4. This is not strictly correct. A claim for Delay Damages can be made under Sub-Clause 8.8 if the Time for Completion expired before the date of termination, subject to having issued a Notice of Claim within 28 days of the expiry of the Time for Completion. If termination occurs before the expiry of the Time for Completion, the Employer has no right to claim Delay Damages.

SB 2017 – SUB-CLAUSES GIVING EXPRESS ENTITLEMENT TO CLAIM			
Clause	**Topic**	**Claiming Party**	**Entitlement to claim**
1.13	Late provision of permits	Contractor	EOT and/or Cost Plus Profit
1.13	Contractor's failure to obtain or comply with permits, etc., or failure to assist Employer	Employer	Costs
2.1	Delay in providing right of access to and possession of Site	Contractor	EOT and/or Cost Plus Profit
4.6	Instructions to cooperate	Contractor	EOT and/or Cost Plus Profit
4.15	Changes to access after Base Date	Contractor	EOT and/or Cost
4.23	Geological or archaeological findings	Contractor	EOT and/or Cost
5.2	Resubmission and review of Contractor's Documents	Employer	Costs
7.4	Delay to testing caused by Engineer	Contractor	EOT and/or Cost Plus Profit
7.4	Delay to testing caused by Contractor	Employer	Costs

SB 2017 – SUB-CLAUSES GIVING EXPRESS ENTITLEMENT TO CLAIM			
Clause	**Topic**	**Claiming Party**	**Entitlement to claim**
7.5	Rejection and retesting	Employer	Costs
7.6	Remedial work attributable to Employer	Contractor	EOT and/or Cost Plus Profit
7.6	Failure to execute remedial work	Employer	Costs
8.5	Extension of Time for Completion	Contractor	EOT
8.6	Delay caused by authorities	Contractor	EOT
8.7	Measures to improve rate of progress	Employer	Costs
8.8	Delay Damages	Employer	Delay Damages
8.10	Employer's suspension	Contractor	EOT and/or Cost Plus Profit
9.2	Delay by Contractor to Tests on Completion	Employer	Costs
9.4	Failure to pass Tests on Completion	Employer	Reduction in Contract Price
10.3	Interference with Tests on Completion	Contractor	EOT and/or Cost Plus Profit
11.3	Extension of DNP	Employer	Extension of DNP
11.4	Failure to remedy defects	Employer	Performance Damages or reduction in Contract Price or reimbursement of sums paid plus demolition costs plus financing costs
11.7	Failure to give right of access after Taking Over	Contractor	Cost Plus Profit
11.8	Search by Contractor for cause of defect shows defect is not Contractor's responsibility	Contractor	Cost Plus Profit
11.8	Search by Employer for cause of defect shows defect is Contractor's responsibility	Employer	Costs of search
11.11	Failure to clear Site	Employer	Costs
12.2	Delay to Tests after Completion attributable to Employer	Contractor	Cost Plus Profit
12.3	Retesting	Employer	Costs
12.4	Failure to pass Tests after Completion	Employer	Performance Damages
12.4	Failure to give right of access for adjustments	Contractor	Cost Plus Profit
13.3	Cost of proposal requested by Engineer but not implemented	Contractor	Cost
13.6	Changes in legislation	Contractor	EOT and/or Cost
13.6	Changes in legislation	Employer	Reduction in Contract Price
14.14	Contractor contests FPC	Contractor	Adjustment to FPC
15.4	Termination for Contractor's default	Employer	Costs plus loss and damage*
16.1	Suspension by Contractor	Contractor	EOT and/or Cost Plus Profit

SB 2017 – SUB-CLAUSES GIVING EXPRESS ENTITLEMENT TO CLAIM			
Clause	**Topic**	**Claiming Party**	**Entitlement to claim**
16.2	Delay and/or costs incurred after Notice of intention to terminate by Contractor	Contractor	EOT and/or Cost Plus Profit
16.3	Contractor's obligations after termination	Contractor	Cost Plus Profit
16.4	Payment after termination by Contractor	Contractor	Loss of profit or other losses and damages
17.2	Loss or damage to the Works, Goods or Contractor's Documents	Contractor	EOT and/or Cost Plus Profit
18.4	Consequences of an Exceptional Event	Contractor	EOT and/or Cost

*The FIDIC 2017 Contracts Guide [2022] at page 502 suggests that the Employer is also entitled to claim Delay Damages under Sub-Clause 15.4. This is not strictly correct. A claim for Delay Damages can be made under Sub-Clause 8.8 if the Time for Completion expired before the date of termination, subject to having issued a Notice of Claim within 28 days of the expiry of the Time for Completion. If termination occurs before the expiry of the Time for Completion, the Employer has no right to claim Delay Damages.

17.1. Notice of Claim

Sub-Clause 20.2.1 [*Notice of Claim*] requires that as soon as practicable, and no later than 28 days after the date on which the claiming Party became aware, or should have become aware, of the event or circumstance giving rise to the Claim, the claiming Party must give a Notice to the Engineer (to the other Party under SB 2017), describing the event or circumstance (the '*Notice of Claim*').

If the claiming Party fails to give such a Notice of Claim within this period of 28 days, the other Party will be discharged from any liability in connection with the event or circumstance giving rise to the Claim. In such a case, the claiming Party will not be entitled to any additional payment, the Contract Price will not be reduced (in the case of the Employer as the claiming Party), and neither the Time for Completion (in the case of the Contractor as the claiming Party) nor the DNP (in the case of the Employer as the claiming Party) will be extended.

At first glance, the requirement for timely submission of the Notice of Claim appears to be clear and decisive. However, whereas the date when the claiming Party became aware of the event or circumstance giving rise to the claim may be relatively easy to establish, the date when the claiming Party should have become aware of the event or circumstance is likely to be more open to disagreement. Moreover, in recent years there has been some support for the argument that the 28-day period for giving Notice of Claim begins when the claiming Party should have become aware that the event or circumstance would give rise to an entitlement to claim. Supporters of this viewpoint consider that it is not knowledge of the occurrence of the event or circumstance which triggers the requirement for Notice but an awareness of the entitlement to claim due to the event or circumstance, which will not arise until the impact of the event or circumstance is felt.

Figure 17.1 Notice of Claim

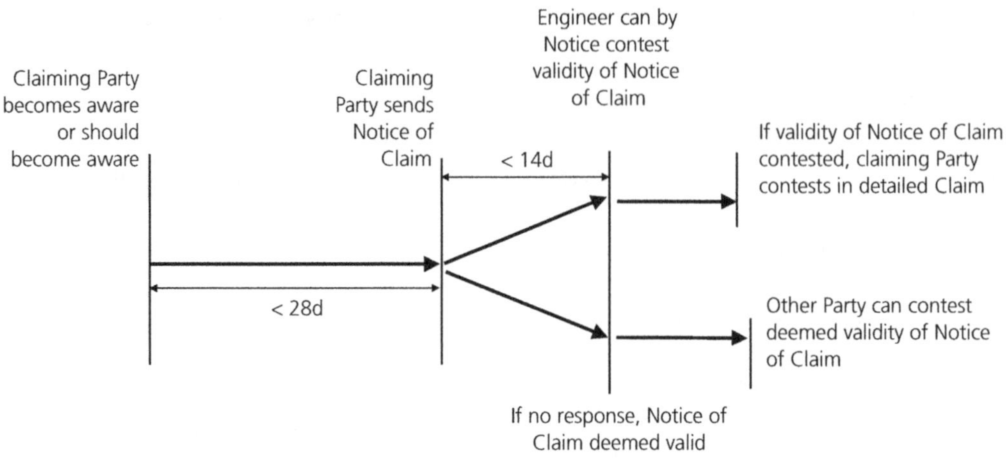

17.2. Response to the Notice of Claim

Within 14 days of receiving the Notice of Claim, if the Engineer (the other Party under SB 2017) considers that the Notice of Claim was not given within the period of 28 days, the Engineer (the other Party under SB 2017) must give a Notice, with reasons, to the claiming Party under Sub-Clause 20.2.2 [*Engineer's initial response*].

If the Engineer (the other Party under SB 2017) does not give such a Notice within this period of 14 days, the Notice of Claim is deemed to be a valid Notice.

If the other Party disagrees with such deemed valid Notice of Claim, the other Party must give a Notice to the Engineer with details of the disagreement. Thereafter, the agreement or determination of the Claim under Sub-Clause 20.2.5 [*Agreement or determination of the Claim*] must include a review by the Engineer of such disagreement.

If the claiming Party receives a Notice under Sub-Clause 20.2.2 from the Engineer (the other Party under SB 2017) and disagrees with the Engineer (the other Party under SB 2017) or considers there are circumstances which justify late submission of the Notice of Claim, the claiming Party must include in its fully detailed Claim under Sub-Clause 20.2.4 [*Fully detailed Claim*] details of the disagreement or why such late submission is justified.

17.3. Contemporary Records

Under Sub-Clause 20.2.3 [*Contemporary Records*], 'contemporary records' are described as 'records that are prepared or generated at the same time, or immediately after, the event or circumstance giving rise to the Claim'.

Such records carry more weight in the eyes of the Engineer, the DAAB and any arbitral tribunal than records compiled long after the event or circumstance. The success of any Claim will depend upon the accuracy, the completeness and the timeliness of records.

The claiming Party must keep those contemporary records that are necessary to substantiate the Claim.

Without admitting the Employer's liability, the Engineer (the Employer under SB 2017) may monitor the Contractor's contemporary records and/or instruct the Contractor to keep additional contemporary records. The Contractor must permit the Engineer (the Employer under SB 2017) to inspect all these records during normal working hours (or at other times agreed by the Contractor) and, if instructed, must submit copies to the Engineer (the Employer under SB 2017). Such monitoring, inspection or instruction (if any) does not imply acceptance of the accuracy or completeness of the Contractor's contemporary records.

Note:

There is a noticeable absence from Sub-Clause 20.2.3 [*Contemporary Records*] in that there is no mention of the Engineer or the Contractor being given an opportunity to monitor or inspect the records prepared by the Employer to substantiate its Claims. However, notwithstanding the absence of an express obligation, it is recommended that the Contractor should be given the opportunity to examine and agree such records and should be given copies if requested.

Although the Engineer (the Employer under SB 2017) is not obliged to monitor the Contractor's records or to specify any additional records that they might require, the Engineer (the Employer under SB 2017) should not hesitate to do so. If the Engineer (the Employer under SB 2017) does not contest the accuracy of the records at the time (or does not compile its own records which permit the Contractor's records to be contested at a later date), the DAAB or an arbitral tribunal will be entitled to rely on the Contractor's uncontested records. Similarly, the Engineer (the Employer under SB 2017) should not simply reject the Contractor's records as insufficient without using their power to specify the records required to be compiled.

In this respect, it should be borne in mind that an agreement on the facts as stated in the records, does not signify that the Engineer (the Employer under SB 2017) accepts that the Claim is valid, and that the Employer is liable. A simple way of ensuring this is for the Engineer to stamp or mark the agreed records as '*Agreed for record purposes only*'.

17.4. Fully detailed Claim

Sub-Clause 20.2.4 [*Fully detailed Claim*] states that a '*fully detailed Claim*' is a submission which includes:

(*a*) a detailed description of the event or circumstance giving rise to the Claim;
(*b*) a statement of the contractual and/or other legal basis of the Claim;
(*c*) all contemporary records on which the claiming Party relies; and
(*d*) detailed supporting particulars of the amount of additional payment claimed (or amount of reduction of the Contract Price in the case of the Employer as the claiming Party), and/or EOT claimed (in the case of the Contractor) or extension of the DNP claimed (in the case of the Employer).

Sub-paragraphs (a) and (b) address the 'principle' of the Claim, whereas (c) and (d) address the quantum of the Claim (that is, the amount of compensation sought).

Within 84 days of the date on which the claiming Party became aware, or should have become aware, of the event or circumstance giving rise to the Claim, the claiming Party must submit a fully detailed Claim to the Engineer (the Employer's Representative under SB 2017), unless a different period is proposed by the claiming Party and agreed by the Engineer (the other Party under SB 2017).

If, within this 84-day period (or other agreed period), the claiming Party fails to submit the statement under sub-paragraph (b) of Sub-Clause 20.2.4, the Notice of Claim is deemed to have lapsed and is no longer considered as a valid Notice.

Within 14 days of the expiry of the 84-day period (or other agreed period), the Engineer (the Employer's Representative under SB 2017) must give a Notice to the claiming Party recording the late submission. If the Engineer (the Employer's Representative under SB 2017) does not give such a Notice within this period of 14 days, the Notice of Claim is deemed not to have lapsed and continues to be a valid Notice.

If the other Party contests the deemed validity of the Notice of Claim, it must give a Notice to the Engineer (the Employer's Representative under SB 2017), giving details of the disagreement. Thereafter, the agreement or determination of the Claim under Sub-Clause 20.2.5 [*Agreement or determination of the Claim*] must include a review by the Engineer (the Employer's Representative under SB 2017) of the disagreement.

If the claiming Party receives from the Engineer (the Employer's Representative under SB 2017) such a Notice of late submission of the statement of contractual and/or legal basis, and if the claiming Party disagrees with the Notice or considers there are circumstances which justify late submission of the statement under sub-paragraph (b) of Sub-Clause 20.2.4, the fully detailed Claim, when submitted, must include details of the claiming Party's disagreement or why such late submission is considered justified.

This assumes that the fully detailed Claim is submitted separately from and after the submission of the statement under sub-paragraph (b). If the statement had been submitted as part of the fully detailed Claim, which is declared by Notice from the Engineer/Employer's Representative to have been submitted late, clearly it will be difficult, if not impossible, to include in the fully detailed Claim details of the claiming Party's disagreement with the Notice.

Note:

The Engineer/Employer's Representative can only issue a Notice of late submission if no statement under sub-paragraph (b) of Sub-Clause 20.2.4 is received within the 84 days or such other agreed period. If the statement is submitted on time, but without the information required under sub-paragraphs (a), (c) or (d) of Sub-Clause 20.2.4, the Engineer/Employer's Representative is not entitled to issue such a Notice.

Figure 17.2 Submission of Claim

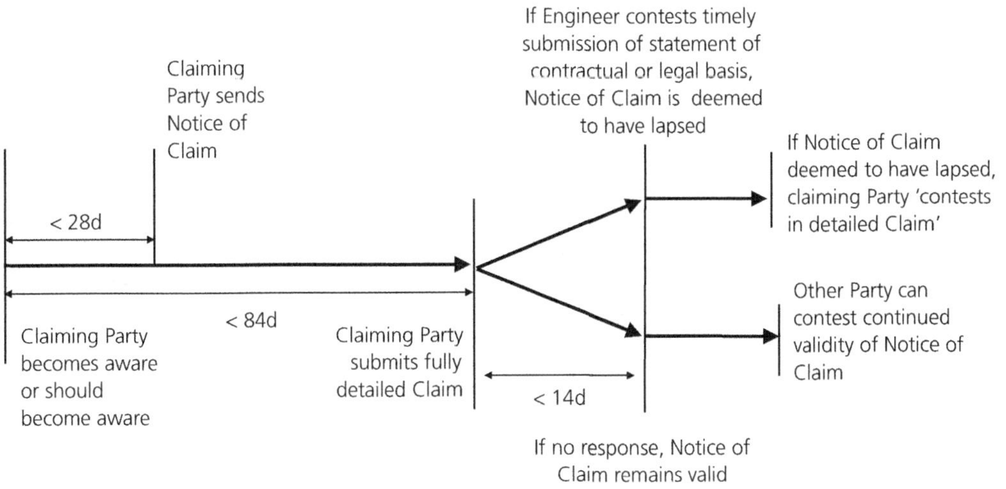

17.5. Claims of continuing effect

If the event or circumstance giving rise to a Claim is of the type set out under sub-paragraphs (a) or (b) of Sub-Clause 20.1 [*Claims*] and has a continuing effect, Sub-Clause 20.2.6 [*Claims of continuing effect*] states:

> (*a*) *the fully detailed Claim submitted under Sub-Clause 20.2.4 [Fully detailed Claim] is to be considered as interim;*
> (*b*) *in respect of this first interim fully detailed Claim, the Engineer (Employer's Representative under SB 2017) must give their response on the contractual or other legal basis of the Claim, by giving a Notice to the claiming Party. This Notice must be served within the time limit for agreement under Sub-Clause 3.7.3 [Time limits];*
> (*c*) *after submitting the first interim fully detailed Claim, the claiming Party must submit further interim fully detailed Claims at monthly intervals, giving the accumulated amount of additional payment or time claimed; and*
> (*d*) *the claiming Party must submit a final fully detailed Claim within 28 days of the end of the effects resulting from the event or circumstance, or within such other period as may be proposed by the claiming Party and agreed by the Engineer. This final fully detailed Claim shall give the total amount of additional payment or time claimed.*

Note:

Sub-Clause 20.2.6 [*Claims of continuing effect*] relates to events or circumstances which have a continuing effect and is not limited to events or circumstances which continue for a long period. An event might last for only one day, but the effects might be felt for years, in which case Sub-Clause 20.2.6 will apply. However, another event might last for four weeks but the impact might end as soon as the event ends, in which case, Sub-Clause 20.2.6 will not apply.

It is also to be noted that the period for the submission of the final Claim is calculated from the end of the effects of the event or circumstance, not from the end of the event or circumstance. In many cases, the effects will only end at a very late stage in the project. For example, the effects of an event which causes delay to completion of the Works will be felt up to the Taking Over of the Works, but it will also delay the release of the second half of the Retention, which will only fully occur after expiry of the DNP and the remedying of all defects notified to the Contractor during the DNP.

Finally, it should be noted that there is no mention of the claiming Party's Notice of Claim being deemed to have lapsed if the claiming Party fails to update the initial Claim at monthly intervals or to submit the final Claim within 28 days of the end of the effects of the event or circumstance.

Figure 17.3 Submission of Claim (ongoing effect)

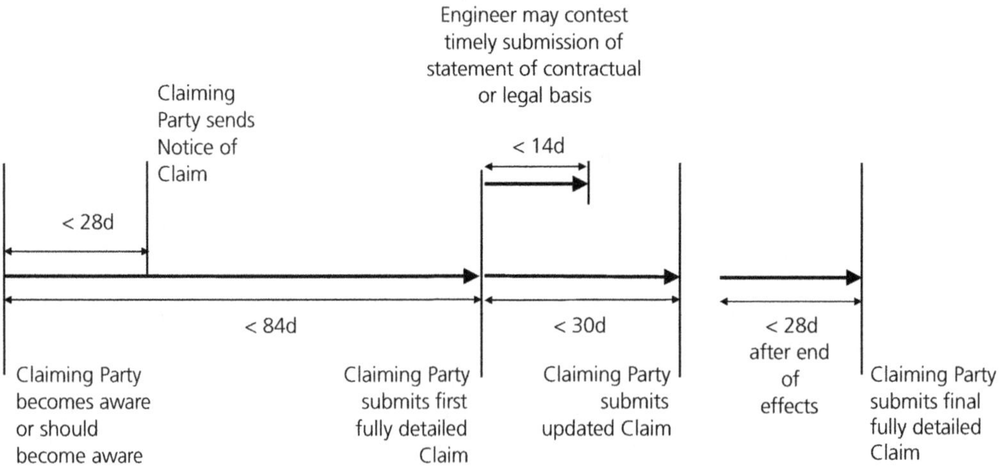

17.6. Agreement or determination of the Claim

After receiving a fully detailed Claim either under Sub-Clause 20.2.4 [*Fully detailed Claim*] or any interim or final fully detailed Claim under Sub-Clause 20.2.6 [*Claims of continuing effect*], the Engineer must proceed under Sub-Clause 20.2.5 [*Agreement or determination of the Claim*] and Sub-Clause 3.7 [*Agreement or Determination*] to agree or determine if the claiming Party is entitled to additional payment or time and, if so, how much. Under SB 2017, the Employer's Representative has this responsibility.

Such agreement or determination of the Claim is notwithstanding any Notice of late submission that might have been given under Sub-Clause 20.2.2 [*Engineer's initial response*] and/or under Sub-Clause 20.2.4 [*Fully detailed Claim*]. The agreement or determination of the Claim must include whether the Notice of Claim is to be treated as valid taking account of the details of the claiming Party's disagreement with the Notice(s) issued under Sub-Clause 20.2.2 [*Engineer's initial response*] and/or under Sub-Clause 20.2.4 [*Fully detailed Claim*] or why late submission is justified.

The circumstances which may be considered when assessing whether the Notice of Claim should be taken as valid or whether late submission is justified, may include:

- whether or to what extent the other Party would be prejudiced by acceptance of the late submission;
- in the case of the 28-day period under Sub-Clause 20.2.1 [*Notice of Claim*], any evidence of the other Party's prior knowledge of the event or circumstance giving rise to the Claim, which the claiming Party may include in its supporting particulars; and/or
- in the case of the 84-day period or other agreed period under Sub-Clause 20.2.4 [*Fully detailed Claim*], any evidence of the other Party's prior knowledge of the contractual and/or other legal basis of the Claim, which the claiming Party may include in its supporting particulars.

If, having received the fully detailed Claim under Sub-Clause 20.2.4 [*Fully detailed Claim*], or in the case of a Claim under Sub-Clause 20.2.6 [*Claims of continuing effect*] an interim or final fully detailed Claim (as the case may be), the Engineer requires additional particulars:

(*a*) they must promptly give a Notice to the claiming Party, describing the additional particulars and the reasons for requiring them;
(*b*) they must nevertheless give their response on the contractual or other legal basis of the Claim, by giving a Notice to the claiming Party, within the time limit for agreement under Sub-Clause 3.7.3 [*Time limits*] (Sub-Clause 3.5.3 under SB 2017), that is, within 42 days of receipt of the Claim;
(*c*) as soon as practicable after receiving the Notice under sub-paragraph (a) above, the claiming Party must submit the additional particulars; and
(*d*) the Engineer (Employer's Representative under SB 2017) must then proceed under Sub-Clause 3.7 [*Agreement or Determination*] to agree or determine the amount of any payment or time to which the claiming Party is entitled. For the purpose of Sub-Clause 3.7.3 [*Time limits*] (Sub-Clause 3.5.3 under SB 2017), the date the Engineer (Employer's Representative under SB 2017) receives the additional particulars from the claiming Party is the date of commencement of the time limit for agreement.

Figure 17.4 Engineer's processing of Claim

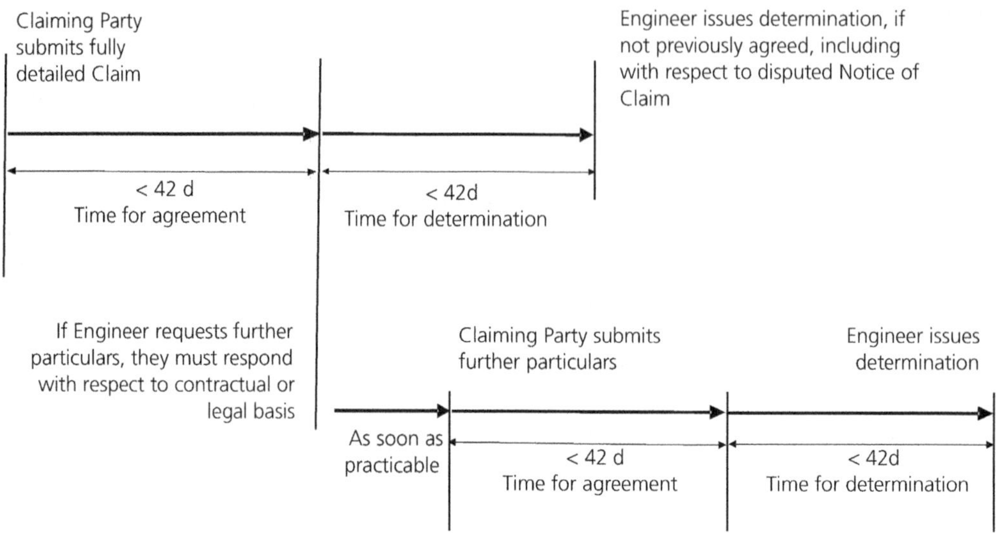

Figure 17.5 Engineer's processing of Claim (ongoing effect)

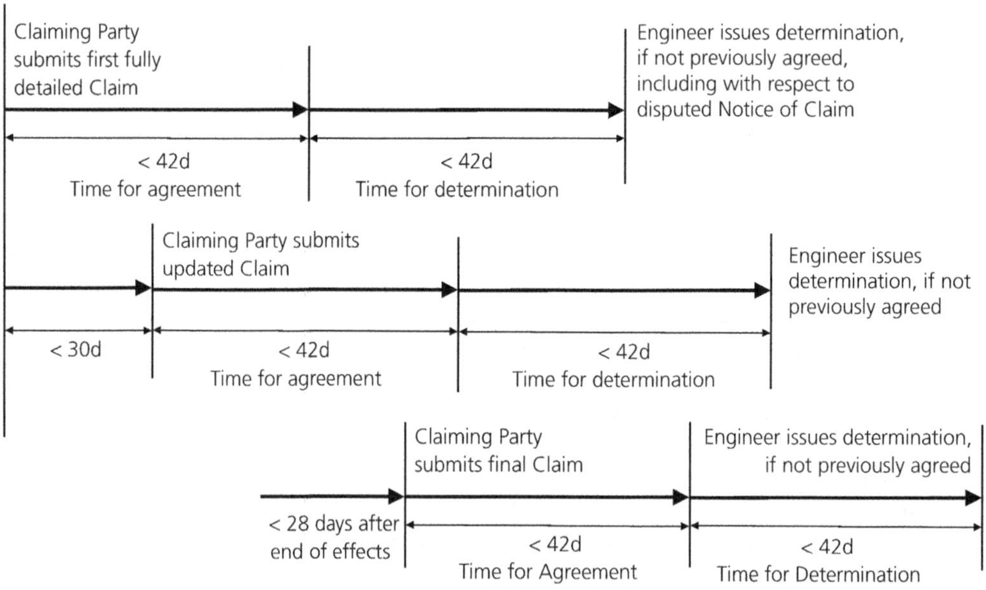

17.7. General requirements

After receiving the Notice of Claim, and until the Claim is agreed or determined under Sub-Clause 20.2.5 [*Agreement or determination of the Claim*], the Engineer must include in each Payment Certificate such amounts for any Claim as have been reasonably substantiated as due to the claiming Party. Similarly, under SB 2017, in each payment under Sub-Clause 14.7 [*Payment*], the Employer must include such amounts for any Claim as have been reasonably substantiated as due.

Thus, payment of amounts that have been reasonably substantiated as due cannot be postponed until the final amount of the Claim is agreed or determined. This obligation can be difficult for the Employer to respect in some countries where the Laws or the system require that any claim must be incorporated in a Contract Amendment before payment can be made.

The Employer is not entitled to claim any payment from the Contractor and/or to extend the DNP or set off against or make any deduction from any amount due to the Contractor, except in accordance with Sub-Clause 20.2 [*Claims for Payment and/or EOT*].

The requirements of Sub-Clause 20.2 [*Claims for Payment and/or EOT*] are in addition to those of any other applicable sub-clause. If the claiming Party fails to comply with this or any other sub-clause in relation to the Claim, any additional payment and/or any EOT (in the case of the Contractor as the claiming Party) or extension of the DNP (in the case of the Employer as the claiming Party) must take account of the extent to which the failure has prevented or prejudiced proper investigation of the Claim by the Engineer (Employer's Representative under SB 2017).

Smith G
ISBN 978-0-7277-6652-6
https://doi.org/10.1680/fcmh.66526.247

Chapter 18
Disputes

18.1. Background

Sub-Clause 1.1.29 RB 2017 & YB 2017 Sub-Clause 1.1.26 SB 2017	***'Dispute'*** *means any situation where:* *(a) one Party makes a claim against the other Party (which may be a Claim, as defined in the Conditions of Contract, or a matter to be determined by the Engineer under those Conditions, or otherwise);* *(b) the other Party (or the Engineer under Sub-Clause 3.7.2 [Engineer's Determination]) rejects the claim in whole or in part; and* *(c) the first Party does not acquiesce (by giving a NOD under Sub-Clause 3.7.5 [Dissatisfaction with Engineer's determination] or otherwise)* *provided however that a failure by the other Party (or the Engineer) to oppose or respond to the claim, in whole or in part, may constitute a rejection if, in the circumstances, the DAAB or the arbitrator(s), as the case may be, deem it reasonable for it to do so.*

Thus, in most cases, three requirements must be satisfied before a Dispute crystallises:

(i) one Party must have submitted a Claim or other matter for determination by the Engineer;
(ii) the Engineer's determination with respect to the Claim or other matter must be negative in the view of the claiming Party; and
(iii) the claiming Party must have served a NOD within 28 days of receipt of the Engineer's determination in accordance with Sub-Clause 3.7.5 [*Dissatisfaction with Engineer's determination*].

It should be noted that the 2022 Reprint has slightly modified this definition, to match the revised definition of Claim (see Chapter 19).

Once a Dispute has arisen, Sub-Clause 21.1 [*Constitution of the DAAB*] foresees that it is to be decided by a DAAB in accordance with Sub-Clause 21.4 [*Obtaining DAAB's Decision]*. However, the role of the DAAB is not only to decide on Disputes, but also to assist the Parties to avoid Disputes developing. Indeed '*DAAB*' stands for '*Dispute Avoidance/Adjudication Board*'.

Under the FIDIC 1999 Contracts, the equivalent name was 'Dispute Adjudication Board' (DAB). The DAB was also empowered to provide non-binding opinions as a way of preventing disputes developing, but FIDIC 2017 Contracts place greater emphasis on this dispute avoidance role.

18.2. DAAB appointment

Sub-Clause 21.1 [*Constitution of the DAAB*] requires the Parties to jointly appoint the member(s) of the DAAB within the period stated in the Contract Data after the date the Contractor receives the Letter of Acceptance. If no period is stated in the Contract Data, the appointments are to be made within 28 days of receipt of the Letter of Acceptance.

In practice, this requirement is not always met by the Parties, either because they are unable to identify suitable candidates and reach agreement with them within the specified period, or because one or both Parties decide not to proceed with the appointment. However, a failure to constitute the DAAB within the stated period does not release the Parties from the obligation to appoint a DAAB.

The DAAB must comprise either one member (the 'sole member') or three members (the 'members'). The number must be stated in the Contract Data; if it is not, and the Parties do not agree otherwise, the DAAB must have three members.

The Contract Data should also include a list of six names, from which the sole member or three members must be selected, unless those listed are unable or unwilling to accept appointment to the DAAB, in which case, the Parties must look for potential members outside the list. The Parties must also seek potential members if the Contract Data does not contain the list of names.

If the DAAB is to comprise three members, each Party must select one member for the agreement of the other Party. The person selected does not represent the Party which selected them. Once each Party has proposed a member who has been accepted by the other Party, the Parties are to seek assistance from the two members with respect to the appointment of the third member, who will act as chairperson.

18.3. Replacement of DAAB member

If at any time the Parties so agree, they may replace one or more members of the DAAB. Unless the Parties agree otherwise, a replacement DAAB member must be appointed if a member declines to act or is unable to act as a result of death, illness, disability, resignation or termination of appointment. The replacement member is to be appointed in the same manner as the replaced member, that is, proposed by the same Party which proposed the replaced member, or proposed by the two members if the replaced member was the chairperson.

The appointment of any member may only be terminated by mutual agreement of both Parties, but not by one of the Parties acting alone.

18.4. Failure to agree on appointment

Sub-Clause 21.2 [*Failure to Appoint DAAB Member(s)*] provides that if the Parties are unable to agree on the appointment of the DAAB member(s) or their replacement, the appointing entity or official named in the Contract Data is to appoint the member(s) of the DAAB at the request of either or both Parties and after due consultation with both Parties.

The circumstances which entitle a Party to request the appointing entity or official to act are as follows:

(*a*) if the DAAB is to comprise a sole member, the Parties fail to agree the appointment of this member within the period fixed by the Contract; or (*b*) if the DAAB is to comprise three persons and if within the period fixed by the Contract:

(i) either Party fails to select a member for agreement by the other Party;
(ii) either Party fails to agree a member selected by the other Party; and/or
(iii) the Parties fail to agree the appointment of the third member (to act as chairperson) of the DAAB;

(*c*) the Parties fail to agree the appointment of a replacement within 42 days of the date on which the sole member or one of the three members declines to act or is unable to act as a result of death, illness, disability, resignation or termination of appointment; or
(*d*) if, after the Parties have agreed the appointment of the member(s) or replacement, one Party refuses or fails to sign a DAAB Agreement with any such member or replacement (as the case may be).

The President of FIDIC is often named in the Contract Data as the appointing entity. The FIDIC website has a page dedicated to this service and includes a form to be completed by a Party (or Parties) seeking the appointment of a DAAB member. When submitting the form, payment is to be made of the appointment fee which, in early 2023, was €3000 for the appointment of a member. Each Party is responsible for paying half of the remuneration of the appointing entity or official. If the Contractor pays the amount in full, the Contractor must include one-half of the amount in a Statement and the Employer must then pay the Contractor in accordance with the Contract. If the Employer pays the remuneration in full, the Engineer must include one-half of the amount as a deduction under sub-paragraph (b) of Sub-Clause 14.6.1 [*The IPC*].

If the President of FIDIC is the appointing entity, the member(s) will be chosen from the '*FIDIC President's List of Approved Dispute Adjudicators*', details of which are also available on the FIDIC website.

Such an appointment is final and conclusive. Thereafter, the Parties and the member(s) so appointed are deemed to have signed and are bound by a DAAB Agreement (see below) under which the remuneration shall be as stated in the terms of the appointment, and the law governing the DAAB Agreement will be the governing law of the Contract. This means that the person(s) so named is/are appointed to the DAAB even without a signed DAAB Agreement, and the Parties are obliged to pay the DAAB member(s) at the rates fixed by the appointing entity or official.

It is stated on the FIDIC website that FIDIC does not recommend fee scales for DAAB members. Nevertheless, FIDIC states that when it is required to fix the fees, it will do so taking due regard of all relevant circumstances (including international practices). As an indication of an acceptable level of fees, FIDIC refers to the 'Memorandum on the Fees and Expenses for ICSID Arbitrators' published by the World Bank's International Centre for Settlement of Disputes (ICSID). The 2022 edition of the ICSID document specifies an hourly rate of US$500 for working time, US$250 for travel time, air fares at one class above Economy Class, and a per diem of US$900 when an overnight stay is required.

18.5. DAAB Agreement

After the Parties have agreed on the choice of the sole member or three members, they must both sign a DAAB Agreement with each of the members. A model DAAB Agreement is provided with each of the

FIDIC 2017 Contracts, and it incorporates by reference the General Conditions of DAAB Agreement set out in the Appendix to the General Conditions of Contract.

The DAAB is deemed to be constituted on the date that the Parties and the sole member or the three members of the DAAB have all signed a DAAB Agreement. For a sole-member DAAB, there will be one DAAB Agreement, whereas for a three-member DAAB, there will be three DAAB Agreements.

The remuneration of the member or members (including the remuneration of any expert whom the DAAB consults) is to be mutually agreed by the Parties when agreeing the terms of each DAAB Agreement. Each Party is responsible for paying one-half of this remuneration.

In reality, at the time of signing the DAAB Agreements, it will be unlikely that the DAAB will have decided to consult an expert, as the need for such an expert will only become apparent if and when a Dispute develops.

18.6. DAAB Procedural Rules

A set of DAAB Procedural Rules is annexed to the General Conditions of the DAAB Agreement. The purpose of the Procedural Rules is described under Rule 1 as:

(*a*) to facilitate the avoidance of Disputes that might otherwise arise between the Parties; and
(*b*) to achieve the expeditious, efficient and cost-effective resolution of any Dispute that arises between the Parties.

The Rules are to be interpreted, the DAAB's Activities are to be conducted and the DAAB is to use its powers under the Contract and the Rules, in the manner necessary to achieve these stated objectives.

The DAAB Procedural Rules will be examined in detail below.

18.7. Notice of Effective Date

Although Sub-Clause 21.1 [*Constitution of the DAAB*] and Sub-Clause 2.1 [*General Provisions*] of the DAAB Agreement state that the DAAB is deemed to be constituted and that the DAAB Agreement takes effect on the date that the Parties and the Members all sign a DAAB Agreement (the '*Effective Date*'), one Party or both Parties must give a Notification to each DAAB Member immediately after the Effective Date confirming that the DAAB Agreement has come into force. If no such Notification is received by the DAAB Member within 182 days of the Effective Date, the DAAB Agreement is void and ineffective.

A Notification is defined by Sub-Clause 1.8 of the General Conditions of DAAB Agreement to be a notice in writing which shall either be a paper original signed by the appropriate person or an electronic original transmitted by the system agreed between the Parties and the DAAB to the electronic address assigned to the appropriate person.

In practice, the Parties often forget to issue such a Notification, in which case, the DAAB Member should remind them of the need to do so.

18.8. Warranties/Declaration of Independence

Under Sub-Clause 2.3 of the General Conditions of DAAB Agreement, the employment of the DAAB Member is stated to be a personal appointment and no assignment, subcontracting or delegation of the DAAB Member's rights and obligations is permitted.

Under Sub-Clause 3 [*Warranties*] of the General Conditions of DAAB Agreement, the DAAB Member warrants and agrees that they are and will remain at all times during the life of the DAAB, impartial and independent of the Employer, the Contractor, the Employer's Personnel and the Contractor's Personnel.

If, after the DAAB Agreement is signed, or deemed to be signed, the DAAB Member becomes aware of any fact or circumstance:

- which might lead to questions about their independence or impartiality; and/or
- which might be, or appear to be, inconsistent with their impartiality and independence,

the DAAB Member undertakes to immediately disclose this in writing to the Parties and the Other Members (if any).

When appointing the DAAB Member, each Party relies on the DAAB Member's representations that they are:

(*a*) experienced and/or knowledgeable in the type of work which the Contractor is to carry out under the Contract;
(*b*) experienced in the interpretation of construction and/or engineering contract documentation; and
(*c*) fluent in the language for communications stated in the Contract Data (or the language agreed between the Parties and the DAAB).

In practice, some flexibility is often needed with respect to these three characteristics. For example, if the work to be carried out under the Contract is mainly concerned with the design, fabrication and installation of an automated baggage-handling system in an airport, it might be difficult to find experienced DAAB practitioners with detailed knowledge of such systems and the Parties may have to accept DAAB members who have general experience of conveyor systems or of mechanical plant.

Clause 4 [*Independence and Impartiality*] of the General Conditions of DAAB Agreement details seven obligations with respect to the DAAB Members' independence and impartiality.

Some of the obligations are more onerous than the equivalent provisions contained in the FIDIC 1999 Contracts. For example, in the 10 years before signing the DAAB Agreement, the DAAB Member must not have been employed as a consultant or otherwise by the Employer, the Contractor, the Employer's Personnel or the Contractor's Personnel. The equivalent provision under the FIDIC 1999 Contracts merely required the DAAB Member to disclose any previous employment as consultant or otherwise, before signing the DAAB Agreement. If the 10-year threshold is applied with respect to previous DAAB appointments on projects involving the same Employer, Engineer or Contractor, this could significantly limit the activities of some DAAB Members. Indeed, it is natural for a Party which has been satisfied by the performance of a DAAB Member on one project to wish to use the same person as a DAAB Member for other projects. Fortunately, a separate obligation is to disclose in writing to the Employer, the Contractor and the Other Members of the DAAB, before the DAAB Agreement is signed or deemed to be signed, any existing and/or past professional or personal relationships with any director, officer or employee of the Employer, the Contractor, the Employer's Personnel or the Contractor's Personnel (including as a dispute resolution practitioner on another project). This implies that a previous DAAB appointment even within the 10-year period should not constitute a conflict of interest, provided that it is disclosed.

It should be noted that the 2022 Reprint reduces the 10-year period to 5 years and this period does not apply if the DAAB member discloses to the Parties the previous relationship (see Chapter 19).

18.9. General Obligations

Under Clause 5 [*General Obligations of the DAAB Member*] of the General Conditions of DAAB Agreement, the DAAB Member undertakes:

- to comply with the DAAB Agreement, the DAAB Rules and relevant provisions of the Works Contract;
- not to give advice to either of the Parties except as part of the DAAB role;
- to be available as required, for Site visits, meetings, hearings and so forth;
- to provide Informal Assistance when requested jointly by the Parties; and
- to become and remain knowledgeable about the Contract and the Works.

('Informal Assistance' is defined in the General Conditions of DAAB Agreement as informal assistance given by the DAAB to the Parties when requested jointly by the Parties under Sub-Clause 21.3 [*Avoidance of Disputes*].)

Under Clause 6 [*General Obligations of the Parties*] of the General Conditions of DAAB Agreement, each Party undertakes:

- to comply with the DAAB Agreement, the DAAB Rules and relevant provisions of the Works Contract;
- to cooperate with the other Party in appointing the DAAB without delay;
- to cooperate in good faith with the DAAB; and to fulfil its duties, and exercise any right or entitlement, under the Contract, the General Conditions of DAAB Agreement and the DAAB Rules and/or otherwise, to facilitate the avoidance of Disputes that might otherwise arise;
- to achieve the expeditious, efficient and cost-effective resolution of any Dispute that does arise;
- not to request advice from or consultation with the DAAB Member regarding the Contract, except as required for the DAAB Member to carry out the DAAB's Activities;
- not to compromise the DAAB's warranty of independence and impartiality;
- to ensure that the DAAB Member remains informed as is necessary to enable them to comply with their obligations.

Under Clause 7 [*Confidentiality*] of the General Conditions of DAAB Agreement, the Parties and the DAAB Member (and any replacement DAAB Member) agree that the details of the DAAB's activities and any documents provided to the DAAB must be treated as private and confidential except to the extent that the information contained in the documents was already in the possession of the other Party prior to its receipt under the DAAB Agreement, or that it became generally available to the public by other means, or that it was obtained from any third party not bound by the obligation of confidentiality.

18.10. Indemnities

Under Clause 8 [*The Parties' Undertaking and Indemnity*] of the General Conditions of DAAB Agreement, the Employer and the Contractor agree that the DAAB Member will not:

- be appointed as an arbitrator in any arbitration under the Contract;
- be called as a witness to give evidence concerning any Dispute in any arbitration under the Contract; or
- be liable for any claims for anything done or omitted to be done by the DAAB Member, except in any case of fraud, gross negligence, deliberate default or reckless misconduct by them.

The Employer and the Contractor must jointly and severally indemnify and hold the DAAB Member harmless with respect to any claim from which the DAAB member is so relieved from liability.

18.11. Remuneration

Clause 9 [*Fees and Expenses*] of the General Conditions of DAAB Agreement sets out the terms of payment to the DAAB Member. The payments are composed of four elements:

(i) a monthly fee, sometimes referred to as a 'retainer', which is to cover time spent in studying monthly reports and so forth provided by the Parties, general office expenses and being available on 28 days' notice for Site visits and hearings and on shorter notice in the case of urgency (3 days for virtual meetings and 14 days for Site visits);
(ii) a daily fee for time spent dealing with requests for decisions, Informal Assistance, meetings, Site visits and travelling time. Such travel time is chargeable at the daily rate for each day or part of a day, up to a maximum of two days in each direction, between the home of the DAAB Member and the Site or the location of any meeting;
(iii) travel expenses and other expenses reasonably incurred (such as telephone calls, courier charges, etc.). Such travel expenses include business-class air fares, hotel and subsistence, other direct travel expenses such as taxi fares and visa fees. Receipts must be provided by the DAAB Member for each item;
(iv) reimbursement or allowance for any taxes levied in the Country with respect to the fees and/or expenses, except if the DAAB Member is a national or permanent resident of the Country.

The monthly fee is payable with effect from the last day of the month in which the Effective Date occurs until the end of the month in which the Term of the DAAB expires or the DAAB declines to act or is unable to do so. It should be noted that under RB 1999, the monthly fee is reduced by 50% during the DNP and under the Pink Book by 33.3%, whereas under the FIDIC 2017 Contracts there is no reduction. However, the Parties and the DAAB Member may agree to modify this, particularly if the Works have been completed without any indication of an ongoing or major dispute.

Invoices for the monthly fee and air fares are to be submitted by the DAAB Member quarterly in advance. If air fares are to be invoiced in advance, they must be based on the estimated cost of the air fare and any over-estimate or under-estimate must be considered in the next quarterly invoice. Alternatively, the dates of Site visits should be agreed more than three months in advance, so that the DAAB members may purchase their tickets and include the amount in their quarterly invoices.

Invoices for the daily fee and other expenses are to be submitted after the meeting, Site visit, hearing, issue of a decision or any note with respect to Informal Assistance. Invoices must contain an outline of activities covered by the invoice. They must be addressed to the Contractor who must make payment in full within 28 days of receipt. The Contractor must include one-half of the amounts paid in its Statements submitted under the Contract. In this way, the Employer pays its share of the DAAB costs.

If the DAAB Member does not receive full payment within the 28-day period, the DAAB Member must inform the Employer who is to promptly pay the amount due. The Employer will then be entitled to claim payment of its share, together with any costs reasonably incurred to recover this share as well as financing charges calculated in accordance with Sub-Clause 14.9 [*Delayed Payment*] of the Conditions of Contract.

If the DAAB Member has not received full payment within 56 days of submission of the invoice, they may suspend services after giving 7 days' notice to the Parties and Other Members and/or resign by giving a Notification of at least 28 days.

18.12. Failure to agree fees

If the Parties and the DAAB Member have agreed all other terms of the DAAB Agreement but are unable to agree the amount of the monthly fee or the daily fee to be inserted in the DAAB Agreement:

(*a*) the DAAB Agreement is deemed to have been signed by the Parties and the DAAB Member, except that the '*non-agreed fee*' proposed by the DAAB Member applies temporarily;
(*b*) either Party or the DAAB Member may apply to the appointing entity or official named in the Contract Data to set the amount of the '*non-agreed fee*';
(*c*) as soon as practicable and in any case within 28 days of receiving any such application, the appointing entity or official will fix the amount of the '*non-agreed fee*', which must be reasonable taking account of the complexity of the Works, the experience and qualifications of the DAAB Member, and all other relevant circumstances;
(*d*) once the appointing entity or official has set the amount of the '*non-agreed fee*', this amount will be final and conclusive, will replace the '*non-agreed fee*' applied temporarily and will be applied from the Effective Date; and
(*e*) thereafter, after giving credit to the Parties for all amounts previously paid in respect of the '*non-agreed fee*', the balance (if any) due from the DAAB Member to the Parties or from the Parties to the DAAB Member is to be paid.

18.13. Resignation of the DAAB Member

The DAAB Agreement can be ended under Clause 10 [*Resignation and Termination*] of the General Conditions of DAAB Agreement for convenience or for default.

The DAAB Member may resign at any time for any reason, by giving a Notification of not less than 28 days (or a shorter period if agreed with the Parties) to the Parties and to the Other Members (if any). During this notice period the Parties must appoint, or have appointed by the appointing entity, a replacement DAAB Member.

On expiry of the notice period, the DAAB Agreement of the resigning DAAB Member terminates with immediate effect except if, on the date of the DAAB Member's Notification, the DAAB is dealing with any request for a decision under Sub-Clause 21.4 [*Obtaining DAAB's Decision*]. In this case, the DAAB Member's resignation does not take effect until after the DAAB has given all the corresponding decisions in accordance with the Contract, unless the resignation arises from an inability to act through illness or disability.

If either Party fails, without justifiable excuse, to pay the DAAB Member in accordance with the DAAB Agreement, the DAAB Member may, without prejudice to their other rights or remedies, terminate the DAAB Agreement by giving a Notification to the Parties. This notice takes effect when received by both Parties.

18.14. Termination of the DAAB Agreement

Under Sub-Clause 10.3 of the General Conditions of DAAB Agreement, the Parties may jointly terminate the DAAB Agreement at any time by giving a Notification of not less than 42 days to the DAAB Member. However, neither the Employer nor the Contractor, acting without the other, can terminate the DAAB Agreement.

If the DAAB Member fails, without justifiable reason, to comply with the General Conditions of DAAB Agreement, the DAAB Procedural Rules or the Conditions of Contract that are relevant to the DAAB's Activities, the Parties may jointly terminate their DAAB Agreement by giving a Notification to the DAAB Member. This Notification (which must be sent by recorded delivery) takes effect when it is received by the DAAB Member. It is without prejudice to the other rights or remedies of the Parties.

18.15. Effects of resignation or termination

Subject to sub-paragraph (b) of Sub-Clause 11.5 which deals with a challenge for lack of independence or impartiality (see below), in the event of resignation or termination, the DAAB Member is nevertheless entitled to payment of any fees and/or expenses under the DAAB Agreement that remain outstanding at the date of termination of the DAAB Agreement.

After resignation by the DAAB Member or termination of the DAAB Agreement under this Clause, the DAAB Member:

- remains bound by their obligation of confidentiality under Sub-Clause 7.1 of the General Conditions of DAAB Agreement; and
- must return the original of any document in their possession to the Party who submitted the document, but only at that Party's written request and cost.

Termination of the DAAB Agreement requires no other action by the Parties or the DAAB Member, except as imposed by the governing law of the DAAB Agreement.

18.16. Challenges

New to FIDIC 2017 is a procedure which permits either Party to raise an objection against a DAAB Member for an alleged lack of independence or impartiality or otherwise. In such a case, the procedure is set out in Clause 11 [*Challenge*] of the General Conditions of DAAB Agreement, and Rule 10 [*Objection Procedure*] and Rule 11 [*Challenge Procedure*] of the DAAB Procedural Rules.

Under FIDIC 1999 Contracts, it sometimes happened that a Party who was dissatisfied with a DAB decision would seek to replace one or more members of the DAB but was unable to do so without the agreement of the other Party. Therefore, to limit the possibility for abuse of the new system in which the other Party is not required to agree to an objection, FIDIC has compiled a two-stage process.

The first stage deals with the initial objection and is set out in Procedural Rule 10.

18.16.1 Rule 10 Objection Procedure

The objecting Party must give a Notification to the DAAB Member of its objection within 7 days of becoming aware of the facts and/or events giving rise to the objection.

The Notification must be simultaneously copied to the other Party and the Other Members. It must:

(i) state that it is given under Rule 10;
(ii) state the reason(s) for the objection;
(iii) substantiate the objection by setting out the facts and describing the events, on which the objection is based; and
(iv) provide supporting particulars.

Within 7 days of receiving the Notification, the objected DAAB Member must respond to the objecting Party, with simultaneous copies to the other Party and the Other Members. If no response is given by the DAAB Member within this period of 7 days, the DAAB Member is deemed to have given a response denying the allegation on which the objection is based.

Within 7 days after receiving the response or deemed response, the objecting Party may formally challenge a DAAB Member in accordance with Rule 11 (see below).

If the objecting Party does not raise a formal challenge within this 7-day period after receiving the response or deemed response, the objecting Party is deemed to have agreed to the DAAB Member remaining on the DAAB and shall not be entitled to object to and/or challenge them again thereafter based on the same facts and/or evidence relied upon with respect to this objection.

18.16.2 Rule 11 Challenge Procedure

If the objecting Party challenges a DAAB Member, within 21 days of learning of the facts upon which the challenge is based, the provisions of Rule 11 apply in addition to those set out in Rule 10 (see above). Any challenge is to be decided by the International Chamber of Commerce (ICC) and administered by the ICC International Centre for ADR. Details of the procedure and associated costs are available on the websites of both FIDIC (http://fidic.org) and the ICC (http://iccwbo.org).

Under Sub-Clause 11.2 of the General Conditions of DAAB Agreement, any decision issued under Rule 11 of the DAAB Rules (the '*Decision on the Challenge*') is final and conclusive.

At any time before the Decision on the Challenge is issued, the challenged DAAB Member may resign under Sub-Clause 10.1 of the General Conditions of DAAB Agreement, and the resignation takes effect immediately, even if the DAAB was dealing with a Dispute at the time of the resignation. Following the resignation, the challenging Party must inform the ICC.

However, unless the challenged DAAB Member has resigned, or their DAAB Agreement has been terminated by the Parties acting jointly under Sub-Clause 10.3 of the General Conditions of DAAB Agreement, the DAAB Member and the Other Members (if any) continue with the DAAB's Activities until the Decision on the Challenge is issued.

If the Decision on the Challenge is that the challenge is successful, the appointment and the DAAB Agreement of the challenged DAAB Member are deemed to have been terminated on the date of the notification of the Decision on the Challenge by the ICC.

In such a case, the challenged DAAB Member will not be entitled to any fees or expenses from the date of the notification of the Decision on the Challenge by ICC. Any decision issued by the DAAB under Sub-Clause 21.4.3 [*The DAAB's decision*] of the Conditions of Contract, after the challenge was referred to the ICC and before the resignation of the challenged DAAB Member or termination of their DAAB Agreement, becomes void and without effect.

In the case of a sole-member DAAB, all other DAAB Activities during this period also become void and ineffective. In the case of a three-member DAAB, all other DAAB Activities during this period remain unaffected by the Decision on the Challenge except if there has been a successful challenge to all three members of the DAAB.

The successfully challenged DAAB Member will be removed from the DAAB, and the Parties must appoint without delay a replacement DAAB Member in accordance with Sub-Clause 21.1 [*Constitution of the DAAB*] of the Conditions of Contract.

Figure 18.1 Objection and Challenge Procedures

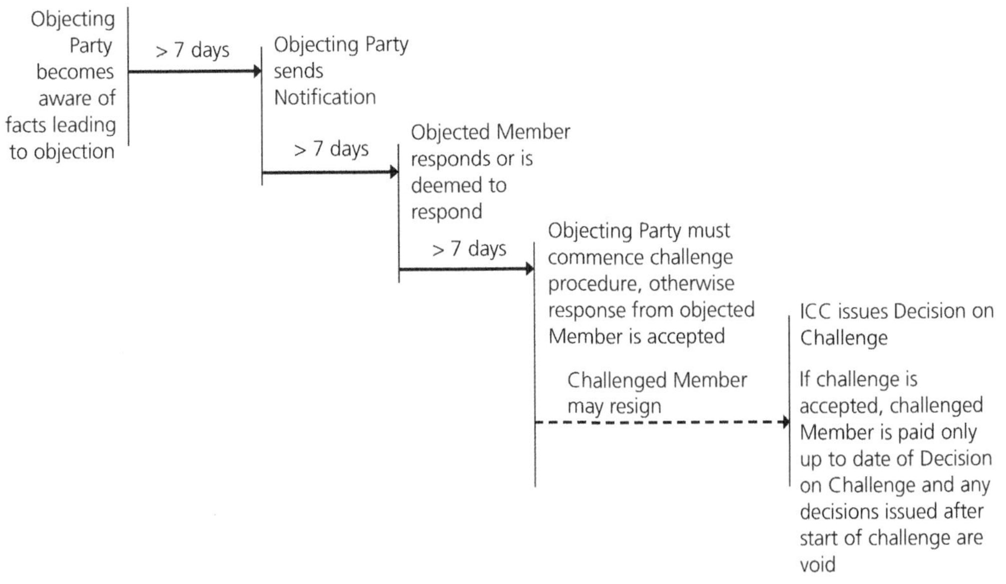

18.17. Duration of DAAB Agreement

Sub-Clause 21.1 [*Constitution of the DAAB*] of the General Conditions states that unless otherwise agreed by the Parties, the role of the DAAB expires (referred to as the '*Term of the DAAB*') at the later of the following dates:

(*a*) the date when the Contractor's discharge takes effect, or is deemed to take effect under Sub-Clause 14.12 [*Discharge*]; or
(*b*) the date 28 days after the DAAB has given its decision on all Disputes, referred to it under Sub-Clause 21.4 [*Obtaining DAAB's Decision*] before the discharge takes effect.

In other words, the DAAB is to be available to decide Disputes until agreement is reached with respect to Final Payment after the end of the DNP and the remedying of notified defects.

However, if the Contract is terminated, the Term of the DAAB (including the appointment of each member) expires 28 days after the first of the following dates:

(*a*) the date when the DAAB has given its decision on all Disputes which were referred to it under Sub-Clause 21.4 [*Obtaining DAAB's Decision*] within 224 days of the date of termination; or
(*b*) the date at which the Parties reach a final agreement on all matters (including payment) in connection with the termination.

The FIDIC 2017 Contracts Guide [2022] explains the period of 224 days as:

- 28 days for submission of the Notice of Claim (starting at the date of termination);
- 56 days for submission of the fully detailed Claim (which is 84 days from the date of termination);
- 28 days for the Engineer to request further particulars;
- 84 days for the Engineer to issue their determination under Sub-Clause 3.7 [*Agreement or Determination*];
- 28 days for submission of a NOD.

Note:

Within this breakdown, there is no mention of the Contractor's response to the Engineer's request for further particulars. It must be assumed that the Engineer's request and the Contractor's response together take no more than 28 days.

18.18. Communications

According to Procedural Rule 4, all written and oral communications between the DAAB and the Parties, including reports and decisions issued by the DAAB and during all Site visits, meetings and hearings relating to the DAAB's Activities, are to be in the language for communications defined in Sub-Clause 1.4 [*Law and Language*] of the Conditions of Contract, unless otherwise agreed jointly by the Parties and the DAAB. This requirement applies equally to communications between the DAAB Members.

All communications and/or documents sent between the DAAB and a Party must simultaneously be copied to the other Party. In the case of a three-member DAAB, the sending Party must send all communications and/or documents to the chairperson of the DAAB, copied simultaneously to the Other Members.

To enable the DAAB to become and remain informed about the performance of the Contract and the progress of the Works, Procedural Rule 4.3 requires the Parties to provide the DAAB with a copy of all documents which the DAAB may request, including:

(*a*) the Contract documents;
(*b*) the Contractor's progress reports issued under Sub-Clause 4.20 [*Progress Reports*] of the Conditions of Contract;
(*c*) the initial programme and each revised programme under Sub-Clause 8.3 [*Programme*] of the Conditions of Contract;
(*d*) relevant instructions given by the Engineer, and Variations under Clause 13.3 [*Variation Procedure*] of the Conditions of Contract;
(*e*) Statements submitted by the Contractor, and all certificates issued by the Engineer under the Contract (Advance Payment Certificate, Interim Payment Certificates, Final Payment Certificate, Taking-Over Certificate(s), Performance Certificate, etc.);
(*f*) Notices;
(*g*) relevant communications between the Parties and between either Party and the Engineer;

and any other relevant document which is necessary for the DAAB to fulfil its obligations under Sub-Clause 5.1 (d) of the General Conditions of DAAB Agreement.

Hint:

The Parties pay the DAAB Members a monthly fee which covers, among other things: '*becoming and remaining knowledgeable about the Contract, informed about the progress of the Works and maintaining a current working file of documents*'. If the Parties do not provide the listed documents regularly and diligently, the DAAB Members are paid for a service which they are prevented from properly providing. Moreover, if the DAAB visits the Site without being given prior knowledge of the situation, the visit will be less efficient and less productive than it would have been if the DAAB was fully informed in advance.

Therefore, from the outset, the Parties should put in place a system which ensures that the required documents are regularly made available to the DAAB, either by a file sharing system or otherwise. Named individuals within the Contractor's and the Engineer's staff should be given the responsibility for ensuring that the documents are made available.

18.19. Site visits and meetings

One of the most important aspects of the DAAB's role is to meet the Parties face to face and/or visit the Site. Such meetings and visits are to take place at regular intervals and/or at the written request of either Party.

Procedural Rule 2 states that the purpose of the meetings and Site visits is to enable the DAAB to:

(*a*) become and remain informed about:

(i) the Parties' performance of the Contract;
(ii) the Site and its surroundings; and
(iii) the progress of the Works (and of any other parts of the project of which the Contract forms part);

(*b*) become aware of, and remain informed about, any actual or potential issue or disagreement between the Parties; and
(*c*) give Informal Assistance if and when jointly requested by the Parties.

The frequency of such meetings and/or Site visits is fixed by Procedural Rule 3.3 to be at intervals of not less than 70 days and not more than 140 days unless otherwise agreed jointly by the Parties and the DAAB, except as required to conduct a hearing in relation to a Dispute or in response to an urgent matter.

According to Procedural Rule 3.2, as soon as practicable after the DAAB has been appointed, the DAAB must convene its first face-to-face meeting with the Parties. At this meeting, the DAAB is to establish a preliminary schedule for future meetings and Site visits in consultation with the Parties.

In addition to the face-to-face meetings, Procedural Rule 3.4 permits the DAAB, with the agreement of the Parties, to also hold meetings by telephone or video conference (in which case, each Party bears the risk of interrupted or faulty telephone or video conference transmission and reception).

Under Procedural Rule 3.7, the date, time and agenda for each meeting and Site visit is to be set by the DAAB in consultation with the Parties. This is slightly different from the equivalent provision in the FIDIC 1999 Contracts (Procedural Rule 2) which empowered the DAB to decide upon the timing and agenda of all visits in the absence of agreement among the Parties and the DAB. In other words, the FIDIC 2017 Contracts have removed the power of the DAAB to fix timing and agenda of visits (and meetings) in the absence of agreement of both Parties. This is unfortunate, as experience under the FIDIC 1999 Contracts shows that the Parties are not always able to agree on the dates of visits and/or the agenda.

At times of critical construction events (which may include periods of suspension of the Works or termination of the Contract), Procedural Rule 3.5 imposes on the DAAB an obligation to visit the Site at the written request of either Party. This request must describe the critical construction event. In addition, if the DAAB becomes aware of a pending critical construction event, it may invite the Parties to make such a request. However, the DAAB has no power to unilaterally impose such a visit (see Procedural Rule 3.7 above).

Under Procedural Rule 3.6, either Party may request an urgent meeting or Site visit by the DAAB. The request must be in writing and must describe the reasons for the urgency of the meeting or Site visit. If the DAAB agrees that such a meeting or Site visit is urgent, the DAAB must use all reasonable endeavours to:

(*a*) hold a meeting with the Parties by telephone or video conference within 3 days of receiving the request; and (*b*) if requested, visit the Site within 14 days of the date of this meeting, provided that, after giving the other Party the opportunity to respond to or oppose this request, the DAAB agrees that a Site visit is necessary.

Under Procedural Rule 3.8, all such meetings with the DAAB and Site visits by the DAAB must be attended by the Employer, the Contractor and the Engineer.

Under Procedural Rule 3.9, the Contractor must coordinate the meetings and Site visits with the co-operation of the Employer and the Engineer. In this respect, the Contractor must provide:

(*a*) personal safety equipment, security controls (if necessary) and site transport for each Site visit;
(*b*) meeting room/conference facilities and secretarial and copying services for each face-to-face meeting; and
(*c*) telephone conference or video conference facilities for each meeting by telephone or video conference.

In hostile political environments, it is usual for the Employer to organise military escorts, bodyguards and so forth.

Under Procedural Rule 3.10, the DAAB must provide the Parties and the Engineer with a report of the meeting or Site visit. If possible, the report must be provided before leaving the venue of the face-to-face meeting or the Site but within 7 days at the latest. The report is not a set of minutes of the meeting or the visit but a summary of the DAAB's activities.

18.20. Avoidance of Disputes

Under Sub-Clause 21.3 [*Avoidance of Disputes*] of the General Conditions of Contract, the Parties may jointly request the DAAB to assist and/or informally discuss and attempt to resolve any issue or disagreement that may have arisen between them. The request must be in writing, with a copy to the Engineer, which was not the case under the FIDIC 1999 Contracts. If the DAAB becomes aware of an issue or disagreement, it may invite the Parties to make such a joint request. Such assistance is referred to in the DAAB Agreement and the Procedural Rules as 'Informal Assistance'.

The joint request may be made at any time, except during the period that the Engineer is carrying out their duties under Sub-Clause 3.7 [*Agreement or Determination*] on the matter at issue or in disagreement, unless the Parties agree otherwise.

Note:

This exception is very significant because for a continuing event or circumstance which results in updated Claims and overlapping determinations by the Engineer, the DAAB may be prevented from playing its dispute avoidance role for many months.

Moreover, once the Engineer has issued the determination, the Parties only have 28 days in which to serve a NOD under Sub-Clause 3.7.5 [*Dissatisfaction with Engineer's determination*] which marks the moment when the Claim becomes a Dispute.

Within 42 days after the NOD being served, the Parties must refer the dispute to the DAAB for its decision (see below). Thus, the DAAB will have a maximum period of only 70 days in which to help the Parties reach an agreement with respect to the contested determination before the DAAB is obliged to begin the procedure for issuing its decision.

Under Procedural Rule 2, the DAAB (the sole member or the three members acting together) may give Informal Assistance during discussions at any meeting with the Parties (whether face-to-face, by telephone or by video conference) or during any Site visit or by an informal written note to the Parties. However, unless the Parties agree otherwise, both Parties must be present during such discussions.

Note:

Most experienced DAAB Members would be very cautious about meeting a Party without the presence of the other Party because of the risk of future complaints that the DAAB is no longer independent, notwithstanding the power of the DAAB under Procedural Rule 5 (l) to proceed without a defaulting Party (see below).

If the DAAB, does decide to meet one Party in the absence of the other, it should ensure that the absent Party agrees in writing to the meeting, or it should record that it is proceeding in accordance with Procedural Rule 5 (l) and set out the circumstances that entitle the DAAB to do so.

The advice given by the DAAB as part of the Informal Assistance will not be binding on the Parties and the DAAB will not be bound in any further Dispute resolution process or decision by any views or advice given as part of the Informal Assistance, regardless of whether it was provided orally or in writing.

18.21. Powers of the DAAB

In addition to the powers granted to the DAAB under the Conditions of Contract, the General Conditions of the DAAB Agreement and elsewhere in the Procedural Rules, under Procedural Rule 5 the Parties empower the DAAB to:

(*a*) establish the procedure to be applied in making Site visits and/or giving Informal Assistance;
(*b*) establish the procedure to be applied in giving decisions under the Conditions of Contract;
(*c*) decide on the DAAB's own jurisdiction, and the scope of any Dispute referred to the DAAB;
(*d*) appoint one or more experts (including legal and technical expert(s)), with the agreement of the Parties;
(*e*) decide whether or not there shall be a hearing (or more than one hearing, if necessary) in respect of any Dispute referred to the DAAB;
(*f*) conduct any meeting with the Parties and/or any hearing as the DAAB thinks fit, not being bound by any rules or procedures for the hearing other than those contained in the Contract and in these Rules;
(*g*) take the initiative in ascertaining the facts and matters required for a DAAB decision;
(*h*) make use of a DAAB Member's own specialist knowledge, if any;
(*i*) decide on the payment of financing charges in accordance with the Contract;
(*j*) decide on any provisional relief such as interim or conservatory measures;

(*k*) open up, review and revise any certificate, decision, determination, instruction, opinion or valuation of (or acceptance, agreement, approval, consent, disapproval, No-objection, permission, or similar act by) the Engineer that is relevant to the Dispute; and
(*l*) proceed with the DAAB's Activities in the absence of a Party who, after receiving a Notification from the DAAB, fails to comply with Sub-Clause 6.3 of the General Conditions of the DAAB Agreement.

The power granted under sub-paragraph (l) is broader than the corresponding power granted under Procedural Rule 7 under the FIDIC 1999 Contracts, which permitted the DAB to proceed with a hearing in the absence of any Party who the DAB was satisfied had received notice of the hearing. Under the FIDIC 2017 Contracts, the DAAB can proceed with any step in the Dispute resolution process without the Party who fails to comply with Sub-Clause 6.3 of the General Conditions of the DAAB Agreement, which requires each Party to:

(*a*) cooperate in good faith with the DAAB; and
(*b*) fulfil its duties, and exercise any right or entitlement, under the Contract, the General Conditions of the DAAB Agreement and the Procedural Rules,

as necessary to achieve the objectives under Procedural Rule 1 (see above).

Nevertheless, the DAAB has discretion to decide whether and to what extent any powers granted to it under the Contract, the General Conditions of the DAAB Agreement or the Procedural Rules may be exercised.

18.22. Reference of a Dispute to the DAAB

If a Dispute arises between the Parties, then either Party may refer the Dispute to the DAAB for its decision under Sub-Clause 21.4 [*Obtaining DAAB's Decision*] of the Conditions of Contract (regardless of whether any informal discussions have been held under Sub-Clause 21.3 [*Avoidance of Disputes*]) and the following provisions shall apply.

The 'reference' of a Dispute to the DAAB must:

(*a*) be made within 42 days of giving or receiving a NOD under Sub-Clause 3.7.5 [*Dissatisfaction with Engineer's determination*] if Sub-Clause 3.7 [*Agreement or Determination*] was applicable to the subject matter of the Dispute. If the Dispute is not referred to the DAAB within this period of 42 days, the NOD is deemed to have lapsed and to be no longer valid;
(*b*) state that it is given under Sub-Clause 21.4 [*Obtaining DAAB's Decision*];
(*c*) set out the referring Party's case relating to the Dispute;
(*d*) be in writing, with copies to the other Party and the Engineer; and
(*e*) for a DAAB of three persons, be deemed to have been received by the DAAB on the date it is received by the chairperson of the DAAB.

Unless prohibited by law, the reference of a Dispute to the DAAB is deemed to interrupt the running of any applicable statute of limitation or prescription period.

After the Dispute has been referred to the DAAB, both Parties must promptly make available to the DAAB all information, access to the Site, and appropriate facilities, that the DAAB requires for the purposes of making its decision on the Dispute.

Unless the Contract has already been abandoned or terminated, the Parties must continue to perform their obligations in accordance with the Contract.

18.23. Hearings

After a Dispute is referred to the DAAB in accordance with Sub-Clause 21.4.1 [*Reference of a Dispute to the DAAB*] of the Conditions of Contract, unless otherwise agreed by the Parties in writing, Procedural Rule 6 requires the DAAB to proceed in accordance with Sub-Clause 21.4 [*Obtaining DAAB's Decision*] of the Conditions of Contract and the DAAB Rules.

In doing so, the DAAB must act fairly and impartially between the Parties and, taking due regard of the 84-day period for the issue of the DAAB decision imposed under Sub-Clause 21.4.3 [*The DAAB's decision*] of the Conditions of Contract, and other relevant circumstances, the DAAB must:

(*a*) give each Party a reasonable opportunity to put forward its case and respond to the other Party's case; and
(*b*) adopt a procedure that is suitable to the Dispute, avoiding unnecessary delay and/or expense.

In these respects, Procedural Rule 7 empowers the DAAB to:

(*a*) decide on the date and place for any hearing, after consulting the Parties;
(*b*) decide on the duration of any hearing;
(*c*) request that written documentation and arguments from the Parties be submitted prior to the hearing;
(*d*) adopt an inquisitorial procedure during any hearing;
(*e*) request the production of documents, and/or oral submissions by the Parties, that the DAAB considers may assist in ascertaining the facts required for the decision;
(*f*) request the attendance at any hearing of persons that the DAAB considers may assist in ascertaining the facts;
(*g*) refuse admission to any hearing, or audience at any hearing, to any persons other than representatives of the Employer, the Contractor and the Engineer;
(*h*) proceed in the absence of any party who the DAAB is satisfied received timely notice of the hearing;
(*i*) adjourn any hearing as and when the DAAB considers further investigation by one Party or both Parties would benefit resolution of the Dispute, for such time as the investigation is carried out, and resume the hearing promptly thereafter.

During any hearing, the DAAB must not express any opinions concerning the merits of any arguments advanced by either Party in respect of the Dispute.

The DAAB must not give any Informal Assistance during a hearing, but if the Parties request Informal Assistance during any hearing:

(*a*) the hearing is to be adjourned for such time as the DAAB is giving Informal Assistance;

(*b*) if the hearing is so adjourned for longer than 2 days, the period under Sub-Clause 21.4.3 [*The DAAB's decision*] is to be temporarily suspended until the date that the hearing is resumed; and
(*c*) the hearing shall be resumed promptly after the DAAB has given such Informal Assistance.

18.24. Decisions

18.24.1 The DAAB's decision

Sub-Clause 21.4.3 [*The DAAB's Decision*] of the Conditions of Contract, and Procedural Rule 8, require that the DAAB issue its decision within 84 days of receiving the reference, or such other period as may be proposed by the DAAB and agreed by the Parties in writing.

In the case of a three-member DAAB:

(*a*) it shall meet in private (after the hearing, if any) in order to have discussions and to start preparation of its decision;
(*b*) the DAAB Members must use all reasonable endeavours to reach a unanimous decision;
(*c*) if it is not possible for the DAAB Members to reach a unanimous decision, the applicable decision is to be made by a majority of the DAAB Members, who may require the minority DAAB Member to prepare a separate written report (with reasons and supporting particulars) which is to be issued to the Parties; and
(*d*) if a DAAB Member fails to:

    (i) attend a hearing (if any) or a DAAB Members' meeting; or
    (ii) fulfil any required function (other than agreeing to a unanimous decision)

the Other Members must nevertheless proceed to make a decision, unless:

- such failure has been caused by exceptional circumstances, of which the Other Members and the Parties have received a Notification from the DAAB Member;
- the DAAB Member has suspended his/her services under sub-paragraph (a) of Sub-Clause 9.7 of the General Conditions of DAAB Agreement, due to non-payment by the due date of a valid invoice; or
- otherwise agreed by the Parties in writing.

In this latter respect, Sub-Clause 21.4.3 [*The DAAB's decision*] states that if at the end of the 84-day period for issuing the decision, any DAAB member's invoice remains unpaid after the due date(s) for payment has passed, the DAAB is not obliged to give its decision until all such outstanding invoices have been paid in full. In this case the DAAB must give its decision as soon as practicable after payment has been received.

Note:

Sub-Clause 21.4.3 [*The DAAB's decision*] entitles the DAAB to refuse to release its decision to the Parties if payment is outstanding. It does not suspend or extend the period of 84 days for making the decision. However, under Sub-Clause 9 of the General Conditions of Dispute Board Agreement, the DAAB member may suspend its services if an invoice has not been paid within 56 days of its submission.

The decision must be given in writing to both Parties with a copy to the Engineer, it must be reasoned and it must state that it is given under Sub-Clause 21.4 [*Obtaining DAAB's Decision*] of the Conditions of Contract.

The decision is immediately binding on both Parties, who must promptly comply with it, even if one or both Parties serves a NOD with respect to the decision under Sub-Clause 21.4.4 [*Dissatisfaction with DAAB's decision*]. The Employer is responsible for and must ensure the Engineer's compliance with the DAAB decision.

If the DAAB decision requires one Party to pay an amount to the other Party:

(*a*) subject to sub-paragraph (b) below, this amount shall be immediately due and payable without any certification or Notice; and
(*b*) the DAAB may (as part of the decision) require the payee to provide an appropriate security (at the DAAB's sole discretion) in respect of such amount. However, the DAAB may do so, only at the request of a Party and only if there are reasonable grounds for the DAAB to believe that the payee will be unable to repay the amount if the DAAB decision is reversed under Sub-Clause 21.6 [*Arbitration*].

The DAAB proceeding is not to be deemed to be an arbitration and the DAAB shall not act as arbitrator(s).

If, after giving a decision, the DAAB finds (and, in the case of a three-member DAAB, the members agree unanimously or by majority) that the decision contained any error of a typographical, clerical or arithmetical nature, Procedural Rule 8.3 provides that the Chairperson or the sole DAAB Member is to advise the Parties of the error within 14 days of giving this decision, and issue an addendum to its original decision in writing to the Parties.

If, within 14 days of receiving a decision from the DAAB, either Party finds a typographical, clerical or arithmetical error in the decision, that Party may request the DAAB to correct such error under Procedural Rule 8.4. The request must be in writing and must clearly identify the error.

If, within 14 days of receiving a decision from the DAAB, either Party believes that the decision contains an ambiguity, that Party may request clarification from the DAAB under Procedural Rule 8.5. The request must be in writing and must clearly identify the ambiguity.

Procedural Rule 8.6 requires the DAAB to respond to a request made under Procedural Rules 8.4 or Rule 8.5 within 14 days of its receipt. The DAAB may decline (at its sole discretion and with no requirement to give reasons) any request for clarification under Rule 8.5. If the DAAB agrees (or in the case of a three-member DAAB the members agree unanimously or by majority) that the decision did contain the error or ambiguity as described in the request, it may correct its decision by issuing in writing to the Parties an addendum to its original decision, together with the DAAB's response under Procedural Rule 8.6.

If the DAAB issues an addendum to its original decision under Procedural Rules 8.3 or 8.6, such an addendum is to form part of the decision and the period for issuing the NOD stated in sub-paragraph (c) of Sub-Clause 21.4.4 [*Dissatisfaction with DAAB's decision*] of the Conditions of Contract is to be calculated from the date the Parties receive this addendum (see below).

18.24.2 Dissatisfaction with DAAB's decision

Under Sub-Clause 21.4.4 [*Dissatisfaction with DAAB's decision*], if either Party is dissatisfied with the DAAB's decision:

(*a*) that Party may give a NOD to the other Party, copied to the DAAB and to the Engineer;
(*b*) this NOD must state that it is a '*Notice of Dissatisfaction with the DAAB's Decision*', must describe the matter in Dispute and state the reason(s) for dissatisfaction; and
(*c*) this NOD must be given within 28 days of receiving the DAAB's decision.

If the dissatisfied Party is dissatisfied with only part(s) of the DAAB's decision:

(i) this part(s) must be clearly identified in the NOD;
(ii) this part(s), and any other parts of the decision that are affected by such part(s) or rely on such part(s) for completeness, are deemed to be severable from the remainder of the decision; and
(iii) the remainder of the decision becomes final and binding on both Parties as if the NOD had not been given.

An example of such a situation could be where the Contractor agrees with the DAAB decision with respect to EOT but is dissatisfied with respect to the evaluation of additional Costs.

If the DAAB fails to give its decision within the 84-day period or other period agreed with the Parties, then either Party may, within 28 days of this period expiring, give a NOD to the other Party in accordance with sub-paragraphs (a) and (b) above.

If the DAAB has given its decision to both Parties, and no NOD under Sub-Clause 21.4.4 [*Dissatisfaction with DAAB's decision*] has been given by either Party within 28 days of receiving the DAAB's decision, then the decision becomes final and binding on both Parties and the Dispute cannot be made the subject of arbitral or court proceedings.

18.24.3 Impact of termination of the DAAB Agreement on the decision process

If the DAAB is dealing with a Dispute under Sub-Clause 21.4 [*Obtaining DAAB's Decision*] of the Conditions of Contract on the date of termination of a member's DAAB Agreement arising from resignation or termination under Clause 10 of the General Conditions of DAAB Agreement:

(*a*) the 84-day period for issuing the decision must be temporarily suspended; and
(*b*) following the appointment of a replacement DAAB Member, the full period of 84 days applies from the date of this appointment.

If the DAAB is composed of three members and the DAAB Agreement of only one DAAB Member is terminated as a result of resignation or termination under Clause 10 of the GCs, the Other Members continue as members of the DAAB but they are not to conduct any hearing or make any decision prior to the replacement of the DAAB Member unless otherwise agreed jointly by the Parties and the Other Members.

Figure 18.2 The DAAB decision

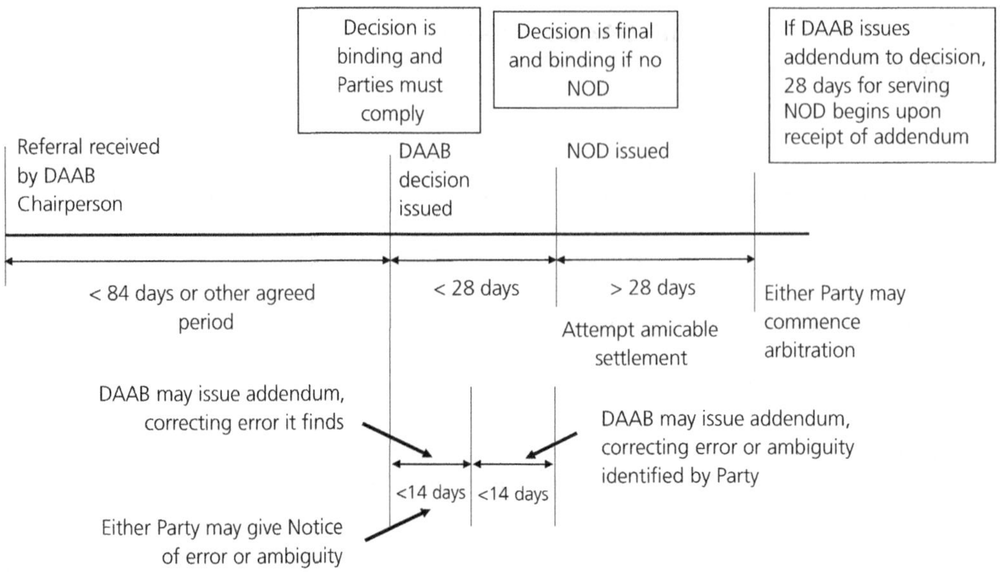

18.25. Amicable Settlement

Where a NOD has been given under Sub-Clause 21.4 [*Obtaining DAAB's Decision*], both Parties are to attempt to settle the Dispute amicably before the commencement of arbitration. However, unless both Parties agree otherwise and even if no attempt at amicable settlement has been made, arbitration may be commenced on or after the twenty-eighth (28th) day after the date on which the NOD was given.

However, there is no obligation to commence arbitration immediately upon expiry of the 28-day period, and the Parties can continue their attempt(s) at amicable settlement for as long as they wish before resorting to arbitration, subject to the applicable limitation or prescription periods.

18.26. Compliance

If a Party fails to comply with any decision of the DAAB, whether binding or final and binding, then the other Party may, without prejudice to any other rights it may have, refer the failure itself directly to arbitration under Sub-Clause 21.6 [*Arbitration*].

In such a case, Sub-Clause 21.4 [*Obtaining DAAB's Decision*] and Sub-Clause 21.5 [*Amicable Settlement*] do not apply to this reference. In other words, it is not necessary to seek a decision from the DAAB with respect to the failure to comply. Nor is it necessary to attempt amicable settlement before referring the failure to arbitration.

The arbitral tribunal shall have the power to order the enforcement of the DAAB decision, by way of summary or other expedited procedure and by means of an interim or provisional measure or an award (as may be appropriate under applicable law or otherwise).

In the case of a binding but not final decision of the DAAB (that is, a decision in relation to which a NOD has been served), such interim or provisional measure or award will be subject to the express reservation that the rights of the Parties as to the merits of the Dispute are reserved until they are resolved by an award.

Any interim or provisional measure or award enforcing compliance with a decision of the DAAB, whether such decision is binding or final and binding, may also include an order or award of damages or other relief.

18.27. Arbitration

Under Sub-Clause 21.6 [*Arbitration*], any Dispute in respect of which the DAAB's decision has not become final and binding is to be finally settled by arbitration, unless settled amicably, and subject always to Sub-Clause 3.7.5 [*Dissatisfaction with Engineer's determination*], Sub-Clause 21.4.4 [*Dissatisfaction with DAAB's decision*], Sub-Clause 21.7 [*Failure to Comply with DAAB's Decision*] and Sub-Clause 21.8 [*No DAAB In Place*].

Unless otherwise agreed by both Parties:

(*a*) the Dispute shall be finally settled under the Rules of Arbitration of the International Chamber of Commerce;
(*b*) the Dispute shall be settled by one or three arbitrators appointed in accordance with these Rules; and
(*c*) the arbitration shall be conducted in the ruling language defined in Sub-Clause 1.4 [*Law and Language*].

The World Bank and some other MDBs have modified these requirements for projects which they finance and there are separate provisions for foreign contractors and for domestic contractors:

'Arbitration shall be conducted as follows:

(a) *if the contract is with foreign contractors, unless otherwise specified in the Contract Data; the dispute shall be finally settled under the Rules of Arbitration of the International Chamber of Commerce; by one or three arbitrators appointed in accordance with these Rules. The place of arbitration shall be the neutral location specified in the Contract Data; and the arbitration shall be conducted in the ruling language defined in Sub-Clause 1.4 [Law and Language].*
(b) *If the Contract is with domestic contractors, arbitration with proceedings conducted in accordance with the laws of the Employer's country.'*

It should be noted that for projects financed by the Asian Development Bank, final settlement of the Dispute involving a foreign Contractor (or a foreign leader of a Joint Venture) is to be in accordance with the Rules of Arbitration of the Singapore International Arbitration Centre and the place of arbitration must be Singapore or any other neutral place mutually agreed by both the Employer and the Contractor.

FIDIC 2017 does not address the place of arbitration but relies on the Rules of Arbitration of the ICC which foresee that the place of arbitration shall be fixed by the ICC Court, unless agreed by the Parties.

The place of arbitration is significant because it is the law of arbitration of that place which will apply to the conduct of the arbitration. Moreover, if the place of arbitration is in a country which has signed the 'Convention on the Recognition and Enforcement of Foreign Arbitral Awards' commonly known as the New York Convention of 1958, the arbitral award can be enforced in any other country which has signed the convention.

The arbitral tribunal shall have full power to open up, review and revise any certificate, determination (other than a final and binding determination), instruction, opinion or valuation of the Engineer, and any decision of the DAAB (other than a final and binding decision) relevant to the Dispute.

Nothing shall disqualify the Engineer from being called as a witness and giving evidence before the arbitral tribunal on any matter relevant to the Dispute. The stress of being a witness in international arbitration should not be under-estimated. In accordance with the Parties' undertaking given under Clause 8.1 (b) of the General Conditions of DAAB Agreement, neither Party may call any member of the DAAB as a witness in the proceedings. However, any decision of the DAAB will be admissible in evidence in the arbitration. Statistics show that international arbitral tribunals tend to confirm DAAB decisions.

In any award dealing with the costs of the arbitration, the arbitral tribunal may take account of the extent (if any) to which a Party failed to cooperate with the other Party in constituting a DAAB under Sub-Clause 21.1 [*Constitution of the DAAB*] and/or Sub-Clause 21.2 [*Failure to Appoint DAAB Member(s)*]. In international arbitration, the Parties' costs are often more than 40% of the total amount in dispute (the sum of the claims and counter-claims). In principal, the losing Party must pay the costs of the successful Party plus interest.

Neither Party is limited in the proceedings before the arbitral tribunal to the evidence or arguments previously put before the DAAB to obtain its decision, or to the reasons for dissatisfaction given in the Party's NOD under Sub-Clause 21.4 [*Obtaining DAAB's Decision*]. In other words, the dispute may be argued and heard from the beginning, as though no DAAB referral had been made.

Arbitration may be commenced before or after completion of the Works. The obligations of the Parties, the Engineer and the DAAB are not altered by reason of any arbitration being conducted during the progress of the Works.

If an award requires a payment of an amount by one Party to the other Party, this amount shall be immediately due and payable without any further certification or Notice.

18.28. No DAAB in place

If a Dispute arises between the Parties and there is no DAAB in place or in the process of being constituted, either Party may refer the Dispute directly to arbitration under Sub-Clause 21.6 [*Arbitration*] without prejudice to any other rights the Party may have, and Sub-Clause 21.4 [*Obtaining DAAB's Decision*] and Sub-Clause 21.5 [*Amicable Settlement*] will not apply.

18.29. World Bank Special Provisions

The World Bank and some other MDBs introduce several changes to the DAAB provisions contained in FIDIC 2017 contracts, via the Special Provisions and the Contract Data which they impose for projects that they finance.

Whereas the standard position with respect to the appointment of the DAAB is that it must be within 28 days of the Contractor's receipt of the Letter of Acceptance unless otherwise stated in the Contract Data (Sub-Clause 21.1 [*Constitution of the DAAB*]), the World Bank and several other MDBs[6] state that the appointment shall be within 42 days of signature by both parties of the Contract Agreement. The Inter-American Development Bank and Asian Development Bank state that the appointment shall be within 28 days of the Commencement Date.

Moreover, the World Bank, AfDB, AIIB and IsDB require the DAAB to have been appointed before the issue of the Notice of Commencement Date under Sub-Clause 8.1 [*Commencement of Work*].

Almost all the MDBs fix criteria for the number of DAAB members: '*For a Contract estimated to cost above USD 50 million, the DAAB shall comprise three members. For a Contract estimated to cost between USD 20 million and USD 50 million, the DAAB may comprise three members or a sole member. For a Contract estimated to cost less than USD 20 million, a sole member is recommended.*'

With respect to the qualifications of the DAAB members, many of the MDBs state under Sub-Clause 21.1 [*Constitution of the DAAB*] of the Special Provisions that '*if the Contract is with a foreign Contractor, the DAAB members shall not have the same nationality as the Employer or the Contractor*'.

Moreover, Sub-Clause 3.3 [*Warranties*] of the General Conditions of DAAB Agreement is expanded by many of the MDBs as follows:

> '*When appointing the DAAB Member, each Party relies on the DAAB Member's representations, that he/she:*
>
> (*a*) *has at least a bachelor's degree in relevant disciplines such as law, engineering, construction management or contract management;*
> (*b*) *has at least ten years of experience in contract administration/management and dispute resolution, out of which at least five years of experience as an arbitrator or adjudicator in construction-related disputes;*
> (*c*) *has received formal training as an adjudicator from an internationally recognized organization;*
> (*d*) *has experience and/or is knowledgeable in the type of work which the Contractor is to carry out under the Contract;*
> (*e*) *has experience in the interpretation of construction and/or engineering contract documents;*

6 African Development Bank (AfDB), Asian Infrastructure Investment Bank (AIIB), Islamic Development Bank (IsDB).

> (*f*) *has familiarity with the forms of contract published by FIDIC since 1999, and an understanding of the dispute resolution procedures contained therein; and*
> (g) *is fluent in the language for communications stated in the Contract Data (or the language as agreed between the Parties and the DAAB).'*

Almost all the MDBs require the Contract to include a specific Provisional Sum to cover the Employer's share of the DAAB's fees and expenses. Accordingly, Sub-Clause 13.4 [*Provisional Sums*] of the Special Provisions reads:

> *'The Provisional Sum shall be used to cover the Employer's share of the DAAB members' fees and expenses, in accordance with Clause 21. No prior instruction of the Engineer shall be required with respect to the work of the DAAB. The Contractor shall submit the DAAB members' invoices and satisfactory evidence of having paid 100% of such invoices as part of the substantiation of those Statements submitted under Sub-Clause 14.3.'*

In January 2021, the World Bank published a special version of its Standard Bidding Documents for Works which allocates a specific role to the DAAB on projects identified as being at high risk of Sexual Exploitation and Abuse (SEA) and Sexual Harassment (SH). For such projects, the Contract must impose obligations on the Employer and the Contractor to prevent, monitor and handle incidents of SEA/SH. In the event of a suspected non-compliance or an allegation, the Employer must refer the matter to the DAAB for a decision on whether the Contractor had failed to comply with any of its SEA/SH obligations. If the DAAB does find such a failure, WB may disqualify the Contractor and any defaulting subcontractor, from bidding for further WB-financed projects for a period of two years. The contract might also be terminated, if merited by the circumstances.

More information is available at: https://thedocs.worldbank.org/en/doc/844301612216257638-0290022021/original/QAsSEASHpreventionProcurementforborrowerscontractors.pdf

Smith G
ISBN 978-0-7277-6652-6
https://doi.org/10.1680/fcmh.66526.273

Chapter 19
The 2022 reprint

19.1. Background

In November 2022, FIDIC published a reprint of its FIDIC 2017 contracts. Each reprint consolidates three sets of amendments introduced since 2017. The first set of amendments was issued as an errata document in December 2018. A second set of amendments was also issued as an errata document in June 2019. The third set of amendments was published in November 2022 to improve the original document following a review of comments and queries raised by users.

Each reprint contains the fully amended General Conditions, Guidance for the Preparation of Particular Conditions, and Advisory Notes to Users of FIDIC Contracts where the project uses BIM systems and Forms. A list of the combined amendments is available at the back of each reprint, although corrections to minor typographical errors and layout which do not affect the meaning of the text, are not listed.

The list of the amendments is also available free of charge from the FIDIC website. In this case, the three sets of amendments are listed separately in chronological order: December 2018, June 2019 and November 2022.

In a press release at the time of launching the reprints, FIDIC stated that the amendments would be '*effective as of 1 January 2023*'. However, there is no mention of such a date of effect in the documents and FIDIC's purpose in making this announcement is unclear. Certainly, the reprints do not automatically replace the earlier version of the FIDIC 2017 contracts. Nor can existing contracts be modified to include the amendments, without the agreement of the Parties.

On the face of each reprint, the only difference compared to the original 2017 document is the mention 'Reprinted 2022 with amendments'. Therefore, to avoid confusion and disputes arising in relation to new contracts, users must clearly identify which version is to apply and must exercise great care to ensure that they work with the correct document.

With respect to RB 2017, the December 2018 amendments were listed on 1.5 pages. The June 2019 amendments were listed on a single page and consisted wholly of typographical corrections.

The November 2022 amendments are listed on 11 pages and, with respect to RB 2017, impact 87 sub-clauses, DAAB Procedural Rules and the Notes on the Preparation of Special Provisions. The most significant are set out in Annex 2. These cover the following topics.

19.2. The definition of '*Claim*'

Under the revised wording of Sub-Clause 1.1.6, the term '*Claim*' excludes a matter to be agreed or determined under sub-paragraph (a) of Sub-Clause 3.7 [*Agreement or Determination*]. Under the revised wording of Sub-Clause 3.7, the matters to be agreed or determined and which are excluded from the definition of 'Claim' may arise under the following sub-clauses:

Sub-Clause	Subject
4.7.3	Agreement or Determination of rectification measures, delay and/or Cost (due to errors in Setting Out reference items)
10.2	Taking Over Parts
11.2	Cost of Remedying Defects
12.1	Works to be Measured
12.3	Valuation of the Works
13.3.1	Variation by Instruction
13.5	Daywork
14.4	Schedule of Payments
14.5	Plant and Materials intended for the Works
14.6.3	Correction or modification of an IPC
15.3	Valuation after Termination for Contractor's Default
15.6	Valuation after Termination for Employer's Convenience
18.5	Optional Termination

This exclusion is logical because if a Party has received an Engineer's determination with which it is dissatisfied, it should not be required to resubmit the matter to the Engineer as a Claim. Instead, and subject to having served a NOD within 28 days of receipt of the determination, the dissatisfied Party may refer the matter to the DAAB.

As a result of this change in definition, Sub-Clause 20.1 [*Claims*] has also been modified to exclude '*determinations*'. However, under Sub-Clause 20.1 (c) a Claim may still concern a certificate or a valuation of the Engineer. This appears to contradict the exclusion of Sub-Clauses 12.3, 13.5, 14.4, 14.5, 15.3, 15.6 and 18.5.

Although Sub-Clause 4.7.3 [*Agreement or Determination of rectification measures, delay and/or Cost*] in relation to errors in the items of reference requires the Engineer to agree or determine whether an error exists, whether such error would have been discovered by an experienced contractor exercising due care and what measures the Contractor is required to take, it also states that if there are no such measures, but the Contractor suffers delay and/or incurs Costs as a result of the error, the matter is to be dealt with as a Claim under Sub-Clause 20.2.

Similarly, whereas the Engineer is required to agree or determine the reduction in the daily rate of Delay Damages following taking over (or deemed taking over) of a Part, if the Contractor incurs Cost as a result of this taking over or deemed taking over, the matter is to be dealt with as a Claim under Sub-Clause 20.2.

19.3. The definition of '*Dispute*'

The definition of '*Dispute*' under Sub-Clause 1.1.29 has been modified to be consistent with the revised definition of '*Claim*'. Under the revised definition, a Dispute means a situation where:

(*a*) one Party has made a Claim, or there has been a matter to be agreed or determined under sub-paragraph (a) of Sub-Clause 3.7 [*Agreement or Determination*];
(*b*) the Engineer's determination under Sub-Clause 3.7.2 [*Engineer's Determination*] was a rejection (in whole or in part) of:

(i) the Claim (or there was a deemed rejection under sub-paragraph (i) of Sub-Clause 3.7.3 [*Time limits*]); or
(ii) a Party's assertion(s) in respect of the matter

as the case may be; and

(*c*) either Party has given a NOD under Sub-Clause 3.7.5 [*Dissatisfaction with Engineer's determination*].

> **Note:**
>
> *Under this revised definition, if the Engineer's determination was an acceptance of a Claim or a Party's assertion, this acceptance cannot become the subject of a Dispute, even if the other Party gives a NOD.*

This definition under Sub-Clause 1.1.29 is not complete and must be read in conjunction with the revised Sub-Clause 21.4 which states:

'In addition to the situation described in the definition of Dispute under Sub-Clause 1.1.29 above, a Dispute shall be deemed to have arisen if:

(*a*) *there is a failure as referred to under sub-paragraph (b), or a non-payment as referred to under sub-paragraph (c), of Sub-Clause 16.2.1 [Notice];*
(*b*) *the Contractor is entitled to receive financing charges under Sub-Clause 14.8 [Delayed Payment] but does not receive payment thereof from the Employer within 28 days after his request for such payment; or*
(*c*) *a Party has given:*

(i) *a Notice of intention to terminate the Contract under Sub-Clause 15.2.1 [Notice] or Sub-Clause 16.2.1 [Notice] (as the case may be); or*
(ii) *a Notice of termination under Sub-Clause 15.2.2 [Termination], Sub-Clause 16.2.2 [Termination], Sub-Clause 18.5 [Optional Termination] or Sub-Clause 18.6 [Release from Performance under the Law] (as the case may be);*

and the other Party has disagreed with the first Party's entitlement to give such Notice;

which Dispute may be referred by either Party under this Sub-Clause 21.4 without the need for a NOD (and Sub-Clause 3.7 [Agreement or Determination] and sub-paragraph (a) of Sub-Clause 21.4.1 [Reference of a Dispute to the DAAB] shall not apply).'

The references under sub-paragraph (a) concern the Engineer's failure to issue an IPC within 56 days of receipt of the Contractor's Statement and the Employer's failure to pay within 42 days of the expiry of the time for payment.

19.4. Clarifications with respect to '*Part(s)*'

Under Sub-Clause 1.1.58 the revised definition of '*Part*' means a part of the Works or part of a Section (as the case may be) which is taken over by the Employer under the first paragraph or used by the Employer and deemed to have been taken over under the second paragraph of Sub-Clause 10.2 [*Taking Over Parts*]. The previous definition only mentioned deemed taking over following use by the Employer.

Sub-Clause 10.2 has been amended to be consistent with this revised definition.

19.5. Errors in Setting Out reference items

In Chapter 4.5 above, reference was made to the confusing wording of Sub-Clause 4.7.3 [*Agreement or Determination of rectification measures, delay and/or Cost*] with respect to the Contractor's entitlement to compensation arising from errors in the items of reference which (taking account of cost and time) could not have been identified by an experienced contractor exercising due care:

- when examining the Site, the Drawings and the Specification before submitting the Tender; or
- if the items of reference are specified on the Drawings and/or in the Specification and the Contractor's Notice is given after the expiry of the period stated in sub-paragraph (a) of Sub-Clause 4.7.2 (28 days or other specified period after the Commencement Date).

This matter has been successfully resolved by the Reprint 2022 which now states:

'After receiving a Notice from the Contractor under Sub-Clause 4.7.2 [Errors], the Engineer shall proceed under Sub-Clause 3.7 [Agreement or Determination] to agree or determine:

(a) whether or not there is an error in the items of reference;
(b) whether or not (taking account of cost and time) an experienced contractor exercising due care would have discovered such an error

- *when examining the Site, the Drawings and the Specification before submitting the Tender; or*
- *if the items of reference are specified on the Drawings and/or in the Specification and the Contractor's Notice is given after the expiry of the period stated in sub-paragraph (a) of Sub-Clause 4.7.2; and*

(c) what measures (if any) the Contractor is required to take to rectify the error

(and, for the purpose of Sub-Clause 3.7.3 [Time limits], the date the Engineer receives the Contractor's Notice under Sub-Clause 4.7.2 [Errors] shall be the date of commencement of the time limit for agreement under Sub-Clause 3.7.3).

If it is agreed or determined, under sub-paragraphs (a) and (b) above, that there is an error in the items of reference that an experienced contractor would not have discovered:

(*a*) *Sub-Clause 13.3.1 [Variation by Instruction] shall apply to the measures that the Contractor is required to take (if any); and*
(*b*) *if there are no such measures, and therefore no Variation, but the Contractor suffers delay and/or incurs Cost as a result of the error, the Contractor shall be entitled subject to Sub-Clause 20.2 [Claims for Payment and/or EOT] to EOT and/or payment of such Cost Plus Profit.'*

Thus, if an experienced contractor could not have identified the problem at the appropriate time, the Contractor is to be compensated as a Variation for complying with any instruction from the Engineer or is to be compensated in terms of time and/or payment if delay to completion or additional Costs are incurred as a result of the error, notwithstanding that the Engineer does not require the Contractor to take measures to address the error. This latter right is subject always to compliance by the Contractor with Sub-Clause 20.2 [*Claims for Payment and/or EOT*].

19.6. Contractor's Discharge

Under Chapter 12.11 above, reference was made to the inconsistent wording with respect to the Engineer's response to a Partially Agreed Final Statement and the corresponding requirement for a Contractor's discharge. The Reprint 2022 attempts to address this, and in particular, the requirements for a Contractor's discharge.

The revised wording of the last paragraph of Sub-Clause 14.13 [*Issue of FPC*] states:

'If the Contractor has not submitted a discharge under Sub-Clause 14.12 [Discharge] but has either:

(i) *submitted a Partially Agreed Final Statement under Sub-Clause 14.11.2 [Agreed Final Statement]; or*
(ii) *not done so but, to the extent that a draft final Statement submitted by the Contractor is deemed by the Engineer to be a Partially Agreed Final Statement,*

the Engineer shall proceed in accordance with Sub-Clause 14.6 [Issue of IPC] to issue an IPC.'

Thus, a discharge is required only with respect to the FPC to be issued in response to the Final Statement, which by definition is agreed. If no such discharge is provided, it is deemed to have been submitted and to have become effective when the Contractor has received full payment of the amount certified in the Final Payment Certificate (FPC) and the Performance Security has been released.

19.7. The appointing official

An important change introduced by the Reprint 2022 is to delete the item in the Contract Data which permitted the Parties to choose the appointing entity under Sub-Clause 21.2 [*Failure to Appoint DAAB Member(s)*] if the Parties are unable to agree on the appointment of the DAAB members. The wording of Sub-Clause 21.2 [*Failure to Appoint DAAB Member(s)*] has been revised to match this deletion and the relevant portion now reads:

'If any of the following conditions apply, namely:

(*a*) *if the DAAB is to comprise a sole member, the Parties fail to agree the appointment of this member by the date stated in the first paragraph of Sub-Clause 21.1 [Constitution of the DAAB]; or*

(*b*) *if the DAAB is to comprise three persons, and if by the date stated in the first paragraph of Sub-Clause 21.1 [Constitution of the DAAB]:*

(i) *either Party fails to select a member (for agreement by the other Party);*

(ii) *either Party fails to agree a member selected by the other Party; and/or*

(iii) *the Parties fail to agree the appointment of the third member (to act as chairperson) of the DAAB;*

(*c*) *the Parties fail to agree the appointment of a replacement within 42 days after the date on which the sole member or one of the three members declines to act or is unable to act as a result of death, illness, disability, resignation, or termination of appointment; or*

(*d*) *if, after the Parties have agreed the appointment of the member(s) or replacement, such appointment cannot be effected because one Party refuses or fails to sign a DAAB Agreement with any such member or replacement (as the case may be) within 14 days of the other Party's request to do so, or because the terms of the DAAB Agreement (including the amount of the monthly fee or the daily fee) cannot be agreed with the member or replacement within 14 days after he/she has been advised by the Parties that they have agreed to his/her appointment,*

then, unless otherwise agreed by the Parties, either or both Parties may apply to the President of FIDIC or a person appointed by the President, who shall be the appointing official under the Contract. The appointing official shall, after due consultation with both Parties and after consulting the prospective member(s) or replacement:

– *appoint the member(s) of the DAAB or the replacement; and*
– *set the terms of the appointment, including the amounts of the monthly fee and the daily fee for each member or replacement.*

Selection of the member(s) or replacement to be so appointed shall not be limited to those persons named in the list in the Contract Data or, in the case of sub-paragraph (d) above, to the member(s) or replacement agreed by the Parties.

This appointment and its terms shall be final and conclusive.'

Thus, the President of FIDIC or a person appointed by the President, shall be the appointing official under the Contract and selection of the member(s) or replacement to be so appointed shall not be limited to those persons named in the list in the Contract Data or, in the case of sub-paragraph (d) above, to the member(s) or replacement agreed by the Parties.

19.8. Referral to the DAAB of a deemed rejection by the Engineer

The revised text of Sub-Clause 21.4.1 [*Reference of a Dispute to the DAAB*] makes it clear that under Sub-Clause 3.7.3 (ii) no NOD is required with respect to an Engineer's failure to issue a determination within the time limit for determination, before referring the deemed rejection to the DAAB.

19.9. General Conditions of DAAB Agreement and DAAB Procedural Rules

The FIDIC 2017 Contracts introduced a requirement for DAAB members not to have been employed as a consultant or otherwise by the Employer, the Contractor, the Employer's Personnel or the Contractor's Personnel in the period of 10 years prior to signing the DAAB Agreement. The Reprint 2022 reduces this period to 5 years and includes an exception to this constraint:

> *'In the 5 years before signing the DAAB Agreement, not have been employed as a consultant or otherwise by the Employer, the Contractor, the Employer's Personnel or the Contractor's Personnel, except in such circumstances as were disclosed in writing to the Employer and the Contractor before they signed the DAAB Agreement (or are deemed to have done so).'*

This revised requirement is more closely aligned to the '*IBA Guidelines on Conflicts of Interest in International Arbitration*', which are often followed by DAAB practitioners.

The revised text of Sub-Clause 9.1 provides clarification with respect to some of the periods for which DAAB members may charge the daily fee:

> *'[time] spent on preparing and attending hearings (and, in case of a three-member DAAB, attending meeting(s) between the DAAB Members in accordance with sub-paragraph (a) of Rule 8.2 of the DAAB Rules, and communicating with the Other Members); and*
>
> *[time] spent on preparing decisions, including studying written documentation and arguments from the Parties'.*

Strangely, the revised text also excludes the entitlement for DAAB members to be reimbursed the cost of telephone calls. This appears to be related to the multiple modifications to the DAAB Procedural Rules to cater for online meetings as a result of the COVID pandemic. It seems that the drafters now assume that communication by internet is possible anywhere in the world and that it will no longer be necessary for DAAB members to pay for telephone calls. At the time of writing, this is not yet the case.

The text of Sub-Clause 9.3 of the General Conditions of DAAB Agreement has been revised to match the revised text of Sub-Clause 21.2 [*Failure to Appoint DAAB Member(s)*] of the Conditions of Contract and the appointment of the DAAB member(s) by the appointing official.

As mentioned above, the DAAB Procedural Rules have been extensively modified to deal with the increased use of online meetings between the DAAB and the Parties, which developed during the COVID pandemic.

19.10. Notes on the Preparation of Special Provisions

The Guidance for Sub-Clause 21.1 [*Constitution of the DAAB*] has been corrected so that it refers to only one example form of DAAB Agreement. The guidance in the FIDIC 2017 Contracts incorrectly referred to two example forms. This reference was a 'left-over' from the FIDIC 1999 Contracts which did provide two example forms of DAB Agreement: one for a sole member DAB, the other for a three-member DAB. The FIDIC 2017 Contracts contain only one example form which can be used for a sole-member DAAB or a three-member DAAB.

The Guidance for Sub-Clause 21.1 [*Constitution of the DAAB*] has been further revised to emphasise the role of FIDIC under Sub-Clause 21.2 [*Failure to Appoint DAAB Member(s)*] if the Parties are unable to agree on the choice and terms of appointment of the DAAB members:

> *'If the Parties cannot agree on any DAAB Member, or replacement, or cannot agree the terms of the DAAB Agreement with any prospective DAAB Member or replacement, Sub-Clause 21.2 [Failure to Appoint DAAB Member(s)] applies. Therefore, unless otherwise agreed by the Parties, the selection and appointment of the DAAB Member(s) will be made by the President of FIDIC or a person appointed by the President.*
>
> *FIDIC is very knowledgeable about the nature and purpose of a DAAB and is an impartial entity to make such appointments, in circumstances where it has not been possible to appoint the DAAB member or replacement member by the agreement of the Parties.*
>
> *FIDIC has its appointment rules and maintains a list of approved and experienced adjudicators for this specific purpose: The FIDIC President's List of Approved Dispute Adjudicators. This list is available to access on FIDIC's website at fidic.org.'*

On many projects implemented under FIDIC 1999 Contracts, the Pink Book and FIDIC 2017, organisations other than FIDIC were named in the Appendix to Tender/Contract Data as the 'appointing entity', charged with selecting DAB/DB/DAAB members if the Parties could not agree. In some cases, the choice of appointing entity was inappropriate, and this led to the selection of inexperienced members. It seems that with the Reprints 2022, FIDIC expects to be the appointing entity or official in all but exceptional cases and to select members exclusively from the FIDIC President's List of Approved Dispute Adjudicators.

Smith G
ISBN 978-0-7277-6652-6
https://doi.org/10.1680/fcmh.66526.281

Annex 1 – Code of conduct for contractor's personnel (ES) form

Note to the Bidder:

The minimum content of the Code of Conduct form as set out by the Employer shall not be substantially modified. However, the Bidder may add requirements as appropriate, including to take into account Contract-specific issues/risks.

The Bidder shall initial and submit the Code of Conduct form as part of its bid.

Note to the Employer:

The following minimum requirements shall not be modified*. The Employer may add additional requirements to address identified issues, informed by relevant environmental and social assessment.*

The types of issues identified could include risks associated with: labor influx, spread of communicable diseases, Sexual Exploitation and Abuse (SEA), Sexual Harassment (SH) etc.

Delete this Box prior to issuance of the bidding documents.

CODE OF CONDUCT FOR CONTRACTOR'S PERSONNEL

We are the Contractor, [*enter name of Contractor*]. We have signed a contract with [*enter name of Employer*] for [*enter description of the Works*]. These Works will be carried out at [*enter the Site and other locations where the Works will be carried out*]. Our contract requires us to implement measures to address environmental and social risks related to the Works, including the risks of sexual exploitation, sexual abuse and sexual harassment.

This Code of Conduct is part of our measures to deal with environmental and social risks related to the Works. It applies to all our staff, laborers and other employees at the Works Site or other places where the Works are being carried out. It also applies to the personnel of each subcontractor and any other personnel assisting us in the execution of the Works. All such persons are referred to as **'Contractor's Personnel'** and are subject to this Code of Conduct.

This Code of Conduct identifies the behavior that we require from all Contractor's Personnel.

Our workplace is an environment where unsafe, offensive, abusive or violent behavior will not be tolerated and where all persons should feel comfortable raising issues or concerns without fear of retaliation.

REQUIRED CONDUCT

Contractor's Personnel shall:

1. carry out his/her duties competently and diligently;
2. comply with this Code of Conduct and all applicable laws, regulations and other requirements, including requirements to protect the health, safety and well-being of other Contractor's Personnel and any other person;
3. maintain a safe working environment including by:
   a. ensuring that workplaces, machinery, equipment and processes under each person's control are safe and without risk to health;
   b. wearing required personal protective equipment;
   c. using appropriate measures relating to chemical, physical and biological substances and agents; and
   d. following applicable emergency operating procedures.
4. report work situations that he/she believes are not safe or healthy and remove himself/herself from a work situation which he/she reasonably believes presents an imminent and serious danger to his/her life or health;
5. treat other people with respect, and not discriminate against specific groups such as women, people with disabilities, migrant workers or children;
6. not engage in Sexual Harassment, which means unwelcome sexual advances, requests for sexual favors, and other verbal or physical conduct of a sexual nature with other Contractor's or Employer's Personnel;
7. not engage in Sexual Exploitation, which means any actual or attempted abuse of position of vulnerability, differential power or trust, for sexual purposes, including, but not limited to, profiting monetarily, socially or politically from the sexual exploitation of another;
8. not engage in Sexual Abuse, which means the actual or threatened physical intrusion of a sexual nature, whether by force or under unequal or coercive conditions;
9. not engage in any form of sexual activity with individuals under the age of 18, except in case of pre-existing marriage;
10. complete relevant training courses that will be provided related to the environmental and social aspects of the Contract, including on health and safety matters, Sexual Exploitation and Abuse (SEA), and Sexual Harassment (SH);
11. report violations of this Code of Conduct; and
12. not retaliate against any person who reports violations of this Code of Conduct, whether to us or the Employer, or who makes use of the grievance mechanism for Contractor's Personnel or the project's Grievance Redress Mechanism.

RAISING CONCERNS

If any person observes behavior that he/she believes may represent a violation of this Code of Conduct, or that otherwise concerns him/her, he/she should raise the issue promptly. This can be done in either of the following ways:

1. Contact [*enter name of the Contractor's Social Expert with relevant experience in handling gender-based violence, or if such person is not required under the Contract, another individual designated by the Contractor to handle these matters*] in writing at this address [ ] or by telephone at [ ] or in person at [ ]; or
2. Call [ ] to reach the Contractor's hotline *(if any)* and leave a message.

The person's identity will be kept confidential, unless reporting of allegations is mandated by the country law. Anonymous complaints or allegations may also be submitted and will be given all due and appropriate consideration. We take seriously all reports of possible misconduct and will investigate and take appropriate action. We will provide warm referrals to service providers that may help support the person who experienced the alleged incident, as appropriate.

There will be no retaliation against any person who raises a concern in good faith about any behavior prohibited by this Code of Conduct. Such retaliation would be a violation of this Code of Conduct.

CONSEQUENCES OF VIOLATING THE CODE OF CONDUCT

Any violation of this Code of Conduct by Contractor's Personnel may result in serious consequences, up to and including termination and possible referral to legal authorities.

FOR CONTRACTOR'S PERSONNEL:

I have received a copy of this Code of Conduct written in a language that I comprehend. I understand that if I have any questions about this Code of Conduct, I can contact [*enter name of Contractor's contact person(s) with relevant experience*] requesting an explanation.

Name of Contractor's Personnel: [insert name]

Signature: ______________________________

Date: (day month year): ______________________________

Countersignature of authorized representative of the Contractor:

Signature: ______________________________

Date: (day month year): ______________________________

ATTACHMENT 1: Behaviors constituting Sexual Exploitation and Abuse (SEA) and behaviors constituting Sexual Harassment (SH)

ATTACHMENT 1 TO THE CODE OF CONDUCT FORM

BEHAVIORS CONSTITUTING SEXUAL EXPLOITATION AND ABUSE (SEA) AND BEHAVIORS CONSTITUTING SEXUAL HARASSMENT (SH)

The following non-exhaustive list is intended to illustrate types of prohibited behaviors.

(1) **Examples of sexual exploitation and abuse** include, but are not limited to:

- A Contractor's Personnel tells a member of the community that he/she can get them jobs related to the work site (e.g. cooking and cleaning) in exchange for sex.
- A Contractor's Personnel that is connecting electricity input to households says that he can connect women headed households to the grid in exchange for sex.
- A Contractor's Personnel rapes, or otherwise sexually assaults a member of the community.
- A Contractor's Personnel denies a person access to the Site unless he/she performs a sexual favor.
- A Contractor's Personnel tells a person applying for employment under the Contract that he/she will only hire him/her if he/she has sex with him/her.

(2) **Examples of sexual harassment in a work context**

- Contractor's Personnel comment on the appearance of another Contractor's Personnel (either positive or negative) and sexual desirability.
- When a Contractor's Personnel complains about comments made by another Contractor's Personnel on his/her appearance, the other Contractor's Personnel comment that he/she is 'asking for it' because of how he/she dresses.
- Unwelcome touching of a Contractor's or Employer's Personnel by another Contractor's Personnel.
- A Contractor's Personnel tells another Contractor's Personnel that he/she will get him/her a salary raise, or promotion if he/she sends him/her naked photographs of himself/herself.

Smith G
ISBN 978-0-7277-6652-6
https://doi.org/10.1680/fcmh.66526.285

Annex 2 – Significant changes in the 2022 reprint

In the table below, the text of RB 2017 which is modified by the 2022 Reprint is highlighted by grey blocks. The modifications introduced by the 2022 Reprint are underlined in the right-hand column.

Sub-Clause	FIDIC 2017	2022 REPRINT
Sub-Clause 1.1.6	**'Claim'** means a request or assertion by one Party to the other Party for an entitlement or relief under any Clause of these Conditions or otherwise in connection with, or arising out of, the Contract or the execution of the Works.	**'Claim'** means a request or assertion by one Party to the other Party (excluding a matter to be agreed or determined under sub-paragraph (a) of Sub-Clause 3.7 [*Agreement or Determination*]) for an entitlement or relief under any Clause of these Conditions or otherwise in connection with, or arising out of, the Contract or the execution of the Works.
Sub-Clause 1.1.24	**'Date of Completion'** means the date stated in the Taking-Over Certificate issued by the Engineer; or, if the last paragraph of Sub-Clause 10.1 [*Taking Over the Works and Sections*] applies, the date on which the Works or Section are deemed to have been completed in accordance with the Contract; or, if Sub-Clause 10.2 [*Taking Over Parts*] or Sub-Clause 10.3. [*Interference with Tests on Completion*] applies, the date on which the Works orSection or Part are deemed to have been taken over by the Employer.	**'Date of Completion'** means the date stated in the Taking-Over Certificate issued by the Engineer under Sub-Clause 10.1 [*Taking Over the Works and Sections*] or the first paragraph of Sub-Clause 10.2 [*Taking Over Parts*] or, if the last paragraph of Sub-Clause 10.1 [*Taking Over the Works and Sections*] applies, the date on which the Works or Section are deemed to have been completed in accordance with the Contract; or, if the second paragraph of Sub-Clause 10.2 [*Taking Over Parts*] or Sub-Clause 10.3. [*Interference with Tests on Completion*] applies, the date on which the Works or Section or Part are deemed to have been taken over by the Employer.
Sub-Clause 1.1.29	**'Dispute'** means any situation where: (a) one Party makes a claim against the other Party (which may be a Claim, as defined in these Conditions, or a matter to be determined by the Engineer under these Conditions, or otherwise);	**'Dispute'** means any situation where: (a) one Party has made a Claim, or there has been a matter to be agreed or determined under sub-paragraph (a) of Sub-Clause 3.7 [*Agreement or Determination*]; (b) the Engineer's determination under Sub-Clause 3.7.2 [*Engineer's Determination*] was a rejection (in whole or in part) of:

Sub-Clause	FIDIC 2017	2022 REPRINT
	(b) the other Party (or the Engineer under Sub-Clause 3.7.2 [*Engineer's Determination*]) rejects the claim in whole or in part; and (c) the first Party does not acquiesce (by giving a NOD under Sub-Clause 3.7.5 [*Dissatisfaction with Engineer's determination*] or otherwise), provided however that a failure by the other Party (or the Engineer) to oppose or respond to the claim, in whole or in part, may constitute a rejection if, in the circumstances, the DAAB or the arbitrator(s), as the case may be, deem it reasonable for it to do so.	(i) the Claim (or there was a deemed rejection under sub-paragraph (i) of Sub-Clause 3.7.3 [*Time limits*]); or (ii) a Party's assertion(s) in respect of the matter as the case may be; and (c) either Party has given a NOD under Sub-Clause 3.7.5 [*Dissatisfaction with Engineer's determination*].
Sub-Clause 1.1.37	**'Exceptional Event'** means an event or circumstance as defined in Sub-Clause 18.1 [*Exceptional Events*].	**'Exceptional Event'** means an exceptional event or circumstance as defined in Sub-Clause 18.1 [*Exceptional Events*].
Sub-Clause 1.1.58	**'Part'** means a part of the Works or part of a Section (as the case may be) which is used by the Employer and deemed to have been taken over under Sub-Clause 10.2 [*Taking Over Parts*].	**'Part'** means a part of the Works or part of a Section (as the case may be) which is taken over by the Employer under the first paragraph, or used by the Employer and deemed to have been taken over under the second paragraph, of Sub-Clause 10.2 [*Taking Over Parts*].
Sub-Clause 1.15	Neither Party shall be liable to the other Party for loss of use of any Works, loss of profit, loss of any contract or for any indirect or consequential loss or damage which may be suffered by the other Party in connection with the Contract, other than under: (a) Sub-Clause 8.8 [*Delay Damages*]; (b) sub-paragraph (c) of Sub-Clause 13.3.1 [*Variation by Instruction*]; (c) Sub-Clause 15.7 [*Payment after Termination for Employer's Convenience*]; (d) Sub-Clause 16.4 [*Payment after Termination by Contractor*]; (e) Sub-Clause 17.3 [*Intellectual and Industrial Property Rights*]; (f) the first paragraph of Sub-Clause 17.4 [*Indemnities by Contractor*]; and (g) Sub-Clause 17.5 [*Indemnities by Employer*].	Neither Party shall be liable to the other Party for loss of use of any Works, loss of profit, loss of any contract or for any indirect or consequential loss or damage which may be suffered by the other Party in connection with the Contract, other than under: (a) Sub-Clause 8.8 [*Delay Damages*]; (b) sub-paragraph (c) of Sub-Clause 13.3.1 [*Variation by Instruction*]; (c) Sub-Clause 15.7 [*Payment after Termination for Employer's Convenience*]; (d) Sub-Clause 16.4 [*Payment after Termination by Contractor*]; (e) Sub-Clause 17.3 [*Intellectual and Industrial Property Rights*]; (f) the first paragraph of Sub-Clause 17.4 [*Indemnities by Contractor*]; and (g) Sub-Clause 17.5 [*Indemnities by Employer*].

Sub-Clause	FIDIC 2017	2022 REPRINT
	The total liability of the Contractor to the Employer under or in connection with the Contract, other than: (i) under Sub-Clause 2.6 *[Employer-Supplied Materials and Employer's Equipment];* (ii) under Sub-Clause 4.19 [*Temporary Utilities*]; (iii) under Sub-Clause 17.3 [*Intellectual and Industrial Property Rights*]; and (iv) under the first paragraph of Sub-Clause 17.4 [*Indemnities by Contractor*], shall not exceed the sum stated in the Contract Data or (if a sum is not so stated) the Accepted Contract Amount. This Sub-Clause shall not limit liability in any case of fraud, gross negligence, deliberate default or reckless misconduct by the defaulting Party.	The total liability of the Contractor to the Employer under or in connection with the Contract, other than under: (i) Sub-Clause 2.6 *[Employer-Supplied Materials and Employer's Equipment];* (ii) Sub-Clause 4.19 [*Temporary Utilities*]; (iii) Sub-Clause 17.3 [*Intellectual and Industrial Property Rights*]; and (iv) the first paragraph of Sub-Clause 17.4 [*Indemnities by Contractor*], shall not exceed the sum stated in the Contract Data or (if a sum is not so stated) the Accepted Contract Amount. This Sub-Clause shall not limit liability in any case of fraud, gross negligence, deliberate default or reckless misconduct by the defaulting Party.
Sub-Clause 3.7	When carrying out his/her duties under this Sub-Clause, the Engineer shall act neutrally between the Parties and shall not be deemed to act for the Employer. Whenever these Conditions provide that the Engineer shall proceed under this Sub-Clause to agree or determine any matter or Claim, the following procedure shall apply: .../...	When carrying out his/her duties under this Sub-Clause, the Engineer shall act neutrally between the Parties and shall not be deemed to act for the Employer. Whenever these Conditions provide that the Engineer shall proceed under this Sub-Clause 3.7 to agree or determine either: (a) any matter, as provided for in Sub-Clauses 4.7.3, 10.2, 11.2, 12.1, 12.3, 13.3.1, 13.5, 14.4, 14.5, 14.6.3, 15.3, 15.6 and 18.5, identifying in the same Sub-Clause the date of commencement of the corresponding time limit for agreement under Sub-Clause 3.7.3 [*Time limits*]; or (b) any Claim, the following procedure shall apply: .../...
Sub-Clause 3.7.3	The Engineer shall give the Notice of agreement, if agreement is achieved, within 42 days or within such other time limit as may be proposed by the Engineer and agreed by both	The Engineer shall give the Notice of agreement, if agreement is achieved, within 42 days or within such other time limit as may be proposed by the Engineer and agreed by both

Sub-Clause	FIDIC 2017	2022 REPRINT
	Parties (the 'time limit for agreement' in these Conditions), after:	Parties (the 'time limit for agreement' in these Conditions), after:
	(a) in the case of a matter to be agreed or determined (not a Claim), the date of commencement of the time limit for agreement as stated in the applicable Sub-Clause of these Conditions;	(a) in the case of a matter to be agreed or determined under sub-paragraph (a) of Sub-Clause 3.7 [*Agreement or Determination*], the date of commencement of the time limit for agreement as stated in the applicable Sub-Clause of these Conditions;
	(b) in the case of a Claim under sub-paragraph (c) of Sub-Clause 20.1 [*Claims*], the date the Engineer receives a Notice under Sub-Clause 20.1 from the claiming Party; or	(b) in the case of a Claim under sub-paragraph (c) of Sub-Clause 20.1 [*Claims*], the date the Engineer receives a Notice under Sub-Clause 20.1 from the claiming Party; or
	(c) in the case of a Claim under sub-paragraph (a) or (b) of Sub-Clause 20.1 [*Claims*], the date the Engineer receives: (i) a fully detailed Claim under Sub-Clause 20.2.4 [*Fully detailed Claim*]; or (ii) in the case of a Claim under Sub-Clause 20.2.6 [*Claims of continuing effect*], an interim or final fully detailed Claim (as the case may be).	(c) in the case of a Claim under sub-paragraph (a) or (b) of Sub-Clause 20.1 [*Claims*], the date the Engineer receives: (i) a fully detailed Claim under Sub-Clause 20.2.4 [*Fully detailed Claim*]; or (ii) in the case of a Claim under Sub-Clause 20.2.6 [*Claims of continuing effect*], an interim or final fully detailed Claim (as the case may be).
		(d) in the case of a Claim under sub-paragraph (c) of Sub-Clause 20.1 [*Claims*], the date the Engineer receives a Notice under Sub-Clause 20.1 from the claiming Party; or
	The Engineer shall give the Notice of his/her determination within 42 days or within such other time limit as may be proposed by the Engineer and agreed by both Parties (the 'time limit for determination' in these Conditions), after the date corresponding to his/her obligation to proceed under the last paragraph of Sub-Clause 3.7.1 [*Consultation to reach agreement*].	The Engineer shall give the Notice of his/her determination within 42 days or within such other time limit as may be proposed by the Engineer and agreed by both Parties (the 'time limit for determination' in these Conditions), after the date corresponding to his/her obligation to proceed under the last paragraph of Sub-Clause 3.7.1 [*Consultation to reach agreement*].
	If the Engineer does not give the Notice of agreement or determination within the relevant time limit:	If the Engineer does not give the Notice of determination within the relevant time limit:
	(i) in the case of a Claim, the Engineer shall be deemed to have given a determination rejecting the Claim; or	(i) in the case of a Claim, the Engineer shall be deemed to have given a determination rejecting the Claim; or
	(ii) in the case of a matter to be agreed or determined, the matter shall be deemed to be a Dispute which may be referred by either Party to the DAAB for its decision under Sub-Clause 21.4 [*Obtaining DAAB's*	(ii) in the case of a matter to be agreed or determined under sub-paragraph (a) of

Sub-Clause	FIDIC 2017	2022 REPRINT
	Decision] without the need for a NOD (and Sub-Clause 3.7.5 [*Dissatisfaction with Engineer's determination*] and sub-paragraph (a) of Sub-Clause 21.4.1 [*Reference of a Dispute to the DAAB*] shall not apply).	Sub-Clause 3.7 [*Agreement or Determination*], the matter shall be deemed to be a Dispute which may be referred by either Party to the DAAB for its decision under Sub-Clause 21.4 [*Obtaining DAAB's Decision*] without the need for a NOD (and Sub-Clause 3.7.5 [*Dissatisfaction with Engineer's determination*] and sub-paragraph (a) of Sub-Clause 21.4.1 [*Reference of a Dispute to the DAAB*] shall not apply).
Sub-Clause 4.7.3	After receiving a Notice from the Contractor under Sub-Clause 4.7.2 [*Errors*], the Engineer shall proceed under Sub-Clause 3.7 [*Agreement or Determination*] to agree or determine: (a) whether or not there is an error in the items of reference; (b) whether or not (taking account of cost and time) an experienced contractor exercising due care would have discovered such an error ▪ when examining the Site, the Drawings and the Specification before submitting the Tender; or ▪ if the items of reference are specified on the Drawings and/or in the Specification and the Contractor's Notice is given after the expiry of the period stated in sub-paragraph (a) of Sub-Clause 4.7.2; and (c) what measures (if any) the Contractor is required to take to rectify the error (and, for the purpose of Sub-Clause 3.7.3 [*Time limits*], the date the Engineer receives the Contractor's Notice under Sub-Clause 4.7.2 [*Errors*] shall be the date of commencement of the time limit for agreement under Sub-Clause 3.7.3). If, under sub-paragraph (b) above, an experienced contractor would not have discovered the error:	After receiving a Notice from the Contractor under Sub-Clause 4.7.2 [*Errors*], the Engineer shall proceed under Sub-Clause 3.7 [*Agreement or Determination*] to agree or determine: (a) whether or not there is an error in the items of reference; (b) whether or not (taking account of cost and time) an experienced contractor exercising due care would have discovered such an error ▪ when examining the Site, the Drawings and the Specification before submitting the Tender; or ▪ if the items of reference are specified on the Drawings and/or in the Specification and the Contractor's Notice is given after the expiry of the period stated in sub-paragraph (a) of Sub-Clause 4.7.2; and (c) what measures (if any) the Contractor is required to take to rectify the error (and, for the purpose of Sub-Clause 3.7.3 [*Time limits*], the date the Engineer receives the Contractor's Notice under Sub-Clause 4.7.2 [*Errors*] shall be the date of commencement of the time limit for agreement under Sub-Clause 3.7.3). If it is agreed or determined, under sub-paragraphs (a) and (b) above, that there is an error in the items of reference that an experienced contractor would not have discovered:

Sub-Clause	FIDIC 2017	2022 REPRINT
	(i) Sub-Clause 13.3.1 [*Variation by Instruction*] shall apply to the measures that the Contractor is required to take (if any); and (ii) if the Contractor suffers delay and/or incurs Cost as a result of the error, the Contractor shall be entitled subject to Sub-Clause 20.2 [*Claims for Payment and/or EOT*] to EOT and/or payment of such Cost Plus Profit.	(i) Sub-Clause 13.3.1 [*Variation by Instruction*] shall apply to the measures that the Contractor is required to take (if any); and (ii) if there are no such measures, and therefore no Variation, but the Contractor suffers delay and/or incurs Cost as a result of the error, the Contractor shall be entitled subject to Sub-Clause 20.2 [*Claims for Payment and/or EOT*] to EOT and/or payment of such Cost Plus Profit.
Sub-Clause 8.5	.../... When determining each EOT under Sub-Clause 20.2 [*Claims for Payment and/or EOT*], the Engineer shall review previous determinations under Sub-Clause 3.7 [*Agreement or Determination*] and may increase, but shall not decrease, the total EOT. .../...	.../... When agreeing or determining each EOT, the Engineer shall review previous agreements and determinations of EOT under Sub-Clause 3.7 [*Agreement or Determination*] and may increase, but shall not decrease, the total EOT. .../...
Sub-Clause 10.2	The Engineer may, at the sole discretion of the Employer, issue a Taking-Over Certificate for any part of the Permanent Works. The Employer shall not use any part of the Works (other than as a temporary measure, which is either stated in the Specification or with the prior agreement of the Contractor) unless and until the Engineer has issued a Taking-Over Certificate for this part. However, if the Employer does use any part of the Works before the Taking-Over Certificate is issued the Contractor shall give a Notice to the Engineer identifying such part and describing such use, and: (a) that Part shall be deemed to have been taken over by the Employer as from the date on which it is used; (b) the Contractor shall cease to be liable for the care of such Part as from this date, when responsibility shall pass to the Employer; and	The Engineer may, at the sole discretion of the Employer, issue a Taking-Over Certificate for any part of the Permanent Works. The Employer shall not use any part of the Works (other than as a temporary measure, which is either stated in the Specification or with the prior agreement of the Contractor) unless and until the Engineer has issued a Taking-Over Certificate for this part. However, if the Employer does use any part of the Works before the Taking-Over Certificate is issued the Contractor shall give a Notice to the Engineer identifying such part and describing such use, and: (a) that Part shall be deemed to have been taken over by the Employer as from the date on which it is used; (b) the Contractor shall cease to be liable for the care of such Part as from this date, when responsibility shall pass to the Employer; and

Sub-Clause	FIDIC 2017	2022 REPRINT
	(c) the Engineer shall immediately issue a Taking-Over Certificate for this Part, and any outstanding work to be completed (including Tests on Completion) and/or defects to be remedied shall be listed in this certificate.	(c) the Engineer shall immediately issue a Taking-Over Certificate for this Part, and any outstanding work to be completed (including Tests on Completion) and/or defects to be remedied shall be listed in this certificate.
	After the Engineer has issued a Taking-Over Certificate for a Part, the Contractor shall be given the earliest opportunity to take such steps as may be necessary to carry out the outstanding work (including Tests on Completion) and/or remedial work for any defects listed in the certificate. The Contractor shall carry out these works as soon as practicable and, in any case, before the expiry date of the relevant DNP.	After a Part has been taken over or is deemed to have been taken over, the Contractor shall be given the earliest opportunity to take such steps as may be necessary to carry out the outstanding work (including Tests on Completion) and/or remedial work for any defects listed in the certificate. The Contractor shall carry out these works as soon as practicable and, in any case, before the expiry date of the relevant DNP.
	If the Contractor incurs Cost as a result of the Employer taking over and/or using a Part, the Contractor shall be entitled subject to Sub-Clause 20.2 [*Claims for Payment and/or EOT*] to payment of such Cost Plus Profit.	If the Contractor incurs Cost as a result of the taking over or deemed taking over of a Part, the Contractor shall be entitled subject to Sub-Clause 20.2 [*Claims for Payment and/or EOT*] to payment of such Cost Plus Profit.
	If the Engineer issues a Taking-Over Certificate for any part of the Works, or if the Employer is deemed to have taken over a Part under sub-paragraph (a) above, for any period of delay after the date under sub-paragraph (a) above, the Delay Damages for completion of the remainder of the Works shall be reduced. Similarly, the Delay Damages for the remainder of the Section (if any) in which this Part is included shall also be reduced. This reduction shall be calculated as the proportion which the value of the Part (except the value of any outstanding works and/or defects to be remedied) bears to the value of the Works or Section (as the case may be) as a whole. The Engineer shall proceed under Sub-Clause 3.7 [*Agreement or Determination*] to agree or determine this reduction (and for	For any period of delay after the date that a Part has been taken over or is deemed to have been taken over, the Delay Damages for completion of the Works or the Section (as the case may be) in which this Part is included shall be reduced. This reduction shall be calculated as the proportion which the value of the Part (except the value of any outstanding works and/or defects to be remedied) bears to the value of the Works or Section (as the case may be) as a whole. The Engineer shall proceed under Sub-Clause 3.7 [*Agreement or Determination]* to agree or determine this reduction (and, for the purpose of Sub-Clause 3.7.3 [*Time limits*], the date the Engineer issues the Taking-Over Certificate under the first paragraph of, or receives the Contractor's Notice under the second paragraph of, this Sub-Clause (as the case may be) shall be the date of

Sub-Clause	FIDIC 2017	2022 REPRINT
	the purpose of Sub-Clause 3.7.3 [*Time limits*], the date the Engineer receives the Contractor's Notice under this Sub-Clause shall be the date of commencement of the time limit for agreement under Sub-Clause 3.7.3). The provisions of this paragraph shall only apply to the daily rate of Delay Damages, and shall not affect the maximum amount of these damages.	commencement of the time limit for agreement under Sub-Clause 3.7.3). The provisions of this paragraph shall only apply to the daily rate of Delay Damages, and shall not affect the maximum amount of these damages.
Sub-Clause 14.13	Within 28 days after receiving the Final Statement or the Partially Agreed Final Statement (as the case may be), and the discharge under Sub-Clause 14.12 [*Discharge*], the Engineer shall issue to the Employer (with a copy to the Contractor), the Final Payment Certificate which shall state: (a) the amount which the Engineer fairly considers is finally due, including any additions and/or deductions which have become due under Sub-Clause 3.7 [*Agreement or Determination*] or under the Contract or otherwise; and (b) after giving credit to the Employer for all amounts previously paid by the Employer and for all sums to which the Employer is entitled, and after giving credit to the Contractor for all amounts (if any) previously paid by the Contractor and/or received by the Employer under the Performance Security, the balance (if any) due from the Employer to the Contractor or from the Contractor to the Employer, as the case may be.	Within 28 days after receiving the Final Statement or the Partially Agreed Final Statement (as the case may be), and the discharge under Sub-Clause 14.12 [*Discharge*], the Engineer shall issue to the Employer (with a copy to the Contractor), the Final Payment Certificate which shall state: (a) the amount which the Engineer fairly considers is finally due, including any additions and/or deductions which have become due under Sub-Clause 3.7 [*Agreement or Determination*] or under the Contract or otherwise; and (b) after giving credit to the Employer for all amounts previously paid by the Employer and for all sums to which the Employer is entitled, and after giving credit to the Contractor for all amounts (if any) previously paid by the Contractor and/or received by the Employer under the Performance Security, the balance (if any) due from the Employer to the Contractor or from the Contractor to the Employer, as the case may be.
	If the Contractor has not submitted a draft final Statement within the time specified under Sub-Clause 14.11.1 [*Draft Final Statement*], the Engineer shall request the Contractor to do so. Thereafter, if the Contractor fails to submit a draft final Statement within a period of 28 days, the Engineer shall issue the FPC for such an amount as the Engineer fairly considers to be due.	If the Contractor has not submitted a draft final Statement within the time specified under Sub-Clause 14.11.1 [*Draft Final Statement*], the Engineer shall request the Contractor to do so. Thereafter, if the Contractor fails to submit a draft final Statement within a period of 28 days, the Engineer shall issue the FPC for such an amount as the Engineer fairly considers to be due.

Sub-Clause	FIDIC 2017	2022 REPRINT
	If: (i) the Contractor has submitted a Partially Agreed Final Statement under Sub-Clause 14.11.2 [*Agreed Final Statement*]; or (ii) no Partially Agreed Final Statement has been submitted by the Contractor but, to the extent that a draft final Statement submitted by the Contractor is deemed by the Engineer to be a Partially Agreed Final Statement the Engineer shall proceed in accordance with Sub-Clause 14.6 [*Issue of IPC*] to issue an IPC.	If the Contractor has not submitted a discharge under Sub-Clause 14.12 [*Discharge*] but has either: (i) submitted a Partially Agreed Final Statement under Sub-Clause 14.11.2 [*Agreed Final Statement*]; or (ii) not done so but, to the extent that a draft final Statement submitted by the Contractor is deemed by the Engineer to be a Partially Agreed Final Statement, the Engineer shall proceed in accordance with Sub-Clause 14.6 [*Issue of IPC*] to issue an IPC.
Sub-Clause 17.2	The Contractor shall be liable for any loss or damage caused by the Contractor to the Works, Goods or Contractor's Documents after the issue of a Taking-Over Certificate. The Contractor shall also be liable for any loss or damage, which occurs after the issue of a Taking-Over Certificate and which arose from an event which occurred before the issue of this Taking-Over Certificate, for which the Contractor was liable. The Contractor shall have no liability whatsoever, whether by way of indemnity or otherwise, for loss or damage to the Works, Goods or Contractor's Documents caused by any of the following events (except to the extent that such Works, Goods or Contractor's Documents have been rejected by the Engineer under Sub-Clause 7.5 [*Defects and Rejection*] before the occurrence of any of the following events): (a) interference, whether temporary or permanent, with any right of way, light, air, water or other easement (other than that resulting from the Contractor's method of construction) which is the unavoidable result of the execution of the Works in accordance with the Contract;	The Contractor shall be liable for any loss or damage caused by the Contractor to the Works, Goods or Contractor's Documents after the issue of a Taking-Over Certificate. The Contractor shall also be liable for any loss or damage, which occurs after the issue of a Taking-Over Certificate and which arose from an event which occurred before the issue of this Taking-Over Certificate, for which the Contractor was liable. The Contractor shall have no liability whatsoever, whether by way of indemnity or otherwise, for loss or damage to the Works, Goods or Contractor's Documents caused by any of the following events (except to the extent that such Works, Goods or Contractor's Documents have been rejected by the Engineer under Sub-Clause 7.5 [*Defects and Rejection*] before the occurrence of any of the following events): (a) interference, whether temporary or permanent, with any right of way, light, air, water or other easement (other than that resulting from the Contractor's method of construction) which is the unavoidable result of the execution of the Works in accordance with the Contract;

Sub-Clause	FIDIC 2017	2022 REPRINT
	(b) use or occupation by the Employer of any part of the Permanent Works, except as may be specified in the Contract;	(b) use or occupation by the Employer of any part of the Permanent Works, except as may be specified in the Contract;
	(c) fault, error, defect or omission in any element of the design of the Works by the Employer or which may be contained in the Specification and Drawings (and which an experienced contractor exercising due care would not have discovered when examining the Site and the Specification and Drawings before submitting the Tender), other than design carried out by the Contractor in accordance with the Contractor's obligations under the Contract;	(c) fault, error, defect or omission in any element of the design of the Works by the Employer or which may be contained in the Specification and Drawings (and which an experienced contractor exercising due care would not have discovered when examining the Site and the Specification and Drawings before submitting the Tender), other than design carried out by the Contractor in accordance with the Contractor's obligations under the Contract;
	(d) any operation of the forces of nature (other than those allocated to the Contractor in the Contract Data) which is Unforeseeable or against which an experienced contractor could not reasonably have been expected to have taken adequate preventative precautions;	(d) any operation of the forces of nature (other than those allocated to the Contractor in the Contract Data) which is Unforeseeable or against which an experienced contractor could not reasonably have been expected to have taken adequate preventative precautions;
	(e) any of the events or circumstances listed under sub-paragraphs (a) to (f) of Sub-Clause 18.1 [*Exceptional Events*]; and/or	(e) any Exceptional Event; and/or
	(f) any act or default of the Employer's Personnel or the Employer's other contractors.	(f) any act or default of the Employer, the Employer's Personnel or the Employer's other contractors.
	Subject to Sub-Clause 18.4 [*Consequences of an Exceptional Event*], if any of the events described in sub-paragraphs (a) to (f) above occurs and results in damage to the Works, Goods or Contractor's Documents the Contractor shall promptly give a Notice to the Engineer. Thereafter, the Contractor shall rectify any such loss and/or damage that may arise to the extent instructed by the Engineer. Such instruction shall be deemed to have been given under Sub-Clause 13.3.1 [*Variation by Instruction*].	If any of the events described in sub-paragraphs (a) to (f) above occurs and results in loss and/or damage to the Works, Goods or Contractor's Documents the Contractor shall promptly give a Notice to the Engineer. Thereafter, the Contractor shall rectify any such loss and/or damage that may arise to the extent instructed by the Engineer. Such instruction shall be deemed to have been given under Sub-Clause 13.3.1 [*Variation by Instruction*] and, in the case of sub-paragraph (e) above, shall be without prejudice to any other rights the Contractor may have under Sub-Clause 18.4 [*Consequences of an Exceptional Event*].
	If the loss or damage to the Works or Goods or Contractor's Documents results from a combination of:	
	(i) any of the events described in sub-paragraphs (a) to (f) above, and	

Sub-Clause	FIDIC 2017	2022 REPRINT
	(ii) a cause for which the Contractor is liable, and the Contractor suffers a delay and/or incurs Cost from rectifying the loss and/or damage, the Contractor shall subject to Sub-Clause 20.2 [*Claims for Payment and/or EOT*] be entitled to a proportion of EOT and/or Cost Plus Profit to the extent that any of the above events have contributed to such delays and/or Cost.	If the loss and/or damage to the Works or Goods or Contractor's Documents results from a combination of: (i) any of the events described in sub-paragraphs (a) to (f) above, and (ii) a cause for which the Contractor is liable, and the Contractor suffers a delay and/or incurs Cost from rectifying the loss and/or damage, the Contractor shall subject to Sub-Clause 20.2 [*Claims for Payment and/or EOT*] be entitled to a proportion of EOT and/or Cost Plus Profit to the extent that any of the above events have contributed to such delay and/or Cost.
Sub-Clause 20.1	A Claim may arise: (a) if the Employer considers that the Employer is entitled to any additional payment from the Contractor (or reduction in the Contract Price) and/or to an extension of the DNP; (b) if the Contractor considers that the Contractor is entitled to any additional payment from the Employer and/or to EOT; or (c) if either Party considers that he/she is entitled to another entitlement or relief against the other Party. Such other entitlement or relief may be of any kind whatsoever (including in connection with any certificate, determination, instruction, Notice, opinion or valuation of the Engineer) except to the extent that it involves any entitlement referred to in sub-paragraphs (a) and/or (b) above. In the case of a Claim under sub-paragraph (a) or (b) above, Sub-Clause 20.2 [*Claims for Payment and/or EOT*] shall apply. In the case of a Claim under sub-paragraph (c) above, where the other Party or the Engineer has disagreed with the requested entitlement or relief (or is deemed to have disagreed if he/she does not respond within a reasonable	A Claim may arise: (a) if the Employer considers that the Employer is entitled to any additional payment from the Contractor (or reduction in the Contract Price) and/or to an extension of the DNP; (b) if the Contractor considers that the Contractor is entitled to any additional payment from the Employer and/or to EOT; or (c) if either Party considers that he/she is entitled to another entitlement or relief against the other Party. Such other entitlement or relief may be of any kind whatsoever (including in connection with any certificate, instruction, Notice, opinion or valuation of the Engineer) except to the extent that it involves any entitlement referred to in sub-paragraphs (a) and/or (b) above. In the case of a Claim under sub-paragraph (a) or (b) above, Sub-Clause 20.2 [*Claims for Payment and/or EOT*] shall apply. In the case of a Claim under sub-paragraph (c) above, where the other Party or the Engineer has disagreed with the requested entitlement or relief (or is deemed to have disagreed if he/she does not respond within a reasonable time), a Dispute shall not be deemed to have

Sub-Clause	FIDIC 2017	2022 REPRINT
	time), a Dispute shall not be deemed to have arisen but the claiming Party may, by giving a Notice refer the Claim to the Engineer and Sub-Clause 3.7 [*Agreement or Determination*] shall apply. This Notice shall be given as soon as practicable after the claiming Party becomes aware of the disagreement (or deemed disagreement) and include details of the claiming Party's case and the other Party's or the Engineer's disagreement (or deemed disagreement).	arisen except if any of sub-paragraphs (a) to (c) of Sub-Clause 21.4 [*Obtaining DAAB's Decision*] applies. The claiming Party may, by giving a Notice refer the Claim to the Engineer and Sub-Clause 3.7 [*Agreement or Determination*] shall apply. This Notice shall be given as soon as practicable after the claiming Party becomes aware of the disagreement (or deemed disagreement) and shall include details of the claiming Party's case and the other Party's or the Engineer's disagreement (or deemed disagreement).
Sub-Clause 21.2	If any of the following conditions apply, namely: (a) if the DAAB is to comprise a sole member, the Parties fail to agree the appointment of this member by the date stated in the first paragraph of Sub-Clause 21.1 [*Constitution of the DAAB*]; or (b) if the DAAB is to comprise three persons, and if by the date stated in the first paragraph of Sub-Clause 21.1 [*Constitution of the DAAB*]: (i) either Party fails to select a member (for agreement by the other Party); (ii) either Party fails to agree a member selected by the other Party; and/or (iii) the Parties fail to agree the appointment of the third member (to act as chairperson) of the DAAB; (c) the Parties fail to agree the appointment of a replacement within 42 days after the date on which the sole member or one of the three members declines to act or is unable to act as a result of death, illness, disability, resignation, or termination of appointment; or (d) if, after the Parties have agreed the appointment of the member(s) or replacement, such appointment cannot be effected because one Party refuses or fails to sign a DAAB Agreement with any such member or replacement (as the case may be) within 14 days of the other Party's request to do so,	If any of the following conditions apply, namely: (a) if the DAAB is to comprise a sole member, the Parties fail to agree the appointment of this member by the date stated in the first paragraph of Sub-Clause 21.1 [*Constitution of the DAAB*]; or (b) if the DAAB is to comprise three persons, and if by the date stated in the first paragraph of Sub-Clause 21.1 [*Constitution of the DAAB*]: (iv) either Party fails to select a member (for agreement by the other Party); (v) either Party fails to agree a member selected by the other Party; and/or (vi) the Parties fail to agree the appointment of the third member (to act as chairperson) of the DAAB; (c) the Parties fail to agree the appointment of a replacement within 42 days after the date on which the sole member or one of the three members declines to act or is unable to act as a result of death, illness, disability, resignation, or termination of appointment; or (d) if, after the Parties have agreed the appointment of the member(s) or replacement, such appointment cannot be effected because one Party refuses or fails to sign a DAAB Agreement with any such member or replacement (as the case may be) within 14 days of the other Party's request to do so, or because the terms of

Sub-Clause	FIDIC 2017	2022 REPRINT
	then the appointing entity or official named in the Contract Data shall, at the request of either or both Parties and after due consultation with both Parties, appoint the member(s) of the DAAB (who, in the case of sub-paragraph (d) above, shall be the agreed member(s) or replacement). This appointment shall be final and conclusive.	the DAAB Agreement (including the amount of the monthly fee or the daily fee) cannot be agreed with the member or replacement within 14 days after he/she has been advised by the Parties that they have agreed to his/her appointment, then, unless otherwise agreed by the Parties, either or both Parties may apply to the President of FIDIC or a person appointed by the President, who shall be the appointing official under the Contract. The appointing official shall, after due consultation with both Parties and after consulting the prospective member(s) or replacement: – appoint the member(s) of the DAAB or the replacement; and – set the terms of the appointment, including the amounts of the monthly fee and the daily fee for each member or replacement. Selection of the member(s) or replacement to be so appointed shall not be limited to those persons named in the list in the Contract Data or, in the case of sub-paragraph (d) above, to the member(s) or replacement agreed by the Parties. This appointment and its terms shall be final and conclusive.
	Thereafter, the Parties and the member(s) so appointed shall be deemed to have signed and be bound by a DAAB Agreement under which: (i) the monthly services fee and daily fee shall be as stated in the terms of the appointment; and (ii) the law governing the DAAB Agreement shall be the governing law of the Contract defined in Sub-Clause 1.4 [*Law and Language*]. Each Party shall be responsible for paying one-half of the remuneration of the appointing entity or official. If the Contractor pays the remuneration in full, the Contractor shall include one-half of the amount of such remuneration in a Statement and the Employer shall then pay the Contractor in accordance with the Contract. If the Employer pays the remuneration in full, the Engineer shall include one-half of the amount of such remuneration as a deduction under sub-paragraph (b) of Sub-Clause 14.6.1 [*The IPC*].	Thereafter, the Parties and the member(s) so appointed shall be deemed to have signed and be bound by a DAAB Agreement under which: (i) the monthly services fee and daily fee shall be as stated in the terms of the appointment; and (ii) the law governing the DAAB Agreement shall be the governing law of the Contract defined in Sub-Clause 1.4 [*Law and Language*]. Each Party shall be responsible for paying one-half of the remuneration of the appointing official. If the Contractor pays the remuneration in full, the Contractor shall include one-half of the amount of such remuneration in a Statement and the Employer shall then pay the

Sub-Clause	FIDIC 2017	2022 REPRINT
		Contractor in accordance with the Contract. If the Employer pays the remuneration in full, the Engineer shall include one-half of the amount of such remuneration as a deduction under sub-paragraph (b) of Sub-Clause 14.6.1 [*The IPC*].
Sub-Clause 21.4	If a Dispute arises between the Parties then either Party may refer the Dispute to the DAAB for its decision (whether or not any informal discussions have been held under Sub-Clause 21.3 [*Avoidance of Disputes*]) and the following provisions shall apply.	If a Dispute arises between the Parties then either Party may refer the Dispute to the DAAB for its decision (whether or not any informal discussions have been held under Sub-Clause 21.3 [*Avoidance of Disputes*]). In addition to the situation described in the definition of Dispute under Sub-Clause 1.1.29 above, a Dispute shall be deemed to have arisen if: (a) there is a failure as referred to under sub-paragraph (b), or a non-payment as referred to under sub-paragraph (c), of Sub-Clause 16.2.1 [*Notice*]; (b) the Contractor is entitled to receive financing charges under Sub-Clause 14.8 [*Delayed Payment*] but does not receive payment thereof from the Employer within 28 days after his request for such payment; or (c) a Party has given: (i) a Notice of intention to terminate the Contract under Sub-Clause 15.2.1 [*Notice*] or Sub-Clause 16.2.1 [*Notice*] (as the case may be); or (ii) a Notice of termination under Sub-Clause 15.2.2 [*Termination*], Sub-Clause 16.2.2 [*Termination*], Sub-Clause 18.5 [*Optional Termination*] or Sub-Clause 18.6 [*Release from Performance under the Law*] (as the case may be); and the other Party has disagreed with the first Party's entitlement to give such Notice; which Dispute may be referred by either Party under this Sub-Clause 21.4 without the need for a NOD (and Sub-Clause 3.7 [*Agreement or Determination*] and sub-paragraph (a) of

Sub-Clause	FIDIC 2017	2022 REPRINT
		Sub-Clause 21.4.1 [*Reference of a Dispute to the DAAB*] shall not apply). Where a Dispute is to be referred to the DAAB for its decision, the following provisions shall apply.
Sub-Clause 21.4.1	The reference of a Dispute to the DAAB (the 'reference' in this Sub-Clause 21.4) shall: (a) if Sub-Clause 3.7 [*Agreement or Determination*] applied to the subject matter of the Dispute, be made within 42 days of giving or receiving (as the case may be) a NOD under Sub-Clause 3.7.5 [*Dissatisfaction with Engineer's determination*]. If the Dispute is not referred to the DAAB within this period of 42 days, such NOD shall be deemed to have lapsed and no longer be valid; (b) state that it is given under this Sub-Clause; (c) set out the referring Party's case relating to the Dispute; (d) be in writing, with copies to the other Party and the Engineer; and (e) for a DAAB of three persons, be deemed to have been received by the DAAB on the date it is received by the chairperson of the DAAB. The reference of a Dispute to the DAAB under this Sub-Clause shall, unless prohibited by law, be deemed to interrupt the running of any applicable statute of limitation or prescription period.	The reference of a Dispute to the DAAB (the 'reference' in this Sub-Clause 21.4) shall: (a) subject to sub-paragraph (ii) of Sub-Clause 3.7.3 [*Time limits*] and the provisions of the second paragraph of Sub-Clause 21.4 [*Obtaining DAAB's Decision*], be made within 42 days of giving or receiving (as the case may be) a NOD under Sub-Clause 3.7.5 [*Dissatisfaction with Engineer's determination*]. If the Dispute is not referred to the DAAB within this period of 42 days, such NOD shall be deemed to have lapsed and no longer be valid; (b) state that it is given under this Sub-Clause; (c) set out the referring Party's case relating to the Dispute; (d) be in writing, with copies to the other Party and the Engineer; and (e) for a DAAB of three persons, be deemed to have been received by the DAAB on the date it is received by the chairperson of the DAAB. The reference of a Dispute to the DAAB under this Sub-Clause shall, unless prohibited by law, be deemed to interrupt the running of any applicable statute of limitation or prescription period.
APPENDIX – GENERAL CONDITIONS OF DAAB AGREEMENT		
Sub-Clause 4.1	Further to Sub-Clauses 3.1 and 3.2 above, the DAAB Member shall: (a) have no financial interest in the Contract, or in the project of which the Works are part, except for payment under the DAA Agreement; (b) have no interest whatsoever (financial or otherwise) in the Employer, the Contractor, the Employer's Personnel or the Contractor's Personnel;	Further to Sub-Clauses 3.1 and 3.2 above, the DAAB Member shall: (a) have no financial interest in the Contract, or in the project of which the Works are part, except for payment under the DAA Agreement; (b) have no interest whatsoever (financial or otherwise) in the Employer, the Contractor, the Employer's Personnel or the Contractor's Personnel;

Sub-Clause	FIDIC 2017	2022 REPRINT
	(c) in the ten years before signing the DAA Agreement, not have been employed as a consultant or otherwise by the Employer, the Contractor, the Employer's Personnel or the Contractor's Personnel;	(c) in the 5 years before signing the DAAB Agreement, not have been employed as a consultant or otherwise by the Employer, the Contractor, the Employer's Personnel or the Contractor's Personnel, except in such circumstances as were disclosed in writing to the Employer and the Contractor before they signed the DAAB Agreement (or are deemed to have done so);
	(d) not previously have acted, and shall not act, in any judicial or arbitral capacity in relation to the Contract;	(d) not previously have acted, and shall not act, in any judicial or arbitral capacity in relation to the Contract;
	(e) have disclosed in writing to the Employer, the Contractor and the Other Members (if any), before signing the DAA Agreement (or before he/she is deemed to have signed the DAA Agreement under the Contract) and to his/her best knowledge and recollection, any: (i) existing and/or past professional or personal relationships with any director, officer or employee of the Employer, the Contractor, the Employer's Personnel or the Contractor's Personnel (including as a dispute resolution practitioner on another project), (ii) facts or circumstances which might call into question his/her independence or impartiality, and (iii) previous involvement in the project of which the Contract forms part;	(e) have disclosed in writing to the Employer, the Contractor and the Other Members (if any), before signing the DAA Agreement (or before he/she is deemed to have signed the DAA Agreement under the Contract) and to his/her best knowledge and recollection, any: (i) existing and/or past professional or personal relationships with any director, officer or employee of the Employer, the Contractor, the Employer's Personnel or the Contractor's Personnel (including as a dispute resolution practitioner on another project) (ii) facts or circumstances which might call into question his/her independence or impartiality, and (iii) previous involvement in the project of which the Contract forms part;
	(f) not, while a DAAB Member and for the Term of the DAAB: (i) be employed as a consultant or otherwise by, and/or (ii) enter into discussions or make any agreement regarding future employment with the Employer, the Contractor, the Employer's Personnel or the Contractor's Personnel, except as may be agreed by the Employer, the Contractor and the Other Members (if any); and/or	(f) not, while a DAAB Member and for the Term of the DAAB: (i) be employed as a consultant or otherwise by, and/or (ii) enter into discussions or make any agreement regarding future employment with the Employer, the Contractor, the Employer's Personnel or the Contractor's Personnel, except as may be agreed by the Employer, the Contractor and the Other Members (if any); and/or
	(g) not solicit, accept or receive (directly or indirectly) any gift, gratuity, commission or other thing of value from the Employer,	

Sub-Clause	FIDIC 2017	2022 REPRINT
	the Contractor, the Employer's Personnel or the Contractor's Personnel, except for payment under the DAA Agreement.	(g) not solicit, accept or receive (directly or indirectly) any gift, gratuity, commission or other thing of value from the Employer, the Contractor, the Employer's Personnel or the Contractor's Personnel, except for payment under the DAA Agreement.
Sub-Clause 9.1	The DAAB Member shall be paid as follows, in the currency named in the DAA Agreement: (a) a monthly fee, which shall be a fixed fee as payment in full for: (i) being available on 28-days' notice for all meetings, Site visits and hearings under the DAAB Rules (and, in the event of a request under Rule 3.6 of the DAAB Rules, being available for an urgent meeting or Site visit); (ii) becoming and remaining knowledge-able about the Contract, informed about the progress of the Works and maintaining a current working file of documents, in accordance with sub-paragraph (d) of Clause 5.1 above; (iii) all office and overhead expenses including secretarial services, photo-copying and office supplies incurred in connection with his/her duties; and (iv) all services performed hereunder except those referred to in sub-paragraphs (b) and (c) of this Clause. This fee shall be paid monthly with effect from the last day of the month in which the Effective Date occurs until the end of the month in which the Term of the DAAB expires, or the DAAB Member declines to act or is unable to act as a result of death, illness, disability, resignation or termination of his/her DAA Agreement. If no monthly fee is stated in the DAA Agreement, the matters described in sub-paragraphs (i) to (iv) above shall be deemed to be covered by the daily fee under sub-paragraph (b) below;	The DAAB Member shall be paid as follows, in the currency named in the DAA Agreement: (a) a monthly fee, which shall be a fixed fee as payment in full for: (i) being available on 28-days' notice for all meetings, Site visits and hearings under the DAAB Rules (and, in the event of a request under Rule 3.6 of the DAAB Rules, being available for an urgent meeting or Site visit); (ii) becoming and remaining knowledgeable about the Contract, informed about the progress of the Works and maintaining a current working file of documents, in accordance with sub-paragraph (d) of Clause 5.1 above; (iii) all office and overhead expenses including secretarial services, photo-copying and office supplies incurred in connection with his/her duties; and (iv) all services performed hereunder except those referred to in sub-paragraphs (b) and (c) of this Clause. This fee shall be paid monthly with effect from the last day of the month in which the Effective Date occurs until the end of the month in which the Term of the DAAB expires, or the DAAB Member declines to act or is unable to act as a result of death, illness, disability, resignation or termination of his/her DAA Agreement. If no monthly fee is stated in the DAA Agreement, the matters described in sub-paragraphs (i) to (iv) above shall be deemed to be covered by the daily fee under sub-paragraph (b) below;

Sub-Clause	FIDIC 2017	2022 REPRINT
	(b) a daily fee, which shall be considered as payment in full for each day: (i) or part of a day, up to a maximum of two days' travel time in each direction, for the journey between the DAAB Member's home and the Site, or another location of a meeting with the Parties and/or the Other Members (if any); (ii) spent on attending meetings and making Site visits in accordance with Rule 3 of the DAAB Rules, and writing reports in accordance with Rule 3.10 of the DAAB Rules; (iii) spent on giving Informal Assistance; (iv) spent on attending hearings (and, in case of a three-member DAAB, attending meeting(s) between the DAAB Members in accordance with sub-paragraph (a) of Rule 8.2 of the DAAB Rules, and communicating with the Other Members), and preparing decisions; and (v) spent in preparation for a hearing, and studying written documentation and arguments from the Parties submitted in accordance with sub-paragraph (c) of Rule 7.1 of the DAAB Rules; (c) all reasonable expenses, including necessary travel expenses (air fare in business class or equivalent, hotel and subsistence and other direct travel expenses, including visa charges) incurred in connection with the DAAB Member's duties, as well as the cost of telephone calls (and video conference calls, if any, and internet access), courier charges and faxes. The DAAB Member shall provide the Parties with a receipt for each item of expenses; (d) any taxes properly levied in the Country on payments made to the DAAB Member (unless a national or permanent resident of the Country) under this Sub-Clause 9.1.	(b) a daily fee, which shall be considered as payment in full for each day: (i) or part of a day, up to a maximum of two days' travel time in each direction, for the journey between the DAAB Member's home and the Site, or another location of a meeting with the Parties and/or the Other Members (if any); (ii) spent on attending meetings and making Site visits in accordance with Rule 3 of the DAAB Rules, and writing reports in accordance with Rule 3.10 of the DAAB Rules; (iii) spent on giving Informal Assistance; (iv) spent on preparing and attending hearings (and, in case of a three-member DAAB, attending meeting(s) between the DAAB Members in accordance with sub-paragraph (a) of Rule 8.2 of the DAAB Rules, and communicating with the Other Members); and (v) spent on preparing decisions, including studying written documentation and arguments from the Parties. (c) all reasonable expenses, including necessary travel expenses (air fare in business class or equivalent, hotel and subsistence and other direct travel expenses, including visa charges) incurred in connection with the DAAB Member's duties, as well as the cost of internet access, courier charges and faxes. The DAAB Member shall provide the Parties with a receipt for each item of expenses; (d) any taxes properly levied in the Country on payments made to the DAAB Member (unless a national or permanent resident of the Country) under this Sub-Clause 9.1.

Sub-Clause	FIDIC 2017	2022 REPRINT
Sub-Clause 9.3	If the Parties and the DAAB Member have agreed all other terms of the DAA Agreement but fail to jointly agree the amount of the monthly fee or the daily fee in the DAA Agreement (the 'non-agreed fee' in this Sub-Clause): (a) the DAA Agreement shall nevertheless be deemed to have been signed by the Parties and the DAAB Member, except that the fee proposed by the DAAB Member shall only temporarily apply; (b) either Party or the DAAB Member may apply to the appointing entity or officialnamed in the Contract Data to set the amount of the non-agreed fee; (c) the appointing entity or official shall, as soon as practicable and in any case within 28 days after receiving any such application, set the amount of the non-agreed fee, which amount shall be reasonable taking due regard of the complexity of the Works, the experience and qualifications of the DAAB Member, and all other relevant circumstances; (d) once the appointing entity or official has set the amount of the non-agreed fee, this amount shall be final and conclusive, shall replace the fee under sub-paragraph (a) above, and shall be deemed to have applied from the Effective Date; and (e) thereafter, after giving credit to the Parties for all amounts previously paid in respect of the non-agreed fee, the balance (if any) due from the DAAB Member to the Parties or from the Parties to the DAAB Member, as the case may be, shall be paid.	If the DAAB Member has been appointed by the appointing official, the amounts of the DAAB Member's monthly fee and daily fee, under Sub-Clause 9.1 above, shall be as referred to under sub-paragraph (i) of Sub-Clause 21.2 [*Failure to Appoint DAAB Member(s)*] of the Conditions of Contract.
Sub-Clause 9.7	If the DAAB Member does not receive payment of the amount due within 56 days	If the DAAB Member does not receive payment of the amount due within 56 days after

Sub-Clause	FIDIC 2017	2022 REPRINT
	after submitting a valid invoice, the DAAB Member may: (a) not less than 7 days after giving a Notification to the Parties and the Other Members (if any), suspend his/her services until the payment is received; and/or (b) resign his/her appointment by giving a Notification under Sub-Clause 10.1 below.	submitting a valid invoice, the DAAB Member may: (a) not less than 7 days after giving a Notification to the Parties and the Other Members (if any), suspend his/her services until the payment is received; and/or (b) without prejudice to his/her other rights or remedies, resign his/her appointment by giving a Notification under Sub-Clause 10.1 below.
Clause 12	Any dispute arising out of or in connection with the DAA Agreement, or the breach, termination or invalidity thereof, shall be finally settled by arbitration under the Rules of Arbitration of the International Chamber of Commerce 2017 by one arbitrator appointed in accordance with these Rules of Arbitration, and Article 30 and the Expedited Procedure Rules at Appendix VI of these Rules of Arbitration shall apply.	Any dispute arising out of or in connection with the DAA Agreement, or the breach, termination or invalidity thereof, shall be finally settled by arbitration under the Rules of Arbitration of the International Chamber of Commerce by one arbitrator appointed in accordance with these Rules of Arbitration, and the Expedited Procedure Rules of these Rules of Arbitration shall apply.
ANNEX – DAAB PROCEDURAL RULES		
Rule 3.2	As soon as practicable after the DAAB is appointed, the DAAB shall convene a face-to-face meeting with the Parties. At this meeting, the DAAB shall establish a schedule of planned meetings and Site visits in consultation with the Parties, which schedule shall reflect the requirements of Rule 3.3 below and shall be subject to adjustment by the DAAB in consultation with the Parties.	As soon as practicable after the DAAB is appointed, the DAAB shall convene an introductory meeting with the Parties. The date, time and type (online or in-person) of, and agenda for, the introductory meeting shall be set by the DAAB in consultation with the Parties. At this meeting, the DAAB shall establish a schedule of planned meetings and Site visits in consultation with the Parties, which schedule shall reflect the requirements of Rule 3.3 below and shall be subject to adjustment by the DAAB in consultation with the Parties.
Rule 3.3	The DAAB shall hold face-to-face meetings with the Parties, and/or visit the Site, at regular intervals and/or at the written request of either Party. The frequency of such meetings and/or Site visits shall be: (a) sufficient to achieve the purpose under Rule 3.1 above;	The DAAB shall hold meetings with the Parties, and visit the Site, at regular intervals and/or at the written request of either Party. The frequency of such meetings and Site visits shall be: (a) sufficient to achieve the purpose under Rule 3.1 above; (b) at intervals of not more than 140 days unless otherwise agreed jointly by the Parties and the DAAB; and

Sub-Clause	FIDIC 2017	2022 REPRINT
	(b) at intervals of not more than 140 days unless otherwise agreed jointly by the Parties and the DAAB; and (c) at intervals of not less than 70 days, subject to Rules 3.5 and 3.6 below and except as required to conduct a hearing as described under Rule 7 below, unless otherwise agreed jointly by the Parties and the DAAB.	(c) at intervals of not less than 70 days, subject to Rules 3.5 and 3.6 below and except as required to conduct a hearing as described under Rule 7 below, unless otherwise agreed jointly by the Parties and the DAAB. Each such meeting shall be face-to-face and each Site visit shall be in-person, unless the Parties and the DAAB agree that exceptional circumstances mean that it would be prudent for the meeting and Site visit to be carried out online. The date, time and agenda for each such meeting and Site visit shall be set by the DAAB in consultation with the Parties.
Rule 3.4	In addition to the face-to-face meetings referred to in Rules 3.2 and 3.3 above, the DAAB may also hold meetings with the Parties by telephone or video conference as agreed with the Parties (in which case each Party bears the risk of interrupted or faulty telephone or video conference transmission and reception).	In addition to the meetings referred to in Rules 3.2 and 3.3 above, the DAAB may also hold meetings with the Parties online.
Rule 3.6	Either Party may request an urgent meeting or Site visit by the DAAB. This shall be a written request and shall give reasons for the urgency of the meeting or Site visit. If the DAAB agrees that such a meeting or Site visit is urgent, the DAAB Members shall use all reasonable endeavours to: (a) hold a meeting with the Parties by telephone or video conference (as agreed with the Parties under Rule 3.4 above) within 3 days after receiving the request; and (b) if requested and (having given the other Party opportunity at this meeting to respond to or oppose this request) the DAAB agrees that a Site visit is necessary, visit the Site within 14 days after the date of this meeting.	Either Party may request an urgent meeting or Site visit by the DAAB. This shall be a written request and shall give reasons for the urgency of the meeting or Site visit. If the DAAB agrees that such a meeting or Site visit is urgent, the DAAB Members shall use all reasonable endeavours to: (a) hold a meeting with the Parties online within 3 days after receiving the request; and (b) if requested and (having given the other Party opportunity at this meeting to respond to or oppose this request) the DAAB agrees that an in-person Site visit is necessary, visit the Site within 14 days after the date of this meeting.
Rule 3.9	Each meeting and Site visit shall be co-ordinated by the Contractor in co-operation with the Employer and the Engineer. The	Each meeting and Site visit shall be co-ordinated by the Contractor in co-operation with the Employer and the Engineer. The

Sub-Clause	FIDIC 2017	2022 REPRINT
	Contractor shall ensure the provision of appropriate: (a) personal safety equipment, security controls (if necessary) and site transport for each Site visit; (b) meeting room/conference facilities and secretarial and copying services for eachface-to-face meeting; and (c) telephone conference or video conference facilities for each meeting by telephone or video conference.	Contractor shall ensure the provision of appropriate: (a) personal safety equipment, security controls (if necessary) and site transport for each Site visit; (b) meeting room/conference facilities and secretarial and copying services for each face-to-face meeting; and (c) access to an online video conference platform for each online meeting and Site visit.
CONTRACT DATA		
Page 7: In the entry for 19.2.1 (iv)	List of Exceptional Risks which shall not be excluded from the insurance cover for the Works	List of risks arising from Exceptional Events which shall not be excluded from the insurance cover for the Works
Page 7: In the entry for 21.2	Appointing entity (official) for DAAB members ..... (*Unless otherwise stated here, it shall be the President of FIDIC or a person appointed by the President*)	[Deleted]
NOTES ON THE PREPARATION OF SPECIAL PROVISIONS		
Guidance for Sub-Clause 18.1	In respect of sub-paragraph (f) of this Sub-Clause, it should be noted that any event of 'exceptionally adverse climatic conditions' is excluded from the definition of what constitutes an Exceptional Event. While this means that there is no right for either Party to suspend the Works in the case of an event of 'exceptionally adverse climatic conditions', if this type of event has the effect of delaying completion and taking over of the Works or Section the Contractor shall be entitled to EOT under sub-paragraph (c) of Sub-Clause 8.5 [*Extension of Time*].	In respect of sub-paragraph (f) of this Sub-Clause, it should be noted that any event of 'exceptionally adverse climatic conditions' (as referred to in sub-paragraph (c) of Sub-Clause 8.5 [*Extension of Time for Completion*]) will not constitute an Exceptional Event unless it is of such severity or magnitude that the conditions stated in sub-paragraphs (ii) and (iii) of this Sub-Clause 18.1 [*Exceptional Events*] are fulfilled. Therefore, unless both such conditions are fulfilled, there is no right for either Party to suspend the Works in the case of an event of 'exceptionally adverse climatic conditions', although if this type of event has the effect of delaying completion and taking over of the Works or Section the Contractor shall be entitled to EOT under sub-paragraph (c) of Sub-Clause 8.5 [*Extension of Time*].

Sub-Clause	FIDIC 2017	2022 REPRINT
Guidance for Sub-Clause 21.1	It is generally accepted that construction projects depend for their success on the avoidance of Disputes between the Employer and the Contractor and, if Disputes do arise, the timely resolution of such Disputes. Therefore, the Contract should include the provisions under Clause 21 which, while not discouraging the Parties from reaching their own agreement on Disputes as the Works proceed, allow them to bring contentious matters to an independent and impartial dispute avoidance/adjudication board ('DAAB') for resolution.	It is generally accepted that construction projects depend for their success on the avoidance of Disputes between the Employer and the Contractor and, if Disputes do arise, the timely resolution of such Disputes. Therefore, the Contract should include the provisions under Clause 21 which, while not discouraging the Parties from reaching their own agreement on Disputes as the Works proceed, allow them to bring contentious matters to an independent and impartial dispute avoidance/adjudication board ("DAAB") for resolution.
	The provisions of this Sub-Clause are intended to provide for the appointment of the DAAB, and FIDIC strongly recommends that the DAAB be appointed, as a 'standing DAAB' – that is, a DAAB that is appointed at the start of the Contract who visits the Site on a regular basis and remains in place for the duration of the Contract to assist the Parties: (a) in the avoidance of Disputes, and (b) in the 'real-time' resolution of Disputes if and when they arise to achieve a successful project. It is for this reason that, under the first paragraph of this Sub-Clause, the Parties are under a joint obligation to appoint the member(s) of the DAAB within 28 days after the Contractor receives the Letter of Acceptance if no other time is stated in the Contract Data. That said, it is preferable that the member(s) of the DAAB are appointed before the Letter of Acceptance is issued.	The provisions of this Sub-Clause are intended to provide for the appointment of the DAAB, and FIDIC strongly recommends that the DAAB be appointed, as a 'standing DAAB' – that is, a DAAB that is appointed at the start of the Contract who visits the Site on a regular basis and remains in place for the duration of the Contract to assist the Parties: (a) in the avoidance of Disputes, and (b) in the 'real-time' resolution of Disputes if and when they arise to achieve a successful project. It is for this reason that, under the first paragraph of this Sub-Clause, the Parties are under a joint obligation to appoint the member(s) of the DAAB within 28 days after the Contractor receives the Letter of Acceptance if no other time is stated in the Contract Data. That said, it is preferable that the member(s) of the DAAB are appointed before the Letter of Acceptance is issued.
	At an early stage in the Employer's planning of the project, consideration should be given as to whether a sole-member DAAB or a three-member DAAB is preferable for a particular project, taking account of its size, duration and the fields of expertise which will be involved.	At an early stage in the Employer's planning of the project, consideration should be given as to whether a sole-member DAAB or a three-member DAAB is preferable for a particular project, taking account of its size, duration and the fields of expertise which will be involved.

Sub-Clause	FIDIC 2017	2022 REPRINT
	This Sub-Clause provides for two alternative arrangements for the DAAB: – a sole-member DAAB of one natural person, who has entered into a tripartite agreement with both Parties; or – a three-member DAAB of three natural persons, each of whom has entered into a tripartite agreement with both Parties.	This Sub-Clause provides for two alternative arrangements for the DAAB: – a sole-member DAAB of one natural person, who has entered into a tripartite agreement with both Parties; or – a three-member DAAB of three natural persons, each of whom has entered into a tripartite agreement with both Parties.
	The tripartite agreement above is referred to as the *DAAB Agreement* under the Conditions of Contract. It is recommended that the form of this agreement be one of the two alternative example forms included at the end of this publication (in the section 'Sample Forms'), as appropriate to the arrangement adopted.	The tripartite agreement above is referred to as the *DAAB Agreement* under the Conditions of Contract. It is recommended that the form of this agreement is the example form included at the end of this publication (in the section 'Forms').
	It should be noted that both forms of the DAAB Agreement incorporate (by reference) the *General Conditions of Dispute Avoidance/ Adjudication Agreement* with its Annex *DAAB Procedural Rules*, which are included as the Appendix to the General Conditions in this publication.	It should be noted that (by reference) the *General Conditions of DAAB Agreement* with its Annex *DAAB Procedural Rules* are included as the Appendix to the General Conditions in this publication.
	Under either of these alternative forms of DAAB Agreement, each natural person of the DAAB is referred to as a *DAAB Member*.	Under either of these alternative forms of DAAB Agreement, each natural person of the DAAB is referred to as a *DAAB Member*.
	A very important factor in the success of the dispute avoidance/adjudication procedure is the Parties' confidence in the agreed individual(s) who will serve on the DAAB. Therefore, it is essential that candidates for this position are not imposed by either Party on the other Party. The appointment of the DAAB is facilitated by the provision in the Contract Data for each Party to name potential DAAB Members. It is important that the Employer and the Contractor each avail himself/herself, of the opportunity at the tender stage of the Contract to name potential DAAB Members in the Contract Data.	A very important factor in the success of the dispute avoidance/adjudication procedure is the Parties' confidence in the agreed individual(s) who will serve on the DAAB. Therefore, it is essential that candidates for this position are not imposed by either Party on the other Party. The appointment of the DAAB is facilitated by the provision in the Contract Data for each Party to name potential DAAB Members. It is important that the Employer and the Contractor each avail himself/herself, of the opportunity at the tender stage of the Contract to name potential DAAB Members in the Contract Data.
		The Contract Data includes two lists for potential DAAB Members to be named: one for

Sub-Clause	FIDIC 2017	2022 REPRINT
	The Contract Data includes two lists for potential DAAB Members to be named: one for the Employer to list three names, the other for the Contractor to list three names. This ensures that both Parties have equal opportunity to put forward (the same number of) names for potential DAAB Members, and so avoid any question that DAAB Member(s) may be imposed on one Party by the other Party.	the Employer to list three names, the other for the Contractor to list three names. This ensures that both Parties have equal opportunity to put forward (the same number of) names for potential DAAB Members, and so avoid any question that DAAB Member(s) may be imposed on one Party by the other Party.
	This provides a total of six potential DAAB Members from which the sole member or three members (as the case may be) can be selected by the Parties. If it is considered necessary to have a wider selection of DAAB Member(s) to choose from, then provision may be made for longer lists in the Contract Data for both the Contractor and the Employer to name (the same number of) additional DAAB Members.	This provides a total of six potential DAAB Members from which the sole member or three members (as the case may be) can be selected by the Parties. If it is considered necessary to have a wider selection of DAAB Member(s) to choose from, then provision may be made for longer lists in the Contract Data for both the Contractor and the Employer to name (the same number of) additional DAAB Members.
	If the Parties cannot agree on any DAAB Member, Sub-Clause 21.2 [*Failure to Appoint DAAB Member(s)*] applies and the selection and appointment of the DAAB Member(s) should be made by a wholly impartial entity with an understanding of the nature and purpose of a DAAB. The President of FIDIC is prepared to perform this role if this authority has been delegated in accordance with the example wording in the Contract Data. FIDIC maintains a list of approved and experienced adjudicators for this specific purpose: *The FIDIC President's List of Approved Dispute Adjudicators* (http://fidic.org/president-list).	If the Parties cannot agree on any DAAB Member, *or replacement, or cannot agree the terms of the DAAB Agreement with any prospective DAAB Member or replacement*, Sub-Clause 21.2 [*Failure to Appoint DAAB Member(s)*] applies. Therefore, unless otherwise agreed by the Parties, the selection and appointment of the DAAB Member(s) will be made by the President of FIDIC or a person appointed by the President.
		FIDIC is very knowledgeable about the nature and purpose of a DAAB and is an impartial entity to make such appointments, in circumstances where it has not been possible to appoint the DAAB member or replacement member by the agreement of the Parties.
	If no potential DAAB Members' names are given in the Contract Data, consideration should be given to stating a time period in the Contract Data that is greater than the default period of 28 days stated in the first paragraph of this Sub-Clause.	FIDIC has its appointment rules and maintains a list of approved and experienced adjudicators for this specific purpose: The FIDIC President's List of Approved Dispute Adjudicators. This list is available to access on FIDIC's website at fidic.org.

Sub-Clause	FIDIC 2017	2022 REPRINT
	The period of 224 days stated in sub-paragraph (i) of this Sub-Clause has been arrived at by taking account of certain time allowances, as follows: date of termination + 28 days to give a Notice of Claim under Sub-Clause 20.2.1 [*Notice of Claim*] + 84 days to submit detailed particulars for the Claim under Sub-Clause 20.2.4 [*Fully detailed Claim*] + 84 days for the Engineer's agreement/ determination of the Claim under Sub-Clause 3.7 [*Agreement or Determination*] + 28 days for a NOD under Sub-Clause 3.7 [*Agreement or Determination*] = 224 days after the date of termination. .../...	If no potential DAAB Members' names are given in the Contract Data, consideration should be given to stating a time period in the Contract Data that is greater than the default period of 28 days stated in the first paragraph of this Sub-Clause. The period of 224 days stated in sub-paragraph (i) of this Sub-Clause has been arrived at by taking account of certain time allowances, as follows: date of termination + 28 days to give a Notice of Claim under Sub-Clause 20.2.1 [*Notice of Claim*] + 56 days to submit detailed particulars for the Claim under Sub-Clause 20.2.4 [*Fully detailed Claim*] (the period of '84 days' stated in sub-paragraph (i) of Sub-Clause 20.2.4 is measured from the date that 'the claiming Party became aware, or should have become aware of, the event or circumstance' which date, in this instance, will be the date of termination. Therefore, the period of 56 days here takes account of the above 28-day period for giving a Notice of Claim) + 28 days to allow for the Engineer to ask for further particulars in respect of the Claim, or for the Parties to agree a longer period for the Engineer's agreement/ determination of the Claim than that allowed under Sub-Clause 3.7 [*Agreement or Determination*] as referred to under Sub-Clause 20.2.5 [*Agreement or determination of the Claim*]. + 84 days for the Engineer's agreement/ determination of the Claim under Sub-Clause 3.7 [*Agreement or Determination*] + 28 days for a NOD under Sub-Clause 3.7 [*Agreement or Determination*] = 224 days after the date of termination.

Smith G
ISBN 978-0-7277-6652-6
https://doi.org/10.1680/fcmh.66526.311
Emerald Publishing Limited:

Index

Page numbers in *italics* refer to figures

Printed and bound by CPI Group (UK) Ltd, Croydon, CR0 4YY

26/02/2024

14458437-0001